Student Edition

Textbook Lite | Activities | Study Guide

# EARTH AND SPACE SCIENCES FOR NGSS

# EARTH & SPACE SCIENCES

FOR NGSS

## Meet the writing team

**Kent Pryor**
I have a BSc from Massey University majoring in zoology and ecology and taught secondary school science for 9 years before joining BIOZONE as an author in 2009.

**Tracey Greenwood (Editor)**
I have been writing resources for students since 1993. I have a Ph.D in biology, specialising in lake ecology and I have taught both graduate and undergraduate biology.

**Lissa Bainbridge-Smith**
I worked in industry in a research and development capacity for 8 years before joining BIOZONE in 2006. I have an M.Sc from Waikato University.

**Richard Allan**
I have had 11 years experience teaching senior secondary school biology. I have a Masters degree in biology and founded BIOZONE in the 1980s after developing resources for my own students.



First edition 2016
Seventh printing with change from inverse notation to solidus.

**ISBN 978-1-927309-37-7**





Published by BIOZONE International Ltd

Printed by Replika (INDIA) LIMITED Using FSC paper

## Cover photograph

The International Space Station (ISS) is a permanently inhabited artificial satellite, which maintains an orbit around Earth at an altitude of 330-435 km. It orbits Earth 15.5 times a day, functioning as a microgravity and space environment research laboratory. The ISS program is a joint project among five participating space agencies and the ownership and use of the station is established by intergovernmental treaties and agreements.

PHOTO: © dollarphotoclub.com


## Thanks to:

The staff at BIOZONE, including Nell Travaglia for design and graphics support, Paolo Curray and Malaki Toleafoa for IT support, Debbie Antoniadis and Arahi Hippolite for office handling and logistics, and the BIOZONE sales team.


**Purchases of this workbook may be made direct from the publisher:**

**BIOZONE** Corporation
USA and Canada
FREE phone: 1-855-246-4555
FREE fax: 1-855-935-3555
Email: sales@thebiozone.com
Web: www.thebiozone.com

# Contents



**CODES:** **Activity** is marked: to be done; when completed

# Contents

ESS2.B

**Plate Tectonics**



ESS2.C

**The Roles of Water in Earth's Surface Processes**



ESS2.D&2.E

**Weather, Climate, and Biogeology**



ESS3.A

**Natural Resources**



ESS3.B

**Natural Hazards**



ESS3.C

**Human Impacts on Earth Systems**



ESS3.D

**Global Climate Change**



**CODES**: **Activity** is marked: to be done / when completed

# Using the Student Edition

Activities make up most of this book. These are usually presented as short instructional sequences allowing you to build a deeper understanding of core ideas and the science and engineering practices that accompany them as you progress through each chapter. Each activity is accompanied by questions or specific tasks for you to complete.
A dark blue question denotes a question that is extension or suitable for gifted and talented students.

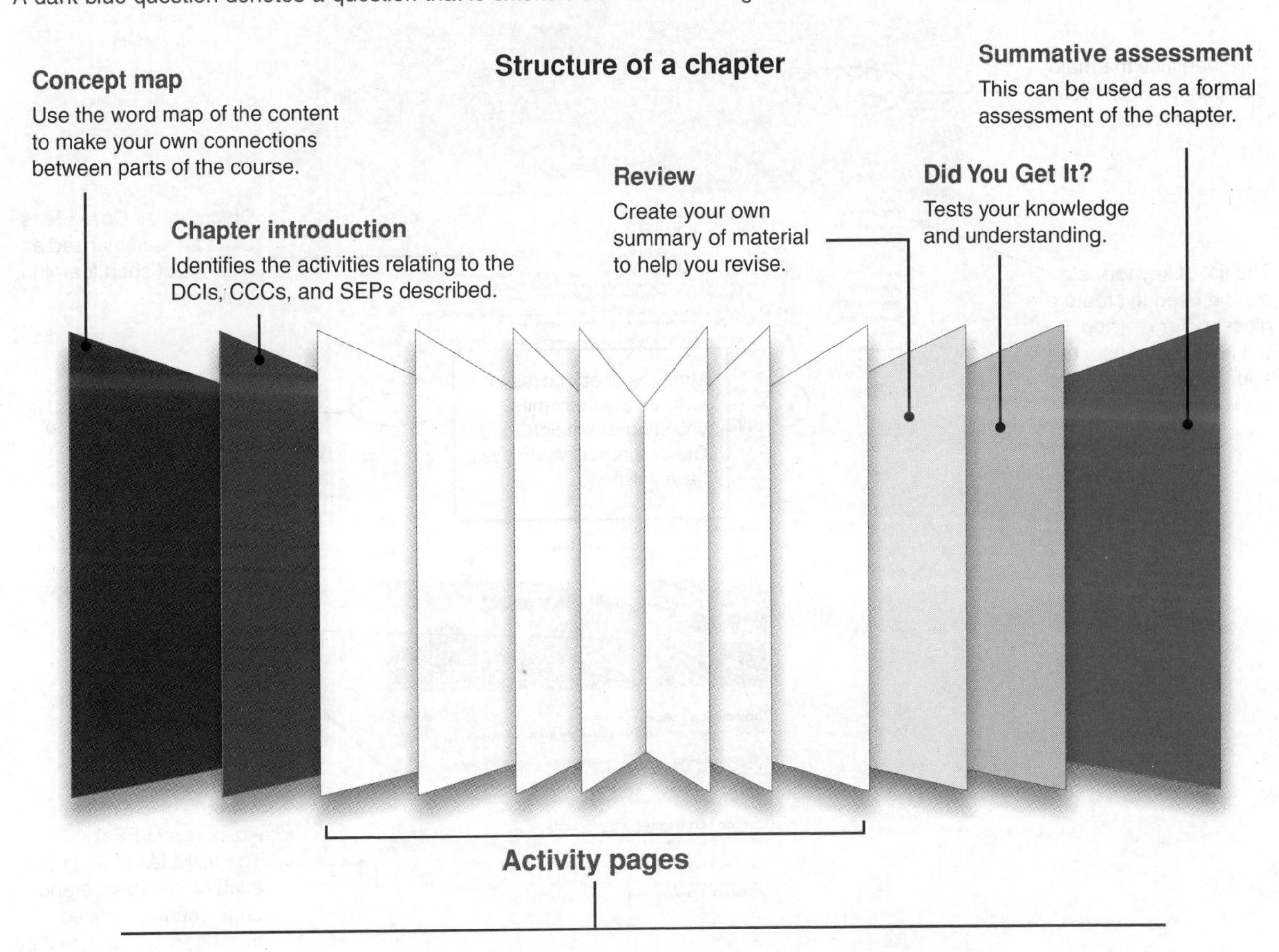


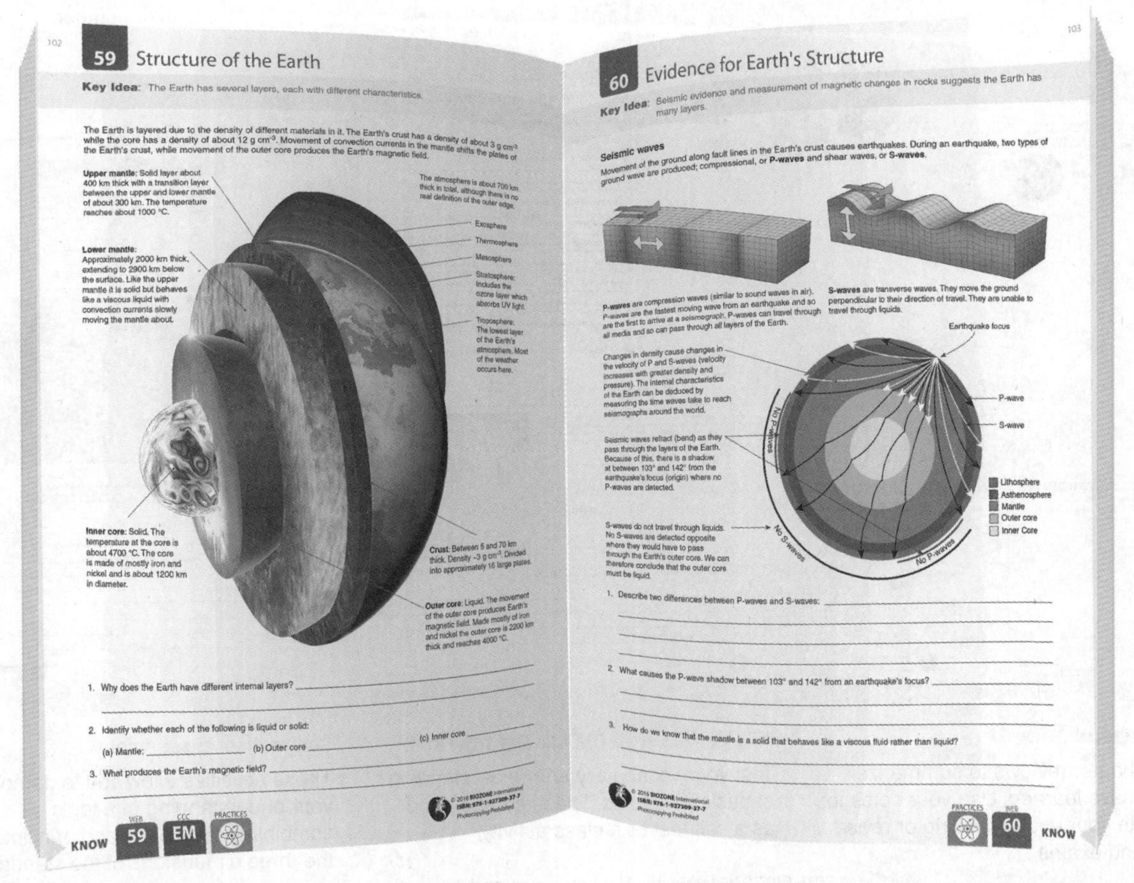

102

## 59 Structure of the Earth

**Key Idea**: The Earth has several layers, each with different characteristics.

The Earth is layered due to the density of different materials in it. The Earth's crust has a density of about 3 g $cm^{-3}$ while the core has a density of about 12 g $cm^{-3}$. Movement of convection currents in the mantle shifts the plates of the Earth's crust, while movement of the outer core produces the Earth's magnetic field.



1. Why does the Earth have different internal layers? ______
2. Identify whether each of the following is liquid or solid:
   (a) Mantle: ______ (b) Outer core ______ (c) Inner core ______
3. What produces the Earth's magnetic field? ______

© 2016 BIOZONE International
ISBN: 978-1-927309-37-7
Photocopying Prohibited

KNOW | WEB 59 | CCC EM | PRACTICES

103

## 60 Evidence for Earth's Structure

**Key Idea**: Seismic evidence and measurement of magnetic changes in rocks suggests the Earth has many layers.

### Seismic waves

Movement of the ground along fault lines in the Earth's crust causes earthquakes. During an earthquake, two types of ground wave are produced; compressional, or **P-waves** and shear waves, or **S-waves**.

**P-waves** are compression waves (similar to sound waves in air). P-waves are the fastest moving wave from an earthquake and so are the first to arrive at a seismograph. P-waves can travel through all media and so can pass through all layers of the Earth.

**S-waves** are transverse waves. They move the ground perpendicular to their direction of travel. They are unable to travel through liquids.



1. Describe two differences between P-waves and S-waves: ______
2. What causes the P-wave shadow between 103° and 142° from an earthquake's focus? ______
3. How do we know that the mantle is a solid that behaves like a viscous fluid rather than liquid? ______

© 2016 BIOZONE International
ISBN: 978-1-927309-37-7
Photocopying Prohibited

PRACTICES | WEB 60 | KNOW

# The Introduction and Activities

The chapter introductions of this book provide a list of learning outcomes for the DCIs, crosscutting concepts, and science and engineering practices. The activities will help you meet these learning outcomes by providing a variety of different ways to build and explore your understanding.

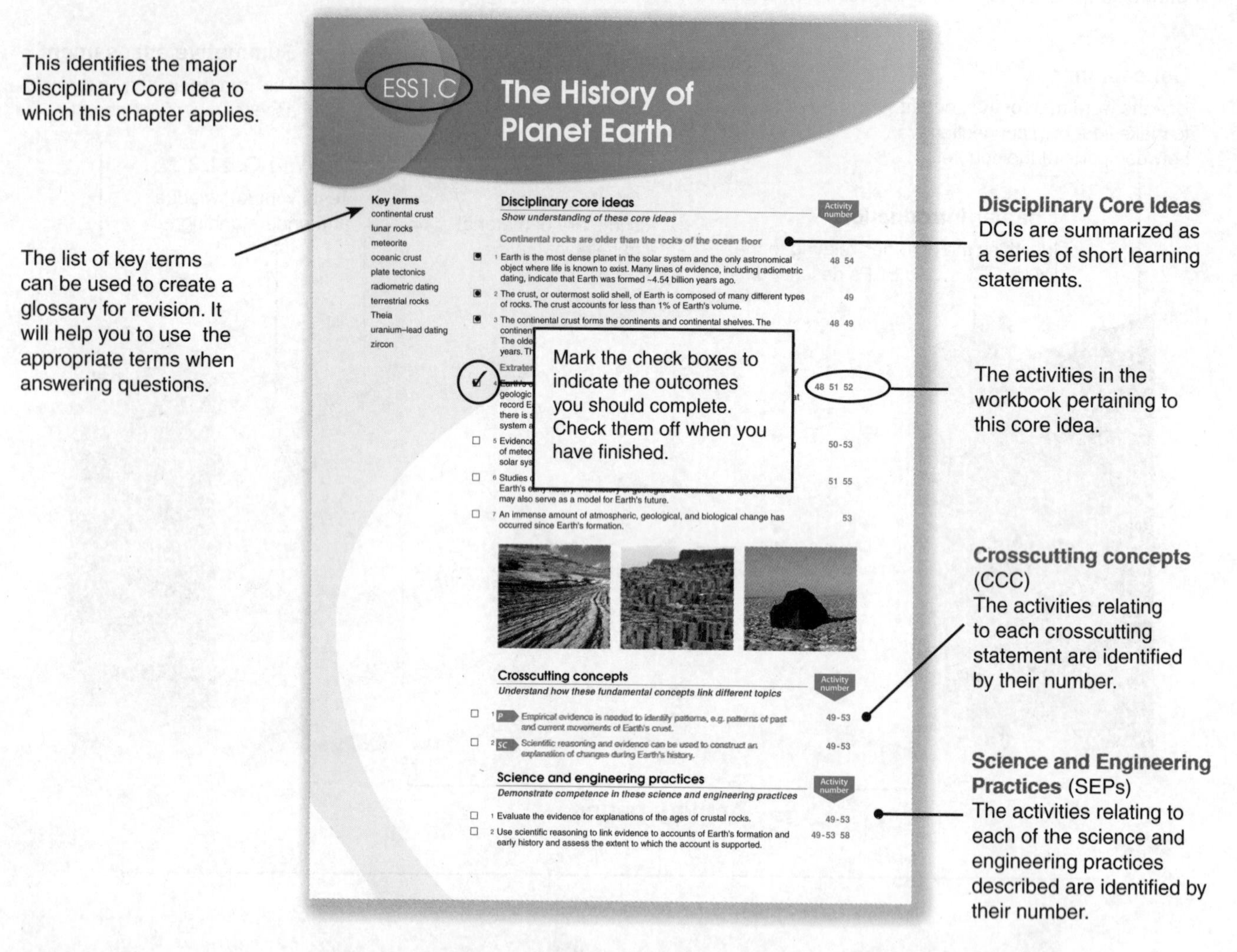


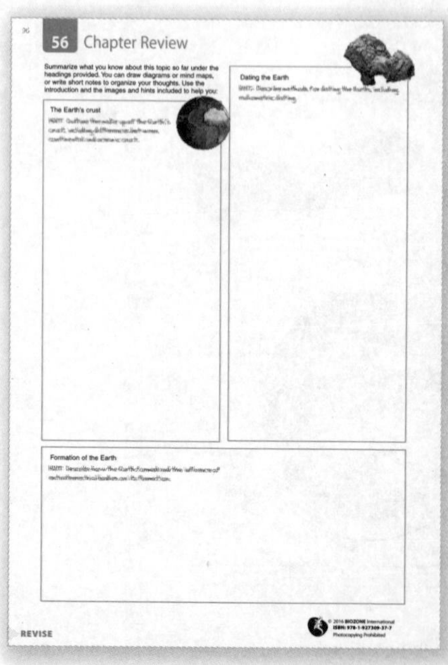


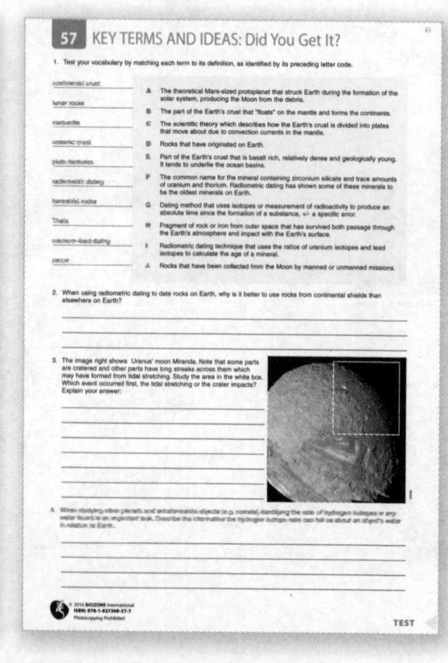


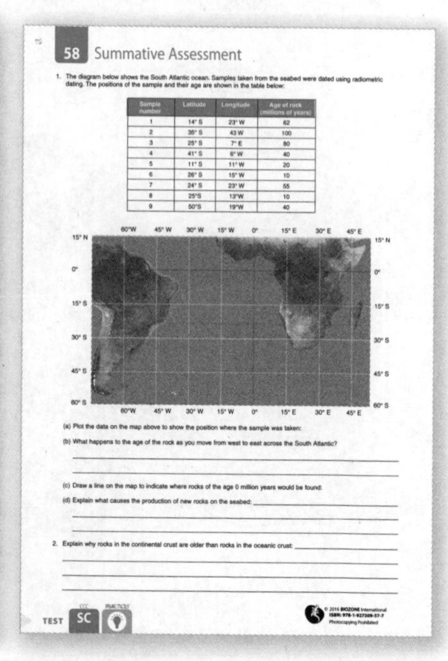


**Chapter Review**

Use the chapter reviews to summarize what you have learned. Use your notes to consolidate your understanding or revise for tests and exams.

**KEY TERMS AND IDEAS: Did You Get It?**

Test your vocabulary and understanding of basic principles. These can be used as a self test or a class activity.

**Summative Assessment**

These activities allow you to demonstrate your understanding of a topic by combining knowledge and principles from the three dimensions of the standards.

# Using the Tab System

The tab system is a useful way to quickly identify crosscutting concepts and science and engineering practices pertaining to the disciplinary core idea of the activity. They also indicate whether or not the activity is supported online.

- The CCC tabs indicate activities (and core ideas) that share the same crosscutting concepts. Not all activities have a crosscutting code and some incorporate more than one. The PRACTICES picture codes identify which science and engineering practices (SEPs) are relevant to the activity (and core ideas). There may be more than one or none.
- The weblinks code is always the same as the activity number on which it is cited. On visiting the weblink page (below), find the number and it will correspond to one or more external websites providing a video or animation of some aspect of the activity's content. Occasionally, the weblink may provide a bank of photographs where images are provided in colour, e.g.. for the appearances of various planetary phenomena.

## TASK CODES

These identify the nature of the activity

**DATA** = data handling and interpretation

**KNOW** = content you need to know

**PRAC** = a paper practical or a practical focus

**REFER** = reference - use this for information

**REVISE** = review the material in the section

**TEST** = test your understanding

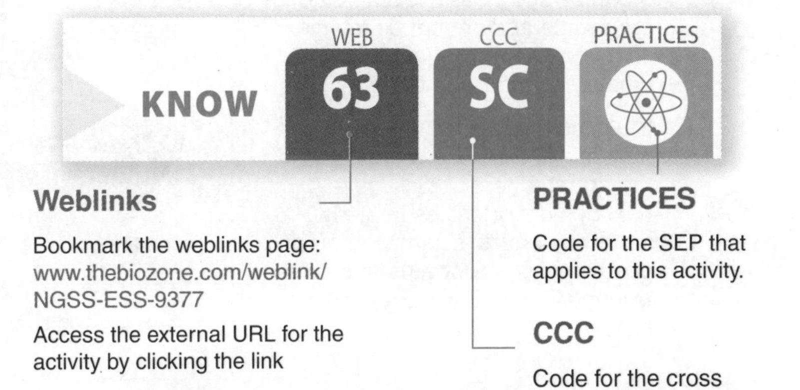


**Weblinks**

Bookmark the weblinks page: www.thebiozone.com/weblink/NGSS-ESS-9377

Access the external URL for the activity by clicking the link

**PRACTICES**

Code for the SEP that applies to this activity.

**CCC**

Code for the cross cutting concept that applies to this activity.

## www.thebiozone.com/weblink/NGSS-ESS-9377

This WEBLINKS page provides links to **external websites** with supporting information for the activities. These sites are distinct from those provided in the BIOLINKS area of BIOZONE's web site. For the most part, they are narrowly focussed animations and video clips directly relevant to some aspect of the activity on which they are cited. They provide great support to help your understanding of basic concepts.

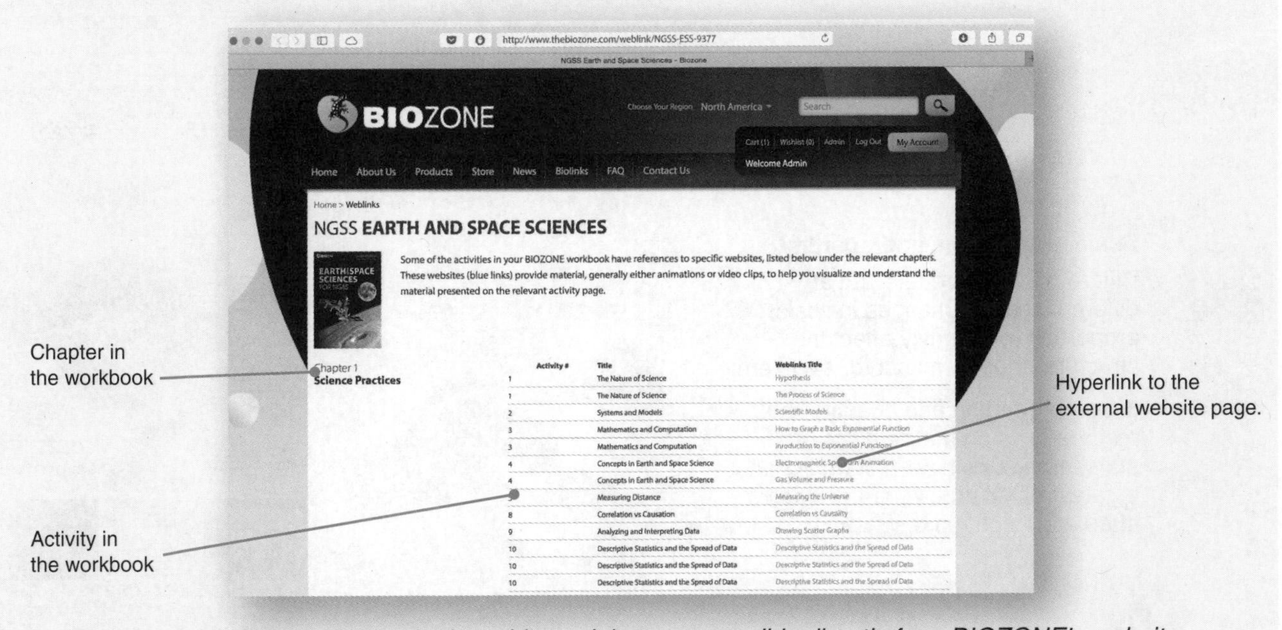


*Bookmark weblinks by typing in the address: it is not accessible directly from BIOZONE's website*

***Corrections and clarifications to current editions are always posted on the weblinks page***

# What Are Crosscutting Concepts?

Crosscutting concepts are ideas that are common to all fields of science. Recognizing the application of crosscutting concepts across different fields of Earth and space sciences, as well as other sciences, will help to deepen your understanding of the core ideas around which this book is structured. You will begin to see that common concepts link different areas of science and understanding this is part of developing a scientifically sound view of the world.

The seven crosscutting concepts are outlined below. Together, they form one the three dimensions of the Framework and the Standards. Each is associated with a number of specific statements. These are identified as they apply under 'Crosscutting Concepts' in each chapter introduction.

### Patterns

We see patterns everywhere in science. These guide how we organize and classify events and organisms and prompt us to ask questions about the factors that create and influence them.

Studying the radioisotopes of minerals in the Earth's crust and the arrangement of rock types provides opportunities to study patterns in Earth science. Analysis of the patterns allows us to produce explanations of Earth's geological history.

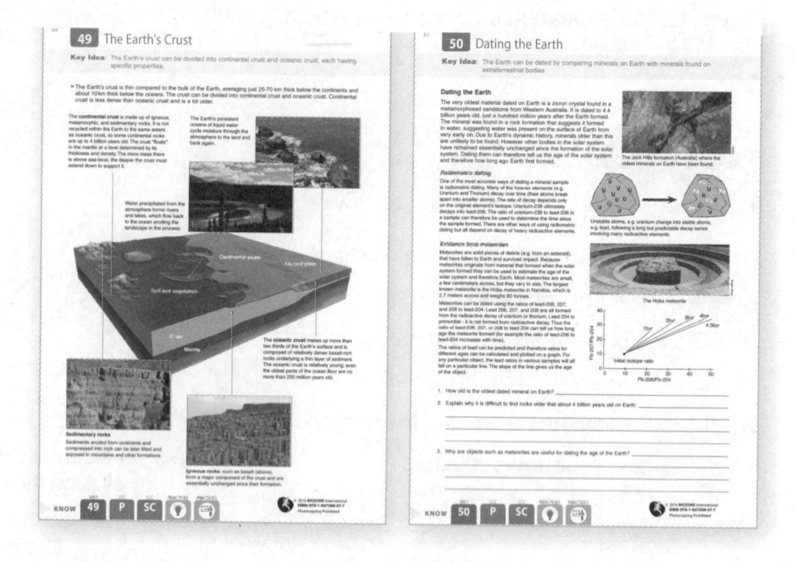
49 The Earth's Crust

50 Dating the Earth

### Cause and effect

A major part of science is investigating and explaining causal relationships. The mechanisms by which they occur can be tested in one context and used to explain and predict events in new contexts.

Studying the energy input from the Sun and reflection from the Earth can help us predict the effects of climate change and help prepare for major climate events such as droughts.

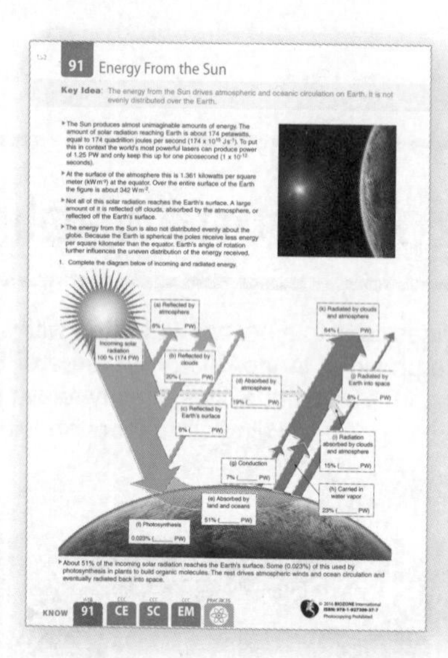
91 Energy From the Sun

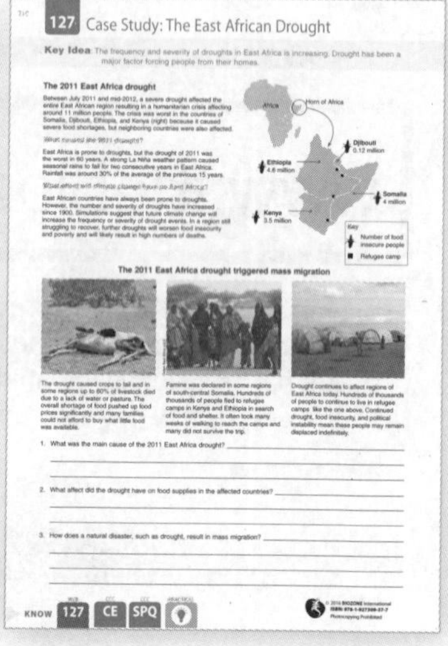
127 Case Study: The East African Drought

### Scale, proportion, and quantity

Different things are relevant at different scales. Changes in scale, proportion, or quantity affect the structure or performance of a system.

Scale, proportion and quantity have major effects on the life cycle of a star. Large stars have a short life and a catastrophic end, whereas smaller stars are much more sedate.

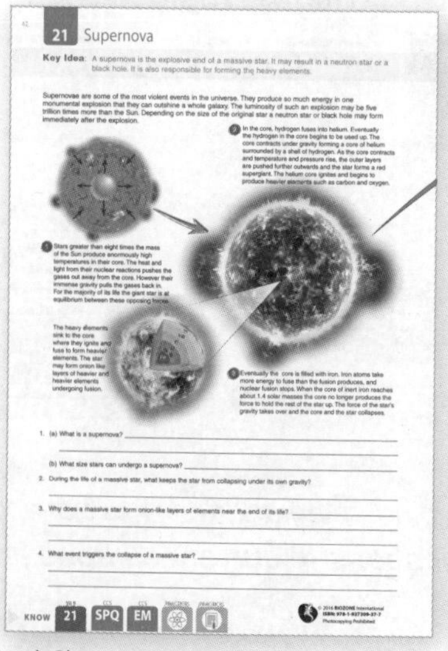
21 Supernova

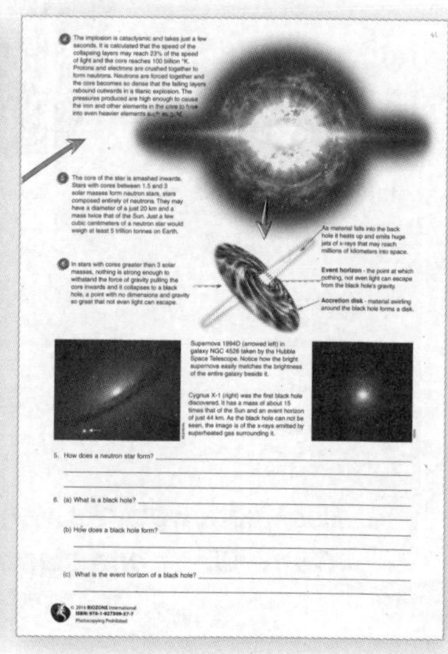

## Stability and change

Science often deals with constructing explanations of how things change or how they remain stable.

How does the Earth's climate remain relatively constant? What causes it to change? Studying feedback loops helps us understand how changes to one part of a system can have significant effects elsewhere.

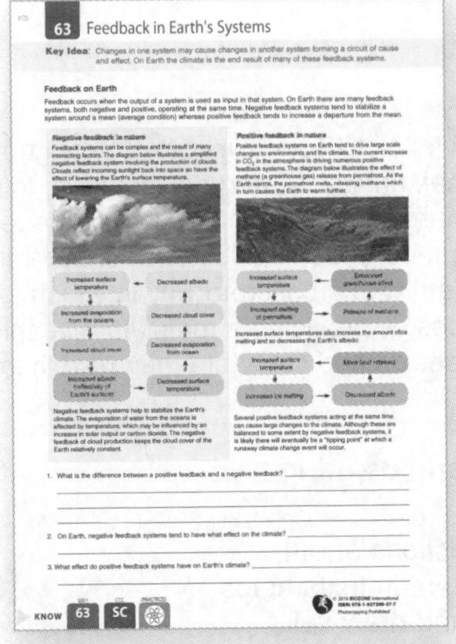
63 Feedback in Earth's Systems

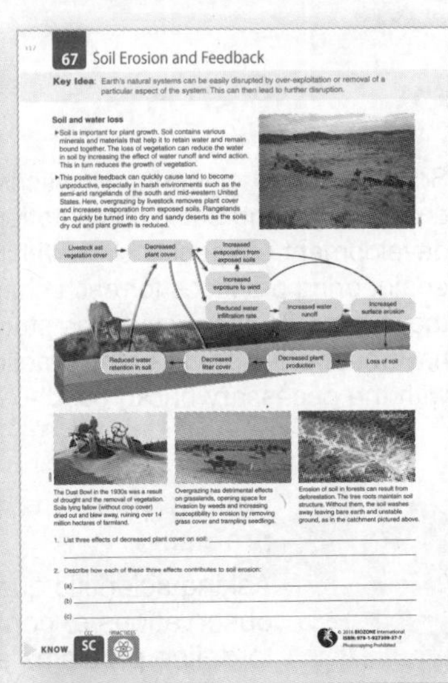
67 Soil Erosion and Feedback

## Energy and matter

Energy cannot be created or destroyed but can be transformed. Nuclear processes convert mass to energy (mass-energy equivalence).

Energy can be transformed between potential energy and kinetic energy. Nuclear fusion is the process that produces the heat and light in stars. During this process matter is converted into energy.

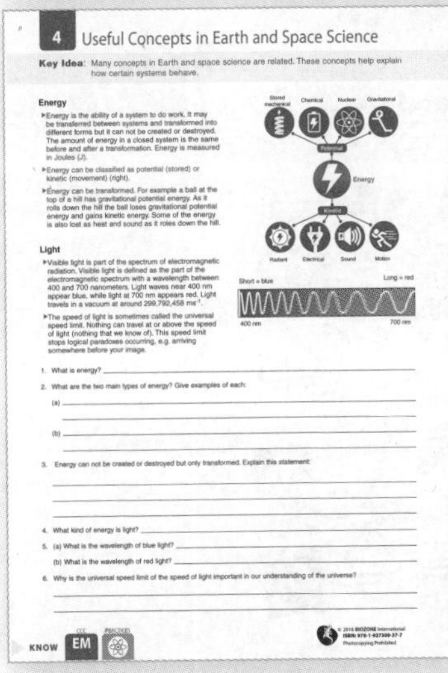
4 Useful Concepts in Earth and Space Science

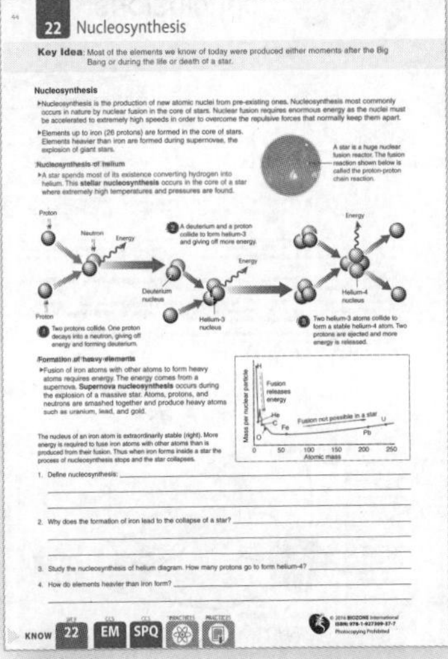
22 Nucleosynthesis

## Systems and system models

Making a model of a system (e.g. physical, mathematical) provides a way to understand and test ideas.

Models can be used to understand how a system works. Climate change models have become increasing sophisticated as new data and modeling techniques have become available.

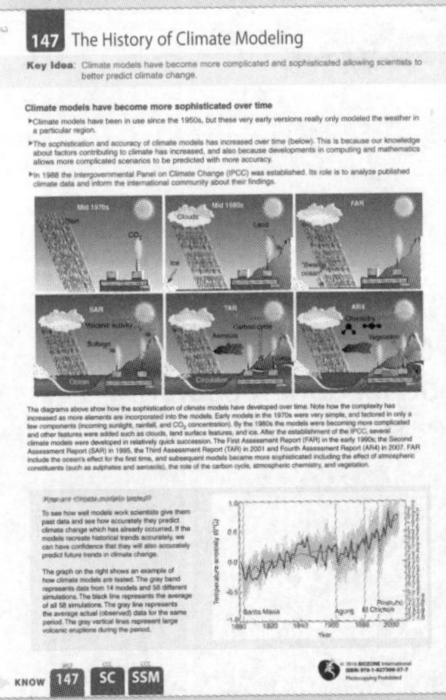
147 The History of Climate Modeling

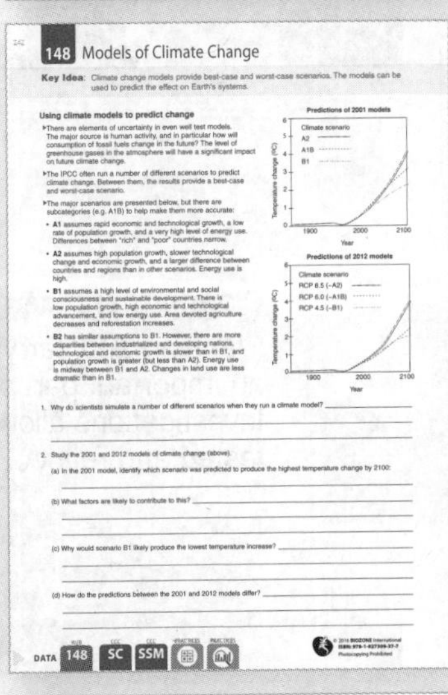
148 Models of Climate Change

CCC
SF

## Structure and function

The functions of objects and systems can be inferred from their structure.

The hydrologic cycle functions to move water through both the biosphere and geosphere. Like many other natural systems (e.g. the rock cycle) it is cyclical and different stages take different amounts of time. Water plays an important role in the structure and stability of soil.

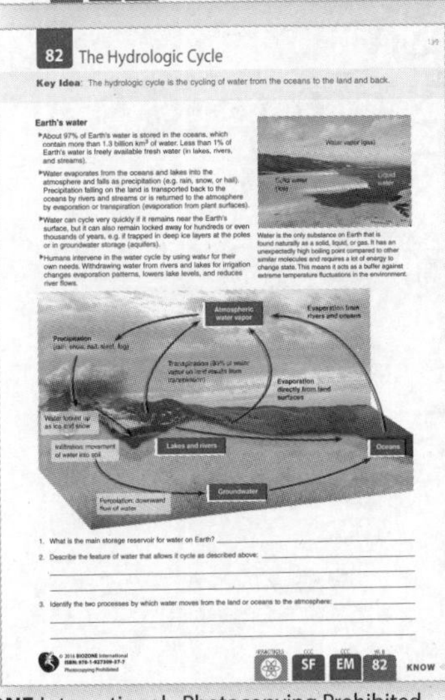
82 The Hydrologic Cycle

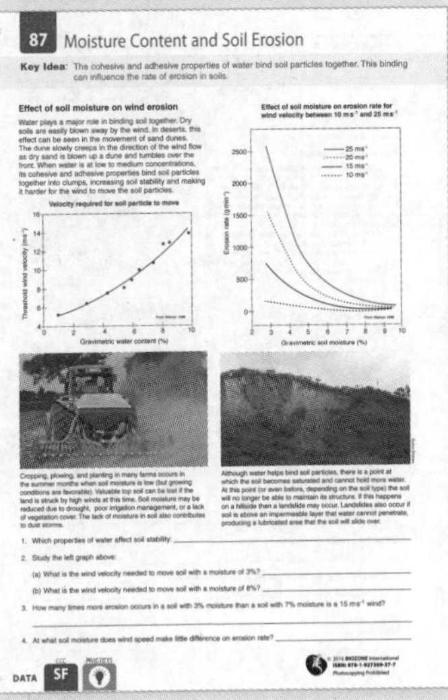
87 Moisture Content and Soil Erosion

# Addressing Science and Engineering Practices

Science and Engineering Practices for NGSS are supported throughout the workbook, beginning with an introductory chapter covering basic computational, analytical, and design skills, to the completion of activities focusing on the development of specific skills within the framework of the DCIs. The learning aims associated with the science and engineering practices for each chapter are identified in each chapter introduction, together with the activities that relate to those aims. These provide the student with a concrete way in which to identify and gain competence in specific skills. We have supported some of the practices by providing a template example for students to work through. This provides them with the necessary background to design and complete their own 'hands-on investigation.

PRACTICES

## Asking questions and defining problems

Asking scientific questions about observations or content in texts helps to define problems and draw valid conclusions.

PRACTICES

## Developing and using models

Models take on many forms. They can be used to represent a system or a part of a system. Using models can help to visualize a structure, process, or design and understand how it works. Models can also be used to improve a design.

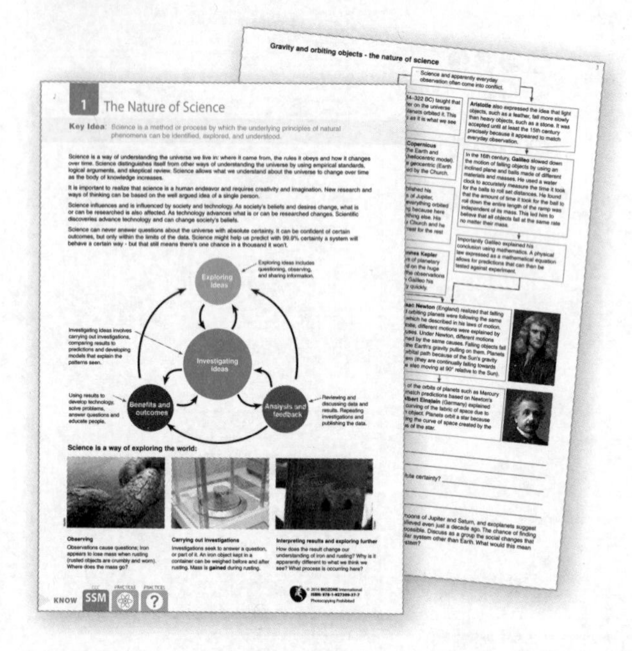

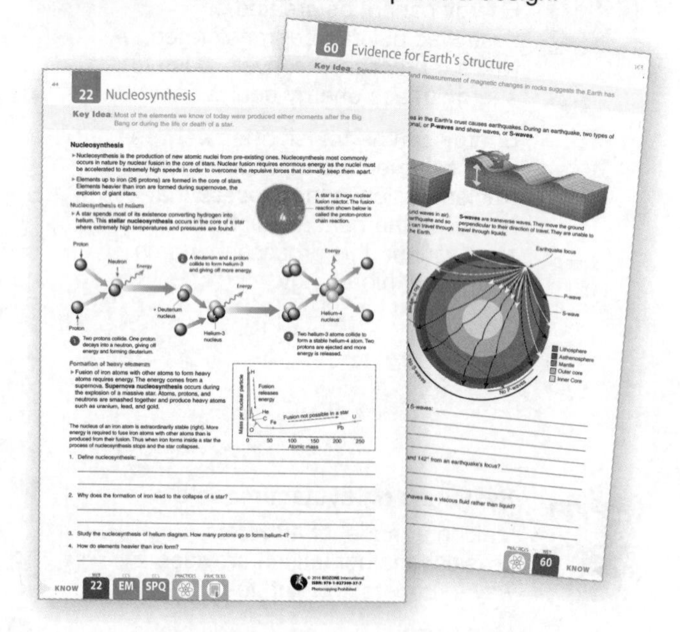

PRACTICES

## Planning and carrying out investigations

Planning and carrying out investigations is an important part of independent research. Investigations allow ideas and models to be tested and refined.

PRACTICES

## Analyzing and interpreting data

Once data is collected it must be analyzed to reveal any patterns or relationships.

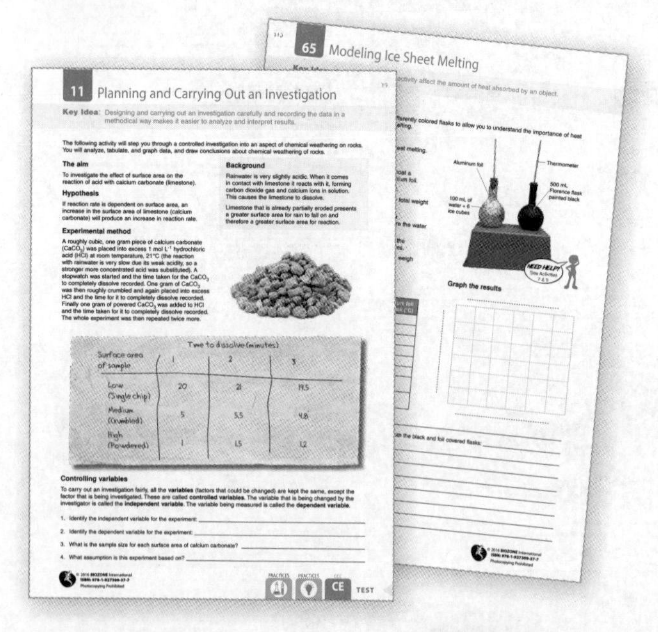

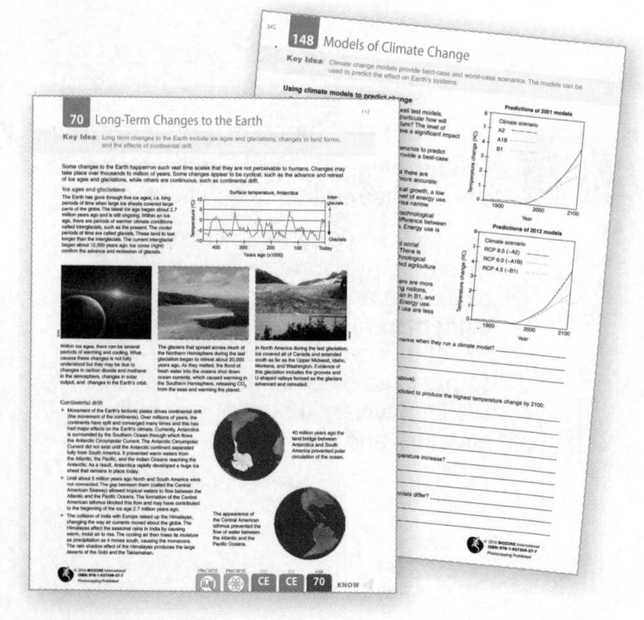

## Using mathematics and computational thinking

Mathematics is a tool for understanding scientific data. Representing the laws of a natural system as a mathematical equation allows us to calculate the properties of many different variations of the system.

## Constructing explanations and designing solutions

Constructing explanations for observations and phenomena is an important part of science. It is a dynamic process and may involve drawing on existing knowledge as well as generating new ideas before an observation is explained or a problem is solved.

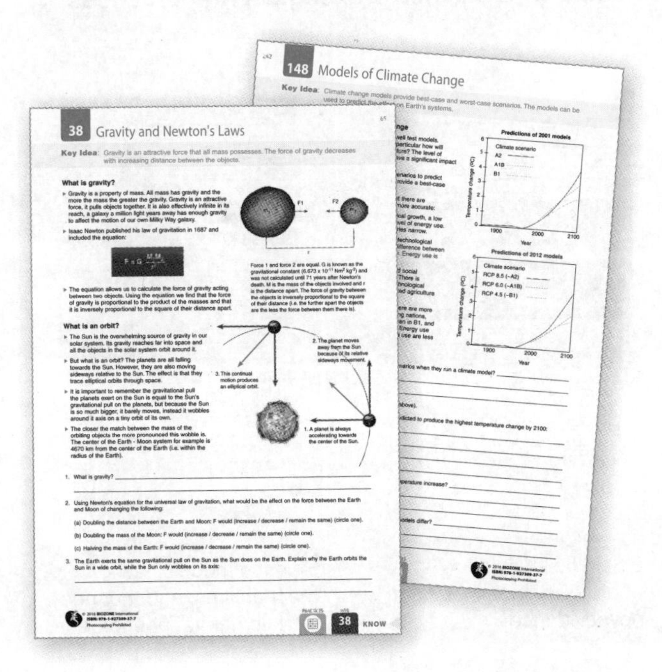

## Engaging in argument from evidence

Scientific argument based on evidence is an important part of gaining acceptance of new ideas in science. Logical reasoning based on empirical evidence is required when considering the merit of new claims or explanations of phenomena.

## Obtaining, evaluating, and communicating information

Evaluating information for scientific accuracy or bias is important in determining its validity and reliability. Communicating information in an effective way includes reports, graphics, oral presentation, and models. Visual models are a useful way of communicating complex scientific ideas.

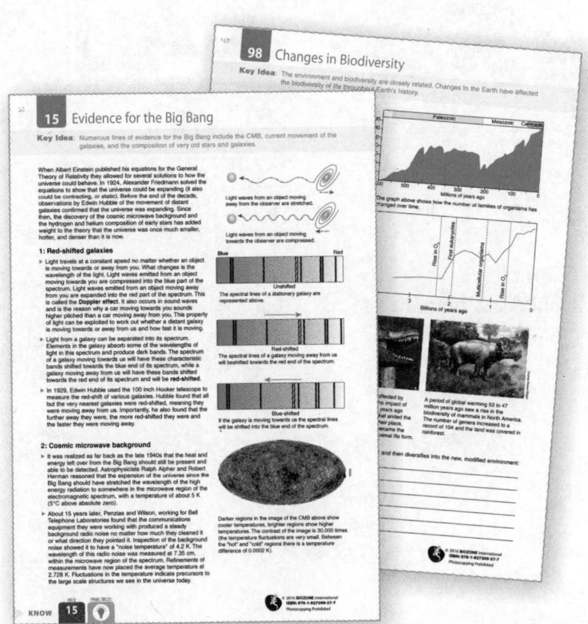


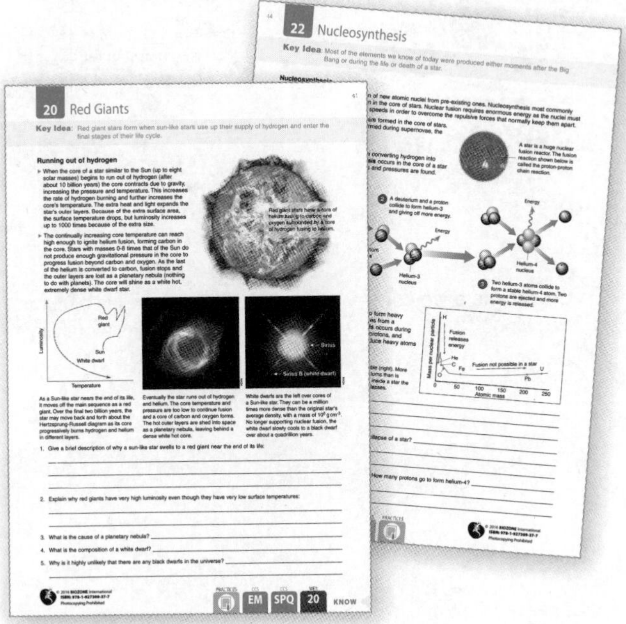

# The Nature of Science

The Nature of Science combines established information with new knowledge to constantly refine what we know about the natural world. Eight Nature of Science understandings are presented in the NGSS document. Four are associated most closely with Science and Engineering Practices, and four with the Crosscutting Concepts. Because the Nature of Science understandings have been incorporated into most activities in the *Earth and Space Sciences for NGSS Student Edition* we have not identified them specifically on the activity page. Some examples of activities relating to the eight Nature of Science understandings are illustrated below.

## Nature of science understandings most closely associated with science and engineering practices

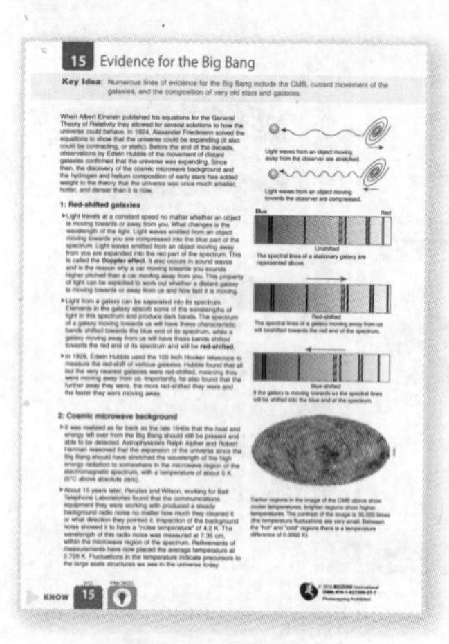
15 Evidence for the Big Bang

Scientific investigations use a variety of methods.

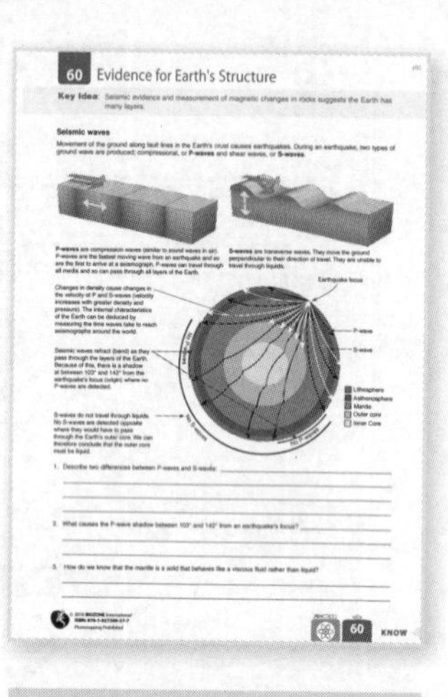
60 Evidence for Earth's Structure

Scientific knowledge is based on empirical evidence.

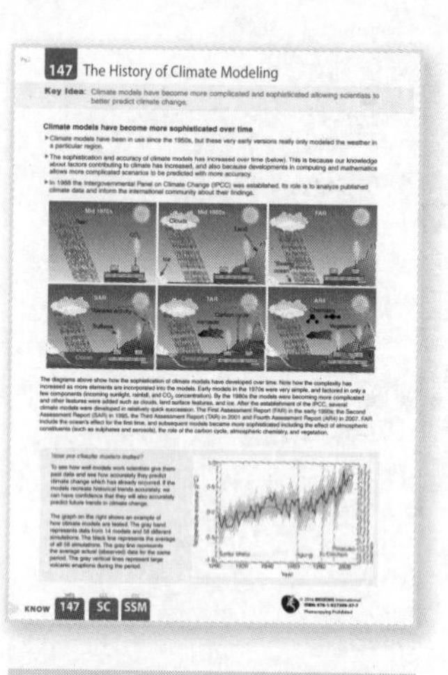
147 The History of Climate Modeling

Scientific knowledge is open to revision in light of new evidence.

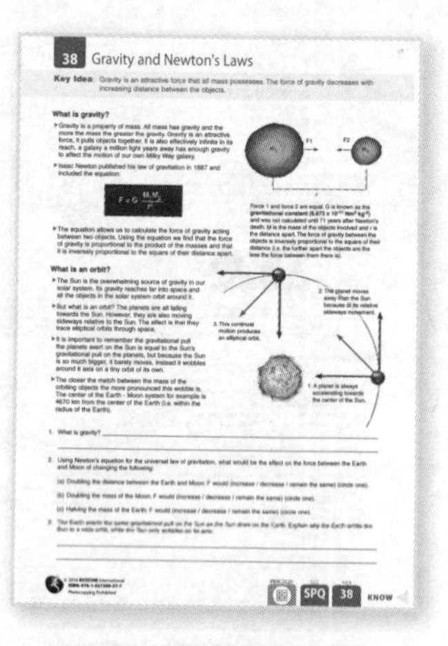
38 Gravity and Newton's Laws

Science models, laws, mechanisms, and theories explain natural phenomena.

## Nature of science understandings most closely associated with crosscutting concepts

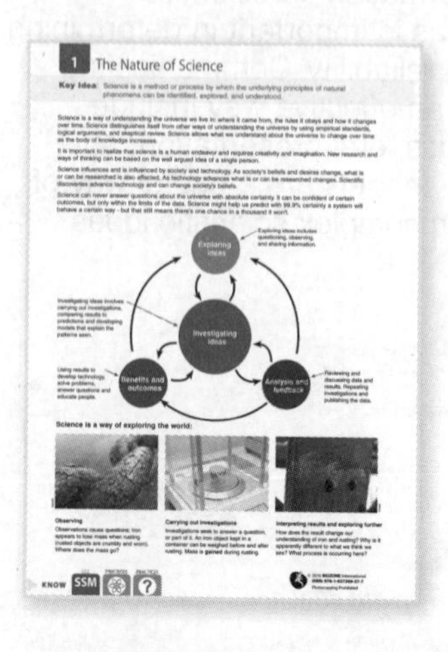
1 The Nature of Science

Science is a way of knowing.

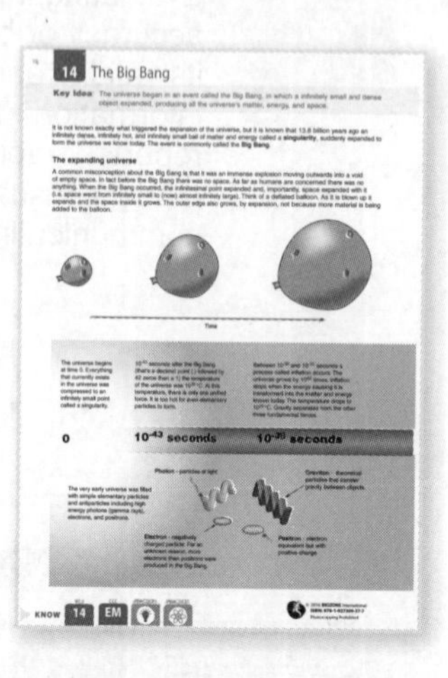
14 The Big Bang

Scientific knowledge assumes an order and consistency in natural systems.

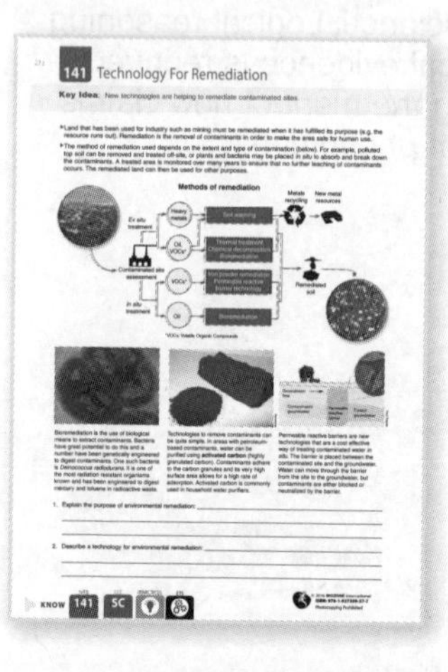
141 Technology For Remediation

Science is a human endeavor.

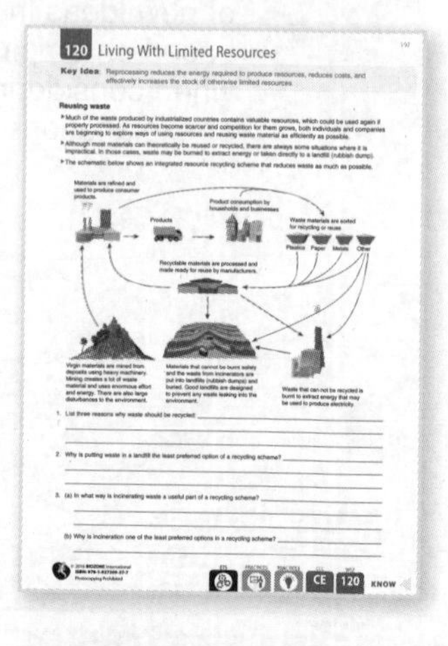
120 Living With Limited Resources

Science addresses questions about the natural and material world.

# Engineering, Technology, and Applications of Science

Activities designed for Engineering, Technology, and Applications of Science (ETS) examine global challenges affecting society. They require you to design and evaluate solutions incorporating knowledge gained through science and engineering. Solutions require consideration of cost, safety, reliability, aesthetics, as well as social, cultural, and environmental impacts. Earth and Space Sciences for NGSS has many opportunities for you to design, model, and evaluate technological solutions to the problems facing humanity today.

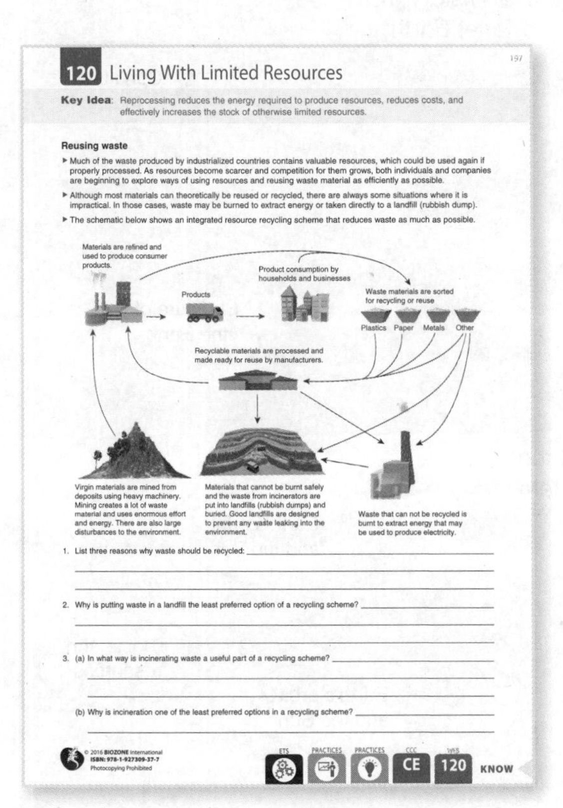

120 Living With Limited Resources

Key Idea: Reprocessing reduces the energy required to produce resources, reduces costs, and effectively increases the stock of otherwise limited resources.

Explore how technology is utilized to make use of the limited resources on Earth.

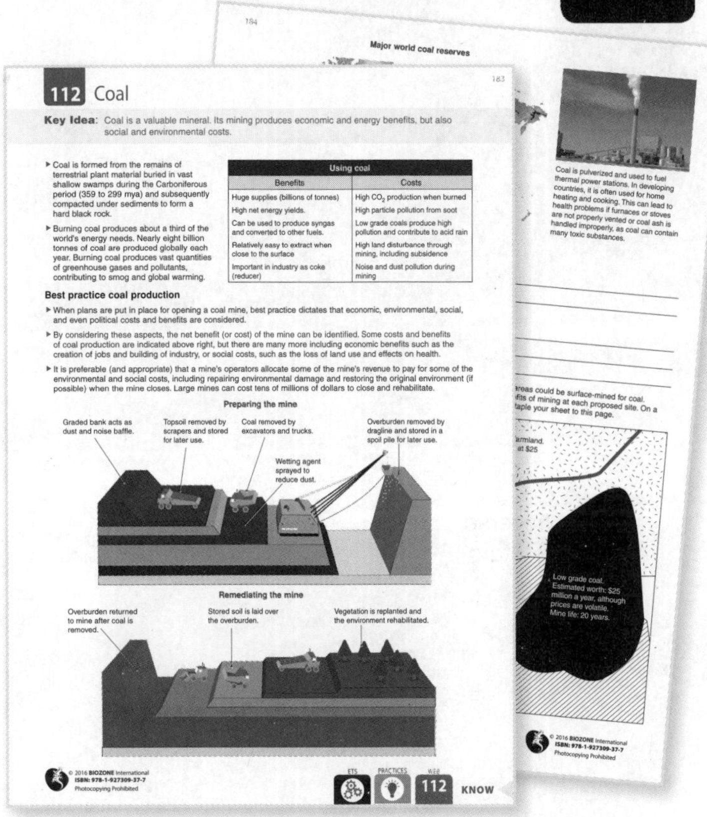

112 Coal

Key Idea: Coal is a valuable mineral. Its mining produces economic and energy benefits, but also social and environmental costs.

Utilize your knowledge of fossil fuels and their extraction to evaluate the benefits and disadvantages of mining coal from two sites.

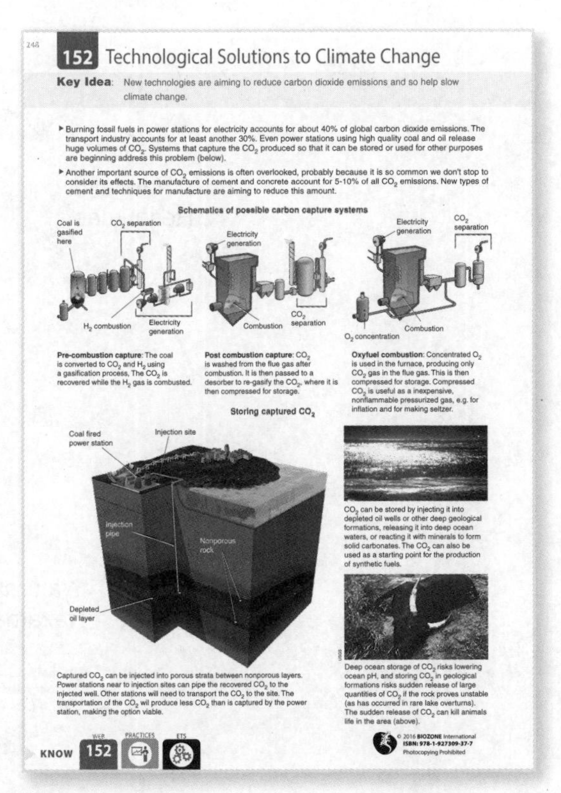

152 Technological Solutions to Climate Change

Key Idea: New technologies are aiming to reduce carbon dioxide emissions and so help slow climate change.

New technologies are aiming to reduce carbon dioxide emissions and so help slow climate change.

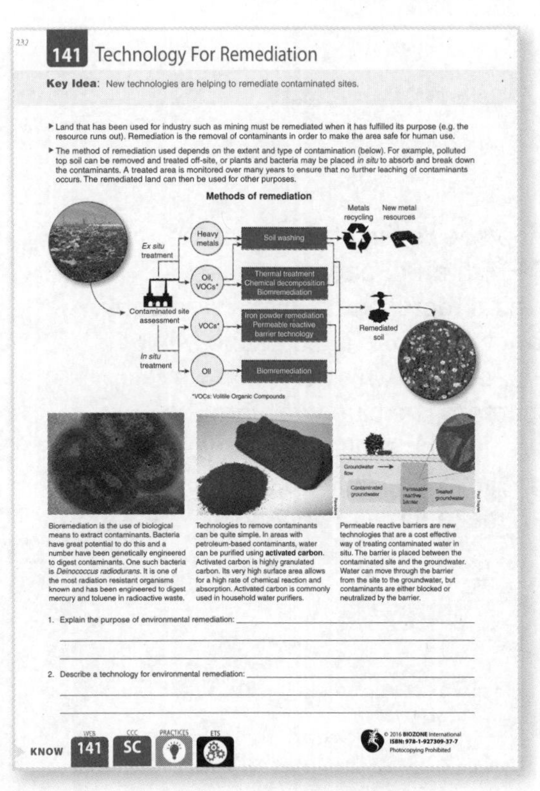

141 Technology For Remediation

Key Idea: New technologies are helping to remediate contaminated sites.

Technological solutions are used to clean up contaminated land so that it can be used again.

# Earth and Space Sciences: Concepts and Connections

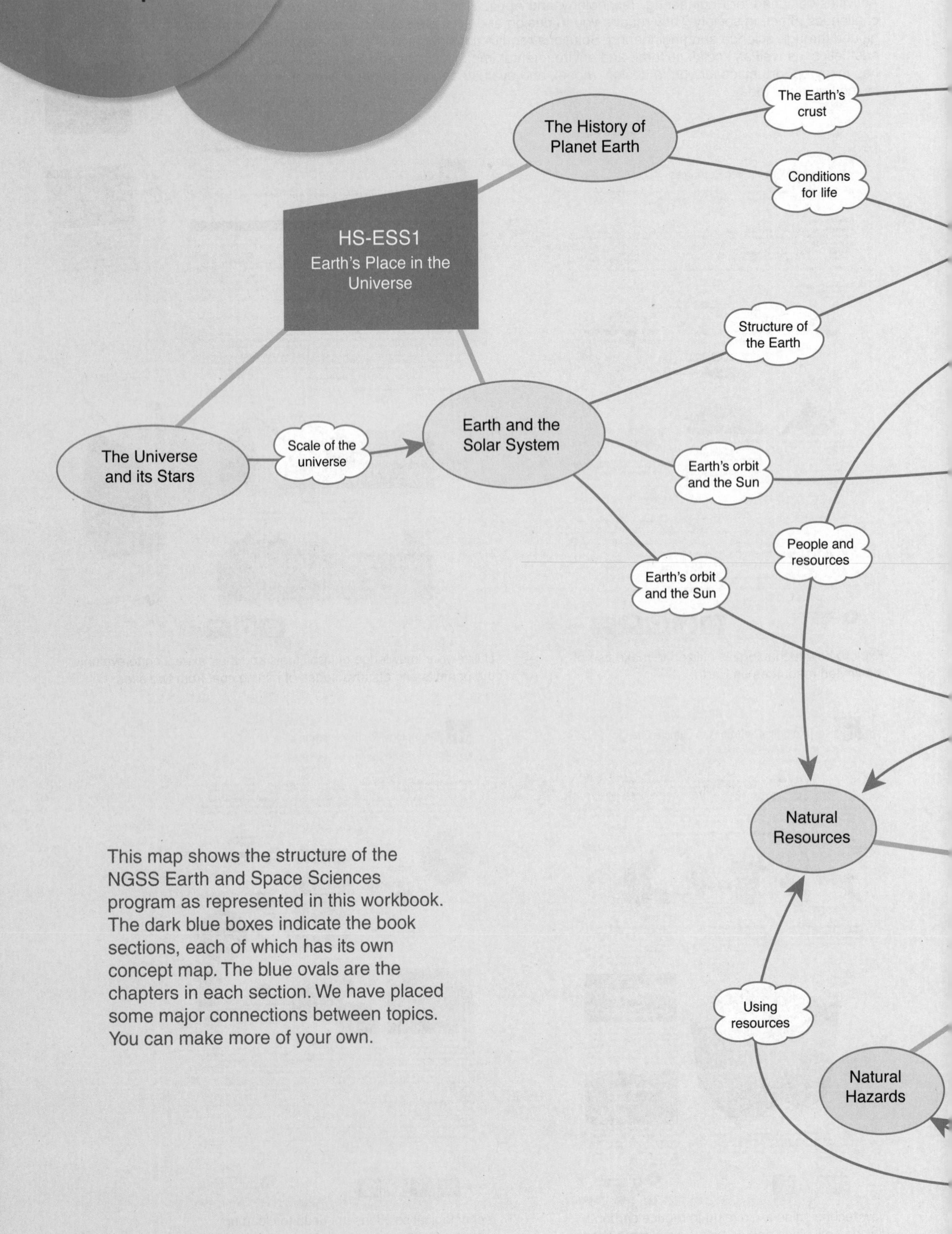


This map shows the structure of the NGSS Earth and Space Sciences program as represented in this workbook. The dark blue boxes indicate the book sections, each of which has its own concept map. The blue ovals are the chapters in each section. We have placed some major connections between topics. You can make more of your own.

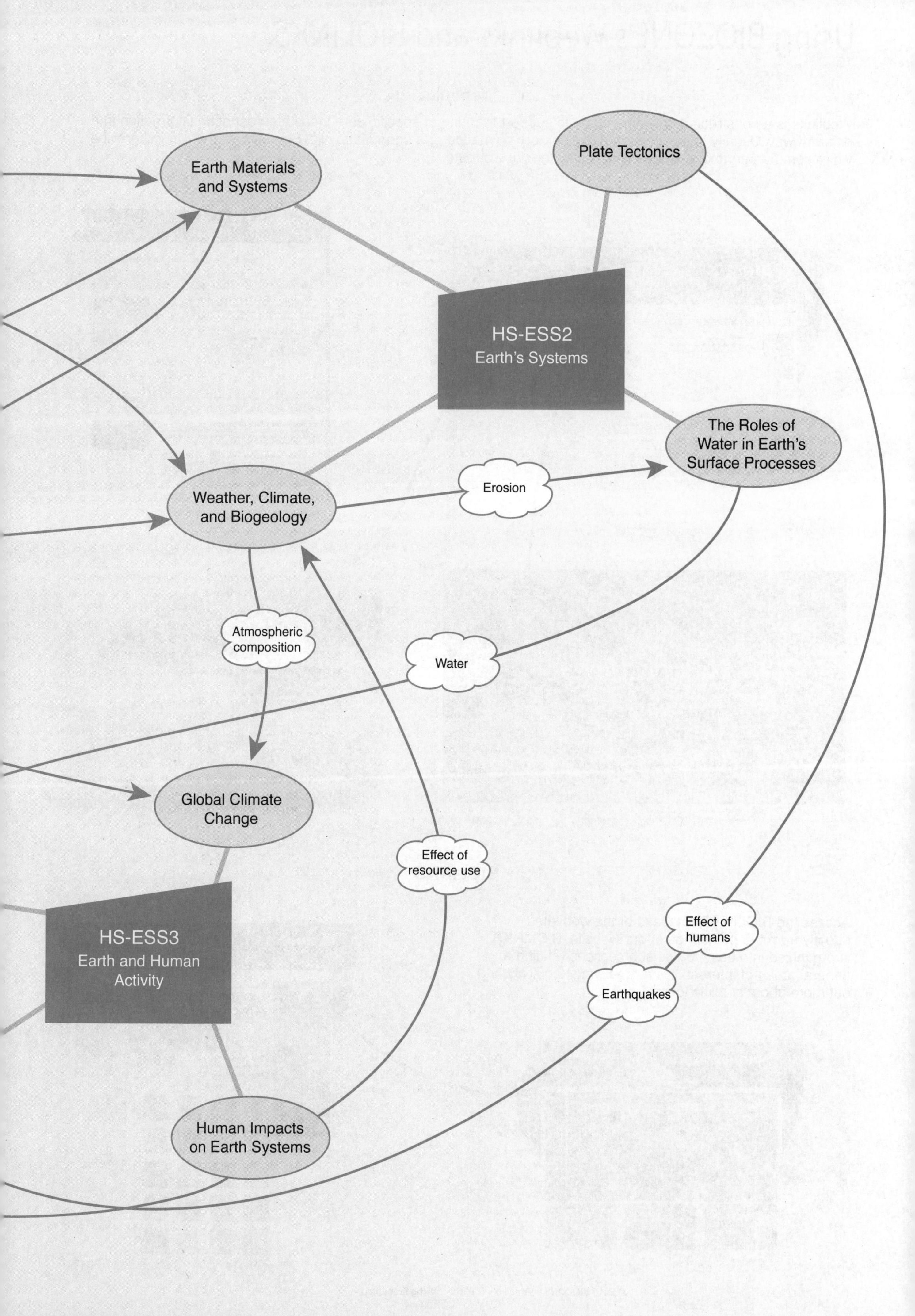

Earth Materials and Systems
Plate Tectonics
HS-ESS2
Earth's Systems
The Roles of Water in Earth's Surface Processes
Erosion
Weather, Climate, and Biogeology
Atmospheric composition
Water
Global Climate Change
Effect of resource use
Effect of humans
HS-ESS3
Earth and Human Activity
Earthquakes
Human Impacts on Earth Systems

# Using BIOZONE's Weblinks and BIOLINKS

## Weblinks

Weblinks is a constructed online resource to support learning of specific core ideas by presenting information in a different way. Usually this is through a explanatory animation or a short video clip. Sometimes, the clip will provide video commentary that provides opportunity for class debate.

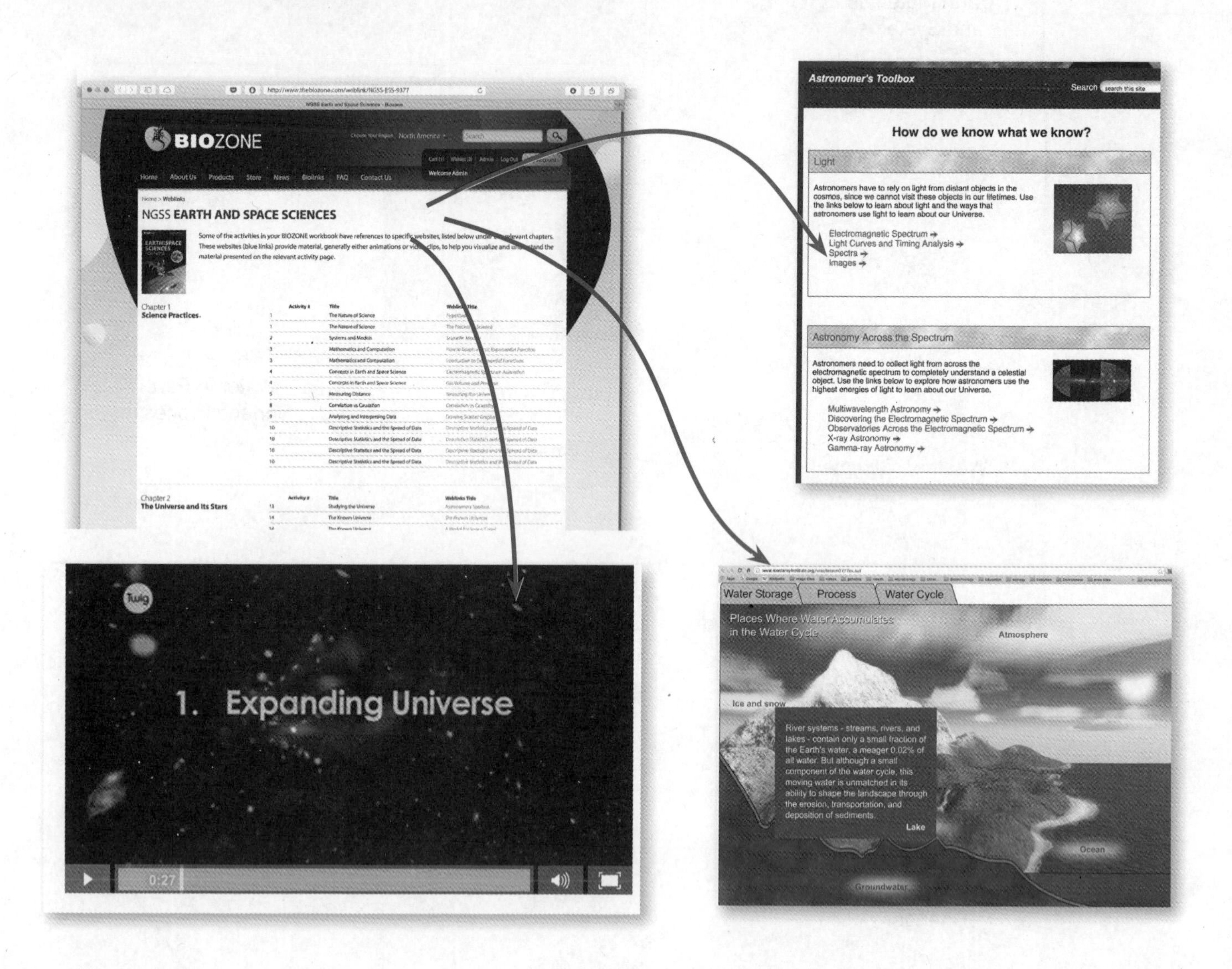


## BIOLINKS

Access the BIOLINKS database of the web sites directly from the homepage of our website. BIOLINKS is organized into easy-to-use sub-sections relating to general areas of interest. It is a great way to quickly find out more about specific topics.

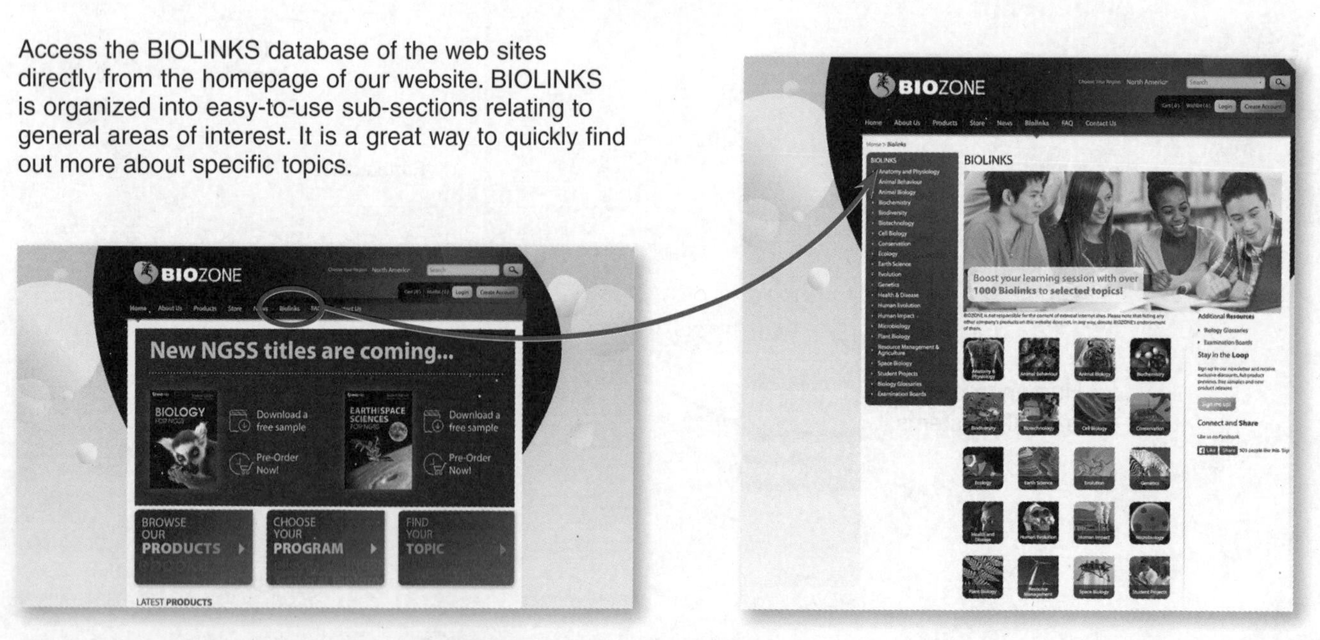

SEPs Background

# Science Practices

## Key terms

accuracy
assumption
control
controlled variable
data
descriptive statistics
dependent variable
graph
hypothesis
independent variable
mean
median
mode
model
observation
precision
prediction
qualitative data
quantitative data
raw data
scientific method
table
variable

## Science and engineering practices

***Supported as noted and throughout subsequent chapters in context***

| | Practice | Activity number |
|---|---|---|
| | **Ask questions and define problems** | |
| ☐ | 1 Demonstrate an understanding of science as inquiry. Appreciate that unexpected results may lead to new hypotheses and to new discoveries. | 1 |
| ☐ | 2 Formulate and evaluate questions that you can feasibly investigate. | 6 |
| | **Develop and use models** | |
| ☐ | 3 Develop and use models based on evidence to describe systems or their components and how they work. | 1 2 4 |
| | **Plan and carry out investigations** | |
| ☐ | 4 Plan and conduct investigations to provide data to test a hypothesis based on observations. Identify any assumptions in the design of your investigation | 11 |
| ☐ | 5 Consider and evaluate the accuracy and precision of the data that you collect. | 10 11 |
| ☐ | 6 Use appropriate tools to collect and record data. Data may be quantitative (continuous or discontinuous), qualitative, or ranked. | 11 |
| ☐ | 7 Variables are factors that can change or be changed in an experiment. Make and test hypotheses about the effect on a dependent variable when an independent variable is manipulated. Understand and use controls appropriately. | 11 |
| | **Analyze and interpret data** | |
| ☐ | 8 Use graphs appropriate to the data to visualize data and identify trends. | 8-11 |
| ☐ | 9 Summarize data and describe its features using descriptive statistics. | 10 |
| ☐ | 10 Apply concepts of statistics and probability to answer questions and solve problems. | 10 |
| | **Use mathematics and computational thinking** | |
| ☐ | 11 Demonstrate an ability to use mathematics and computational tools to analyze, represent, and model data. Recognize and use appropriate units in calculations. | 3 5 7 |
| ☐ | 12 Demonstrate an ability to apply ratios, rates, percentages, and unit conversions. | 3 5 |
| | **Construct explanations and design solutions** | |
| ☐ | 13 Explain results based on evidence and application of scientific ideas and principles. | 9 11 |
| | **Engage in argument from evidence** | |
| ☐ | 14 Use evidence to defend and evaluate claims and explanations about science. | 9 |
| | **Obtain, evaluate, and communicate information** | |
| ☐ | 15 Evaluate the validity and reliability of designs, methods, claims, and evidence. | 11 |

# 1 The Nature of Science

**Key Idea**: Science is a method or process by which the underlying principles of natural phenomena can be identified, explored, and understood.

Science is a way of understanding the universe we live in: where it came from, the rules it obeys and how it changes over time. Science distinguishes itself from other ways of understanding the universe by using empirical standards, logical arguments, and skeptical review. What we understand about the universe changes over time as the body of knowledge increases.

Science is a human endeavor and requires creativity and imagination. New research and ways of thinking can be based on the well argued idea of a single person.

Science influences and is influenced by society and technology. As society's beliefs and desires change, what is or can be researched is also affected. As technology advances, what is or can be researched changes. Scientific discoveries advance technology and can change society's beliefs.

Science can never answer questions about the universe with absolute certainty. It can be confident of certain outcomes, but only within the limits of the data. Science might help us predict with 99.9% certainty a system will behave a certain way, but that still means there's one chance in a thousand it won't.

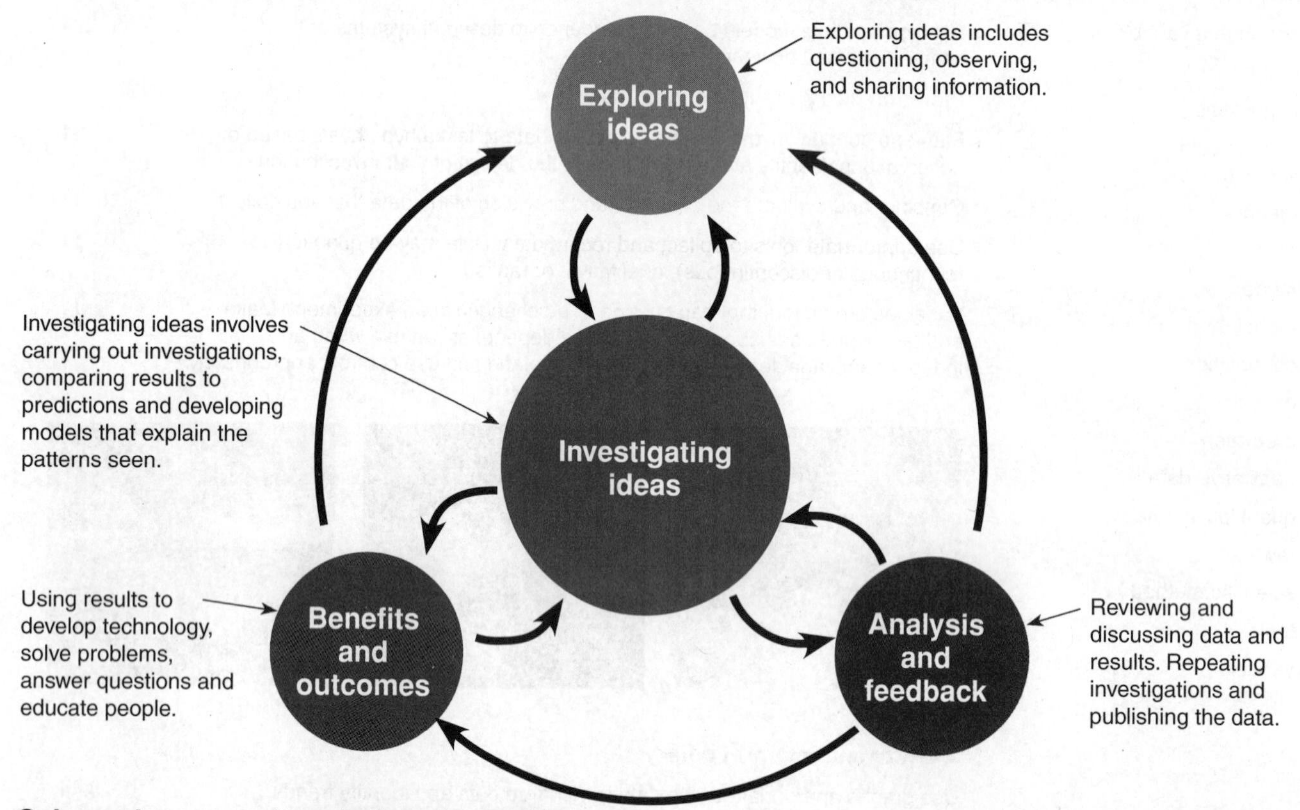


## Science is a way of exploring the world

**Observing**

Observations cause questions; Iron appears to lose mass when rusting (rusted objects are crumbly and worn). Where does the mass go?

**Carrying out investigations**

Investigations seek to answer a question, or part of it. An iron object kept in a container can be weighed before and after rusting. Mass is **gained** during rusting.

**Interpreting results and exploring further**

How does the result change our understanding of iron and rusting? Why is it apparently different to what we think we see? What process is occurring here?

**ISBN: 978-1-927309-37-7**

## Gravity and orbiting objects - the nature of science

Science and apparently everyday observation often come into conflict.

Knowledge was based on observation and experience. Ideas were not tested under controlled conditions.

**Aristotle** (Greece) (384–322 BC) taught that the Earth was the center on the universe and that the Sun and planets orbited it. This seemed plainly obvious as it is what we see standing here on Earth.

Many ideas that were at odds with social and political beliefs were oppressed, often ruthlessly.

Around 1514 **Nicolaus Copernicus** (Poland) proposed that the Earth and planets orbited the Sun (heliocentric model). This was at odds with the geocentric (Earth centered) model supported by the Church.

The testing of ideas under controlled conditions allows the underlying principles to be discovered.

In 1610, **Galileo** (Italy) published his observations of the moons of Jupiter, arguing that the idea that everything orbited the Earth was wrong because here were objects orbiting something else. His ideas were at odds with the Church and he spent the rest of his life under house arrest.

As new ideas based on facts obtained by controlled experiment became more widely accepted, social and political ideas began to change.

Around the same time, **Johannes Kepler** (Germany) published his laws of planetary motion around the Sun based on observational data collected by astronomer Tycho Brahe. In contrast to Galileo, his ideas were accepted relatively quickly.

**Aristotle** also expressed the idea that light objects, such as a feather, fall more slowly than heavy objects, such as a stone. It was accepted until at least the 15th century precisely because it appeared to match everyday observation.

In the 16th century, **Galileo** slowed down the motion of falling objects by using an inclined plane and balls made of different materials and masses. He used a water clock to accurately measure the time it took for the balls to roll set distances. He found that the amount of time it took for the ball to roll down the entire length of the ramp was independent of its mass. This led him to believe that all objects fall at the same rate no matter their mass.

Importantly Galileo explained his conclusion using mathematics. A physical law expressed as a mathematical equation allows for predictions that can then be tested experimentally.

As more information becomes available phenomena that were once thought to be separate e.g. falling objects and planetary orbits, can now be described under one over arching principle, e.g. gravity.

In 1687 **Isaac Newton** (England) realized that falling bodies and orbiting planets were following the same principles, which he described in his laws of motion. Under Aristotle, different motions were explained by different causes. Under Newton, different motions were explained by the same causes. Falling objects fall because of the Earth's gravity pulling on them. Planets follow their orbital path because of the Sun's gravity pulling on them (they are continually falling towards the Sun while also moving at 90° relative to the Sun).

New mathematics, knowledge, and inspiration can redefine whole areas of science and society. Einstein's theory of relativity was revolutionary and has stood up to every scientific test, including gravitational waves.

Observations of the orbits of planets such as Mercury didn't always match predictions based on Newton's law. In 1905 **Albert Einstein** (Germany) explained gravity as the curving of the fabric of space due to the mass of an object. Planets orbit a star because they are following the curve of space created by the enormous mass of the star.

1. Describe how science is a tool for understanding the universe: ______________________________

2. Why can science not predict with absolute certainty how a system will behave? ______________________________

3. New discoveries on other planets and moons including Mars, the moons of Jupiter and Saturn, and exoplanets, suggest the elements and conditions for life are more common than was believed even just a decade ago. The chance of finding life of some kind somewhere other than Earth seems increasingly possible. As a group, discuss the social changes that might occur if life (even simple life) was found somewhere in the solar system other than Earth. What would this mean for us as humans and as members of the population of the solar system?


**ISBN: 978-1-927309-37-7**

# 2 Systems and Models

**Key Idea**: A system is an assemblage of connected components. A model is a way of explaining how the components of a system interact.

**Systems** are assemblages of interrelated components working together by way of a driving force. A simple example of a system is our eight-planet solar system. Each of the planet's orbits represents a single component of the system. The driving force of the system is gravity from the Sun.

Modeling systems helps to understand how they work. A **model** is a representation of an object or system that shares important characteristics with the object or system being studied. A model does not necessarily have to incorporate all the characteristics or be fully accurate to be useful. It depends in the level of understanding required.

## Modeling data

- Models are extremely important when trying to understand how a system operates. Models are useful for breaking complex systems down into manageable parts and often only part of a system is modeled at a time. As understanding of the system progresses, more and more data can be built into the model so that it more closely represents the real world system or object.
- A common example is the use of models to represent atoms. The three illustrations below become more complex from left to right.

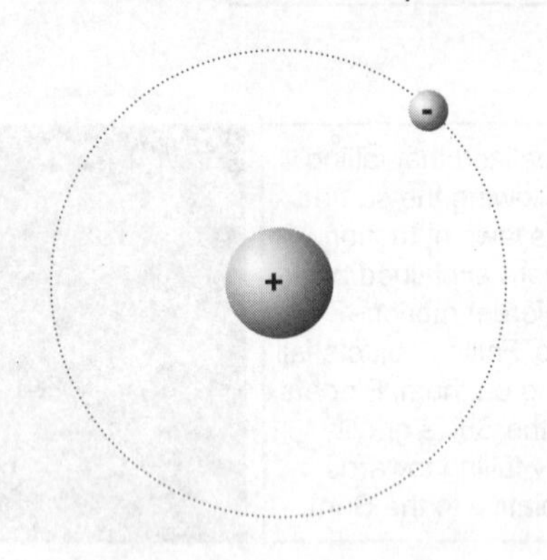

Rutherford's atomic model showing position of charge

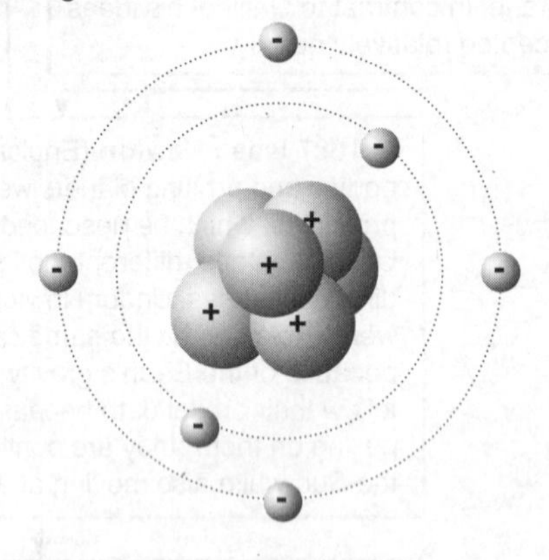

Bohr atomic model showing electron orbitals (2D)

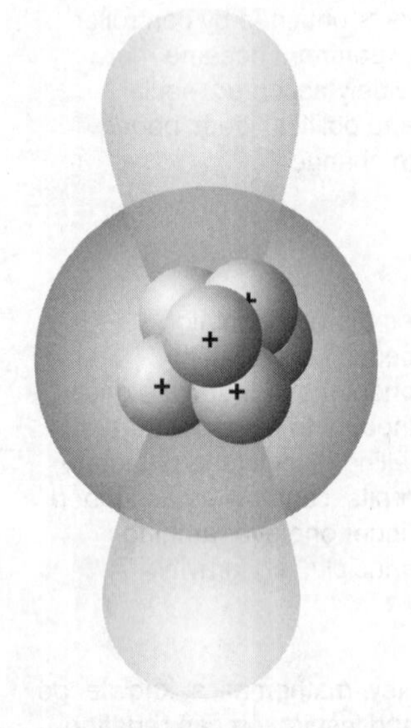

Atomic model using quantum mechanics to show 3D electron clouds.

### Modeling the weather

Weather systems are modeled in order to predict future weather. A weather model has several parts to it where current weather data can be entered. Computers are used to run the model under different scenarios. The outcomes are used to predict the most likely weather pattern in the future.

### Modeling the Earth's structure

The Earth's structure can be modeled using data from earthquakes, Earth's magnetic field, and volcanic activity. The Earth's magnetic field suggests an iron core surrounded by molten material. The bending of earthquake waves suggests a thin solid crust. Volcanic activity shows there is molten material below the crust.

### Modeling the evolution of stars

Models of stellar evolution are based on visual observation of numerous stars, their luminosity, measured surface color, and mass. Models of nuclear fusion are added to this data. Based on these observations, we can predict how a star of a certain mass will behave and develop over millions to billions of years.

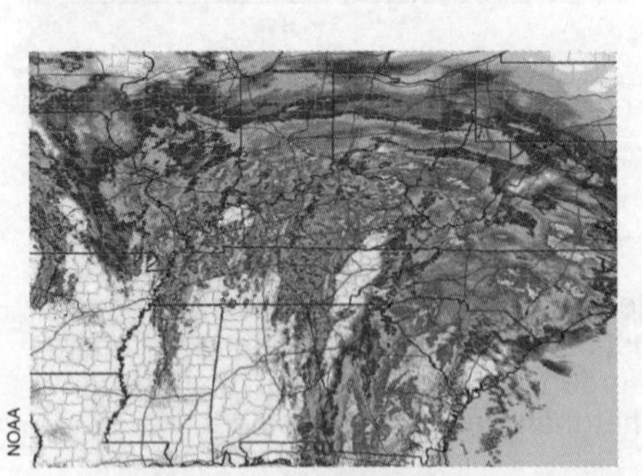


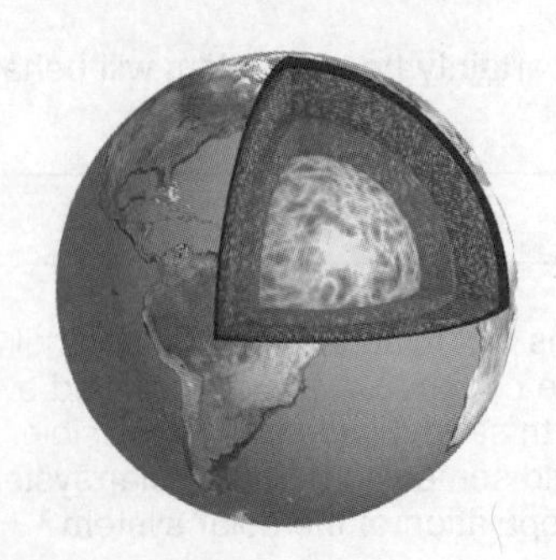

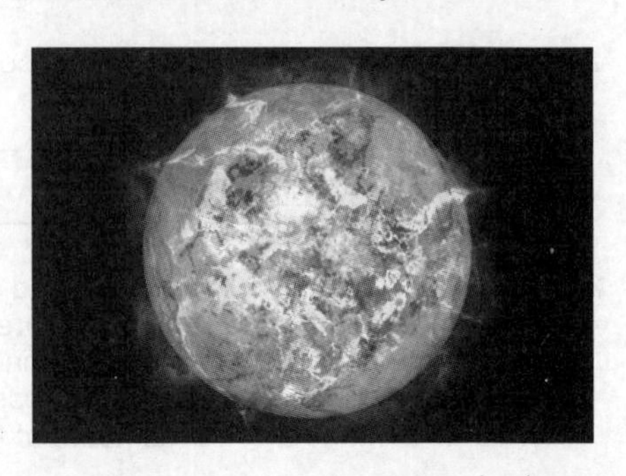

KNOW

CCC SSM

PRACTICES

**ISBN: 978-1-927309-37-7**

## Closed, open, and isolated systems

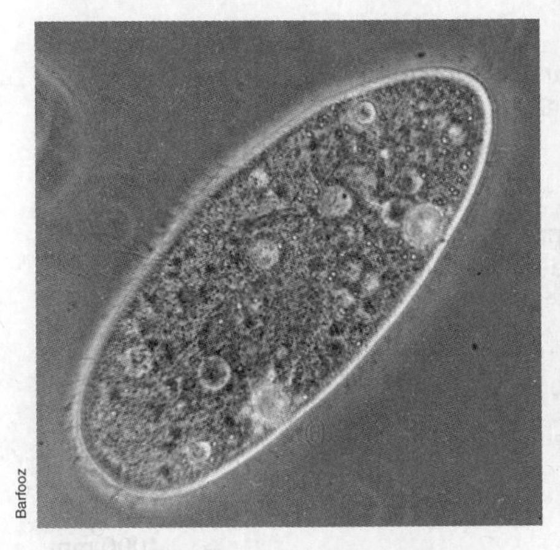

**Open systems** are able to exchange matter, energy and information with their surroundings. This causes them to be constantly changing, although the overall processes and outcomes remain relatively constant. Open systems are the most common type in natural systems. Examples include ecosystems, living organisms, and the ocean.

**Closed systems** exchange energy with their surroundings, but not matter. Closed systems are uncommon on Earth, although the cycling of certain materials, such as water and nitrogen, approximates them. The Earth itself is essentially a closed system. It receives energy from the Sun but exchanges virtually no matter with the universe (apart from the occasional meteorite).

**Isolated systems** exchange no energy, information or matter with their surroundings. No such systems are known to exist (except possibly the entire universe). Some natural systems approximate isolated systems, at least for certain lengths of time. The solar system is essentially isolated, as is the Milky Way galaxy if gravity from nearby stars or galaxies is ignored.

1. (a) What is a system? ______________________

   (b) What is a model? ______________________

   (c) Explain why models are never 100% accurate representations of the system being studied: ______________________

   (d) Discuss the advantages and disadvantages of using models to explain a system: ______________________

   (e) Why is it easier to use a series of simple models to explain a complex system than one complex model? ______________________

2. Identify each of the following as either an open, closed, or isolated system:

   (a) Reef ecosystem: ______________________

   (b) Nitrogen cycle: ______________________

   (c) Earth: ______________________

   (d) Biosphere: ______________________

   (e) Solar system: ______________________

   (f) Digestive system: ______________________

   (g) A national park: ______________________

   (h) A large lake: ______________________


**ISBN: 978-1-927309-37-7**

# 3 Mathematics and Computation

**Key Idea**: Using correct mathematical notation and being able to carry out simple calculations and conversions are fundamental skills in Earth and space sciences.

- Mathematics is used in Earth and space sciences to analyze, interpret, and compare data. It is important that you are familiar with mathematical notation (the language of mathematics) and can confidently apply some basic mathematical principles and calculations to your data.
- Much of our understanding of Earth and space science is based on our ability to use mathematics to interpret the patterns seen in collected data and express laws of the universe in simple notation.

### Commonly used mathematical symbols

In mathematics, universal symbols are used to represent mathematical concepts. They save time and space when writing. Some commonly used symbols are shown below.

$=$ Equal to

$<$ The value on the left is **less than** the value on the right

$>$ The value on the left is **greater than** the value on the right

$\propto$ Proportional to. $A \propto B$ means that A = a constant X B

$\sim$ Approximately equal to

$\infty$ Infinity

$\sqrt{b}$ The square root of b

$b^2$ b squared (b x b)

$b^n$ b to the power of n ( b x b... n times)

$\Delta$ The change in. For example $\Delta T / \Delta d$ = the change in T ÷ the change in d (see rates below right).

### Length

| | |
|---|---|
| Kilometer (km) | 1000 m |
| Meter (m) | 1000 mm |

### Volume

| | |
|---|---|
| Liter (L) | 1000 mL |
| Milliliter (mL) | = 1 $mm^3$ |

### Area

| | |
|---|---|
| Square kilometer | 1,000,000 $m^2$ |
| Hectare | 10,000 $m^2$ |
| Square meter | 1,000,000 $mm^2$ |

### Temperature

0°C = freezing point of pure water

100°C = boiling point of pure water

Kelvin scale (K) and °C have the same magnitude. Kelvin scale starts at absolute zero (–273.15 °C).

### Decimal and standard form

**Decimal form** (also called ordinary form) is the longhand way of writing a number (e.g. 15,000,000). Very large or very small numbers can take up too much space if written in decimal form and are often expressed in a condensed **standard form**. For example, 15,000,000 is written as $1.5 \times 10^7$ in standard form.

In standard form a number is always written as $A \times 10^n$, where A is a number between 1 and 10, and n (the exponent) indicates how many places to move the decimal point. n can be positive or negative.

For the example above, A = 1.5 and n = 7 because the decimal point moved seven places (see below).

$$15\,000\,000 = 1.5 \times 10^7$$

Small numbers can also be written in standard form. The exponent (n) will be negative. For example, 0.00101 is written as $1.01 \times 10^{-3}$.

$$0.00101 = 1.01 \times 10^{-3}$$

### Adding numbers in standard form

Numbers in standard form can be added together so long as they are both raised to the same power of ten. E.g: $1 \times 10^4 + 2 \times 10^3 = 1 \times 10^4 + 0.2 \times 10^4 = 1.2 \times 10^4$

### Rates

**Rates** are expressed as a measure per unit of time and show how a variable changes over time. Rates are used to provide meaningful comparisons of data that may have been recorded over different time periods.

Often rates are expressed as a mean rate over the duration of the measurement period, but it is also useful to calculate the rate at various times to understand how rate changes over time. The table below shows the distance traveled by a rolling ball. A worked example for the rate at 4 seconds is provided below.

| Time (s) | Distance traveled (m) | Rate of movement (speed) (m/s) |
|---|---|---|
| 0 | 0 | 0 |
| 2 | 34 | 17 |
| 4 | 42 | 4* |
| 6 | 48 | 3 |
| 8 | 50 | 1 |
| 10 | 50 | 0 |

* meters moved between 2 - 4 seconds: 42 m – 34 m = 8 m

Rate of movement (speed) between 2 - 4 seconds
8 m ÷ 2 seconds = 4 m/s

1. Use the information above to complete the following calculations:

(a) $\sqrt{9}$: ______________________

(b) $4^3$: ______________________

(c) Write 6,340,000 in standard form: ______________________

(d) Write 0.00103 in standard form: ______________________

(e) Convert 10 cm to millimeters: ______________________

(f) Convert 4 liters to milliliters: ______________________

(g) Write $7.82 \times 10^7$ as a number: ______________________

(h) $4.5 \times 10^4 + 6.45 \times 10^5$: ______________________

PRACTICES

DATA 

ISBN: 978-1-927309-37-7

## Dealing with large numbers

Earth and space sciences often deal with very large numbers or scales. Numerical data indicating scale can often increase or decrease exponentially. Large scale changes in numerical data can be made more manageable by transforming the data using logarithms.

### Exponential function

- Exponential growth or decay occurs at an increasingly rapid rate in proportion to the increasing or decreasing total number or size.
- In an exponential function, the base number is fixed (constant) and the exponent is variable.
- The equation for an exponential function is $y = c^x$.
- An example of exponential decay is radioactive decay. Any radioactive element has a half-life, the amount of time required for its radioactivity to fall to half its original value.

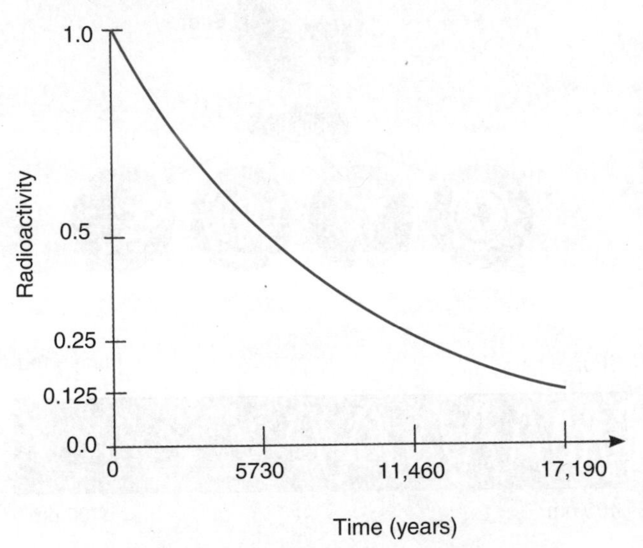


**Example above**: Carbon-14 ($^{14}C$) has a half life of 5730 years. If a sample with a mass of 10 g was left for 5730 years half the sample will have decayed, leaving 5 g of radioactive material. After another 5730 years, 2.5 g of radioactive carbon will be left.

### Log transformations

- A log transformation can make very large numbers easier to work with.
- The log of a number is the exponent to which a fixed value (the base) is raised to get that number. So $\log_{10}(1000) = 3$ because $10^3 = 1000$.
- Both $\log_{10}$ and $\log_e$ (natural logs or *ln*) are commonly used.
- Log transformations are useful for data where there is an exponential increase or decrease in numbers. In this case, the transformation will produce a straight line plot.
- To find the $\log_{10}$ of a number, e.g. 32, using a calculator, key in log 32 = . The answer should be 1.51.
- An example of a log scale is the Moment Magnitude scale used to measure the energy released during earthquakes. Each step of the scale is approximately $10^{1.5}$ times greater than the step below it. Calculating the **difference in energy released** between earthquakes can be done by finding the inverse $\log_{10}$ of the difference in magnitude ($10^{(1.5 \times (m1-m2))}$).
- Also the number of earthquakes around the world at each magnitude follows a negative logarithmic spread (below).

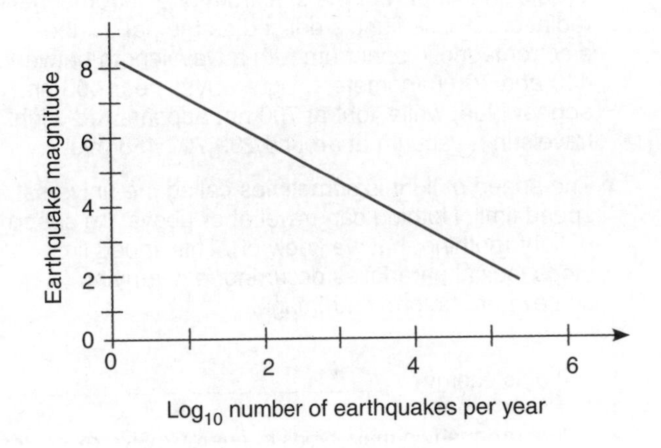


2. The Moment Magnitude scale is a measure of the energy released during an earthquake.

   (a) How many times more energy is released by the magnitude 6 earthquake than a magnitude 4 earthquake?

   (b) How many times more energy is released by the magnitude 7.5 earthquake than a magnitude 4.3 earthquake?

3. The pH scale measures the acidity of a substance. It is a negative logarithmic scale. A pH of 3 has a hydrogen ion concentration (which is responsible for acidity) ten times greater than a pH of 4.

   How many times greater is the hydrogen ion concentration of a pH 2 solution than a pH 6 solution?

4. Carbon-14 ($^{14}C$) is found in living organisms. It has a half life of 5730 years. When an organism dies it stops taking in $^{14}C$ and the ratio of $^{14}C$ to $^{12}C$ changes.
   Using these pieces of information explain how we can calculate how long ago an organism died:


**ISBN: 978-1-927309-37-7**

# 4 Useful Concepts in Earth and Space Science

**Key Idea**: Many concepts in Earth and space science are related. These concepts help explain how certain systems behave.

## Energy

- Energy is the ability of a system to do work. It may be transferred between systems and transformed into different forms but it can not be created or destroyed. The amount of energy in a closed system is the same before and after a transformation. Energy is measured in joules (J).
- Energy can be classified as potential (stored) or kinetic (movement) (right).
- Energy can be transformed. For example a ball at the top of a hill has gravitational potential energy. As it rolls down the hill the ball loses gravitational potential energy and gains kinetic energy. Some of the energy is also lost as heat and sound as it rolls down the hill.

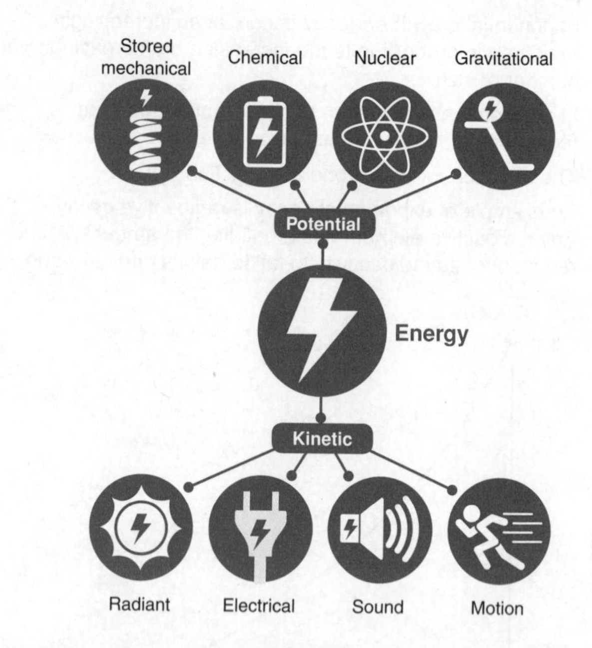


## Light

- Visible light is part of the spectrum of electromagnetic radiation. Visible light is defined as the part of the electromagnetic spectrum with a wavelength between 400 and 700 nanometers. Light waves near 400 nm appear blue, while light at 700 nm appears red. Light travels in a vacuum at around 299,792,458 m/s.
- The speed of light is sometimes called the universal speed limit. Nothing can travel at or above the speed of light (nothing that we know of). This speed limit stops logical paradoxes occurring, e.g. arriving somewhere before your image.

Short = blue | Long = red

400 nm | 700 nm

1. What is energy? ____________________

2. What are the two main types of energy? Give examples of each:

   (a) ____________________

   (b) ____________________

3. Energy can not be created or destroyed but only transformed. Explain this statement:

____________________

4. What kind of energy is light? ____________________

5. (a) What is the wavelength of blue light? ____________________

   (b) What is the wavelength of red light? ____________________

6. Why is the universal speed limit of the speed of light important in our understanding of the universe?

____________________

KNOW | CCC EM | PRACTICES


**ISBN: 978-1-927309-37-7**

## Temperature

- Temperature is a measure of the energy an object has. Molecules constantly vibrate (kinetic energy) and temperature is a measure of these vibrations. The faster and larger the vibrations the higher the object's temperature. Temperature rises as heat is added to a system and lowers when heat is removed.
- Temperature can be inferred from color. A heated piece of iron will initially glow red, then orange-yellow as it gets hotter, then white. The temperature of stars is linked to their color in a similar way (right).
- The coolest stars appear red. Stars like our Sun appear yellow. Even hotter stars emit white light, and the very hottest stars appear blue. Although the color of star can be difficult to see visually we can measure the peak wavelength of light and match that to a known temperature scale.

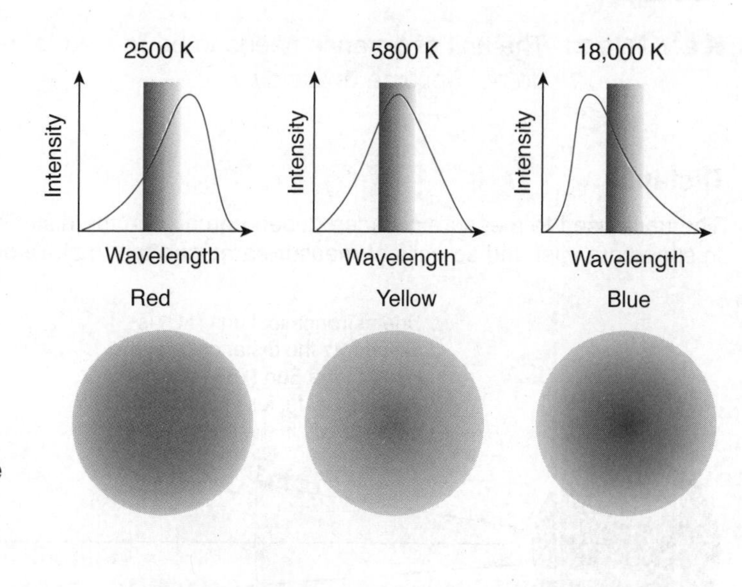


## Pressure

- Pressure is an important concept in Earth and space science. Pressure is a measure of the force being applied per unit surface area. The pressure of the air pressing on the surface of the Earth is 101.3 Pa (pascals). Pressure is important in the formation of rocks, the weather, and stars. Pressure and temperature are related by Gay-Lussac's Law, formulated in 1809, which states P (pressure) $\propto$ T (temperature) for a given mass and volume of gas.

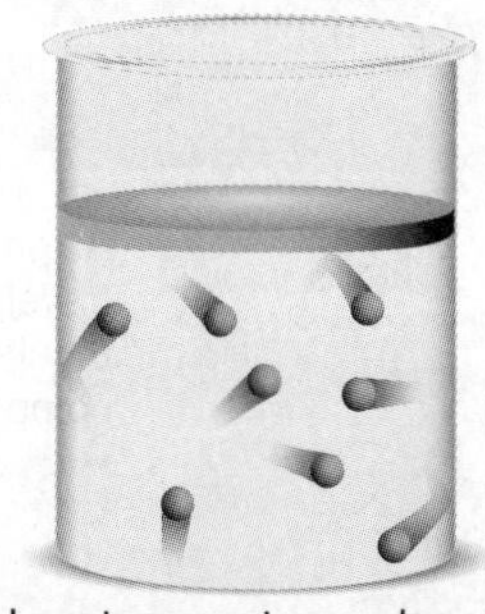

Low temperature - slow movement of gas particles

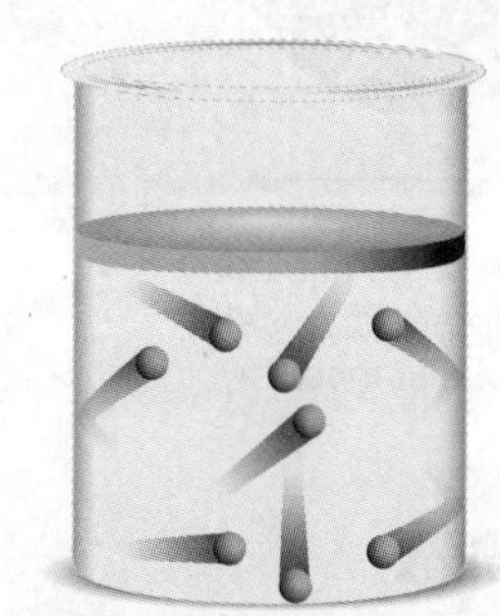

High temperature - fast movement of gas particles

7. What is temperature measuring? ______________________________

8. Explain how a star's temperature can be inferred from its color: ______________________________

9. What is pressure? ______________________________

10. (a) Explain why the pressure in a sealed container of gas increases when heat is added to the system: ______________________________

    (b) Explain why the pressure in a sealed container of gas decreases when the system is cooled: ______________________________


**ISBN: 978-1-927309-37-7**

# 5 Measuring Distance

**Key Idea**: The unit of distance needs to be related to the distance being measured so that numbers do not become unwieldy.

## Distance

The units used to measure distance depend greatly on the distance being measured. The distances between objects in space are vast and so units of measure are used that make working with these distances simpler.

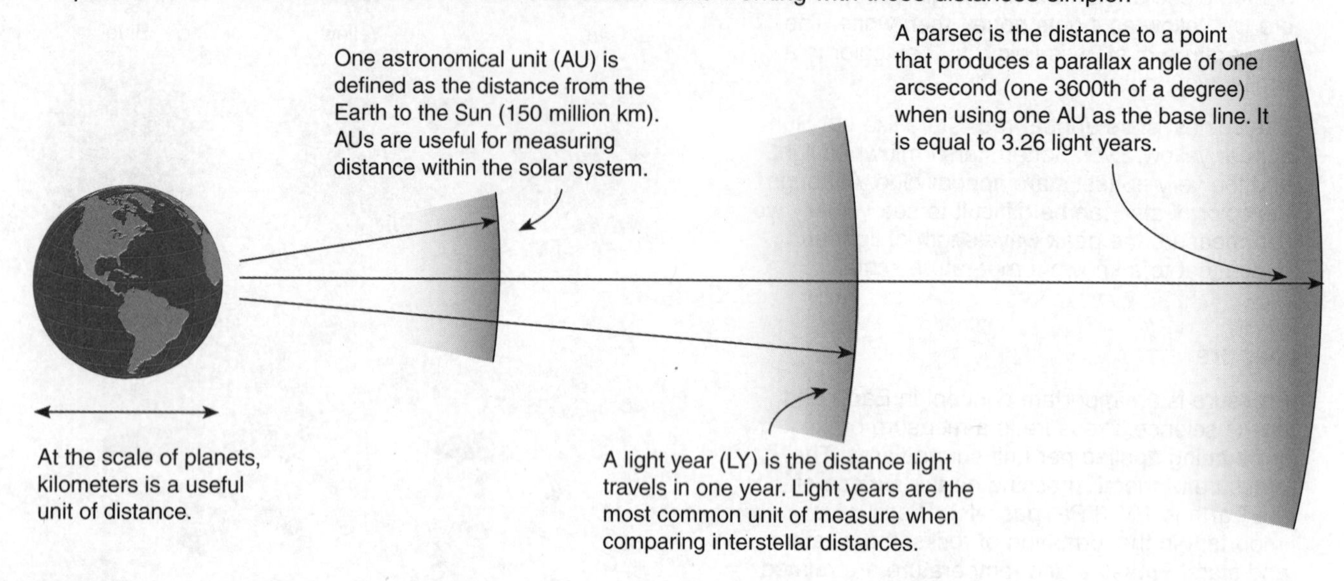


## Parallax

Parallax is the difference in apparent position of an object compared to background objects when viewed from different lines of sight. Parallax is a useful way of measuring the distance to the stars. The distance to a star can be calculated using the parallax angle and the distance from the Earth to the Sun.

The distance to a star can be calculated using the equation $d = 1/p$ where d is the distance in parsecs and p is the parallax angle.

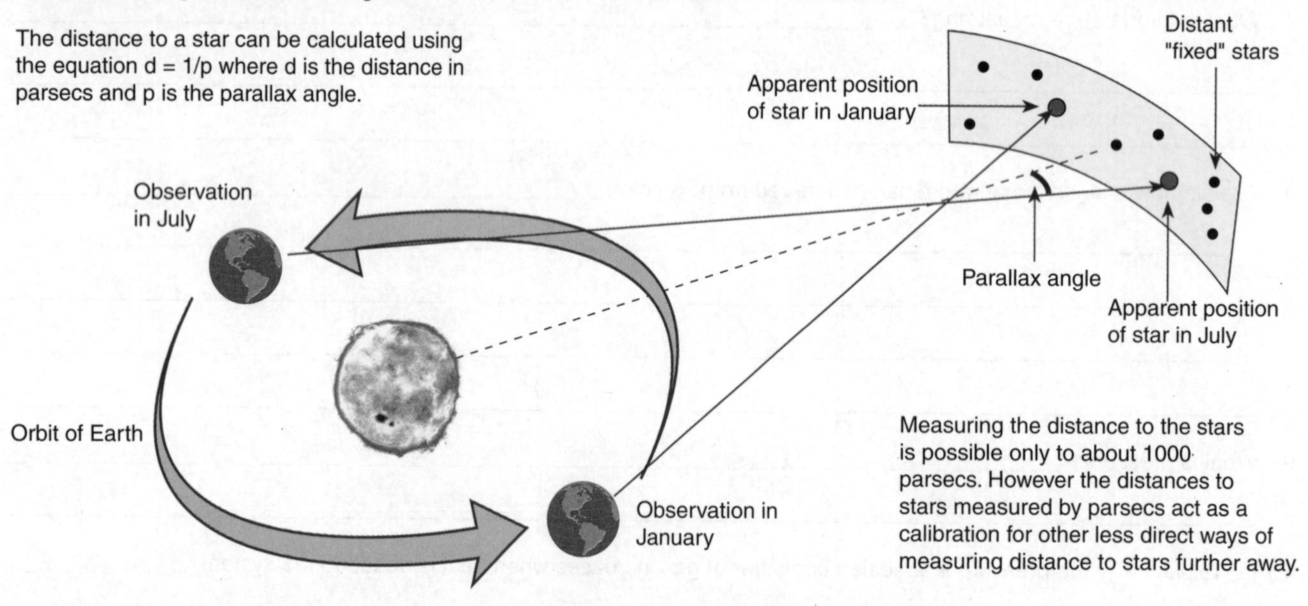


Measuring the distance to the stars is possible only to about 1000 parsecs. However the distances to stars measured by parsecs act as a calibration for other less direct ways of measuring distance to stars further away.

1. Light travels at 299,792.458 kilometers per second. How far in kilometers is one light year? __________

2. What is an astronomical unit? __________

3. The parallax of two stars observed from Earth was determined. Star A has a parallax angle of 0.17 arcseconds. Star B has a parallax angle of 0. 34 arcseconds. Which of these stars is closest to Earth? Explain your decision:

__________

__________

__________

**ISBN: 978-1-927309-37-7**

# 6 Observations and Assumptions

**Key Idea**: Many of the theories developed in Earth and space science depend on the assumption that all physical laws and processes are the same everywhere in the universe.

## Observations and hypotheses

- An observation is watching or recording what is happening. Observation is the basis for forming hypotheses and making predictions. An observation may generate a number of hypotheses (tentative explanations for what we see). Each hypothesis will lead to one or more predictions, which can be tested by investigation.
- A hypothesis can be written as a statement to include the prediction: "If X is true, then if I do Y (the experiment), I expect Z (the prediction)". Hypotheses are accepted, changed, or rejected on the basis of investigations. A hypothesis should have a sound theoretical basis and should be testable.

**Observation 1**: Sediments appear in layers, the oldest layers at the bottom the youngest layers on top.

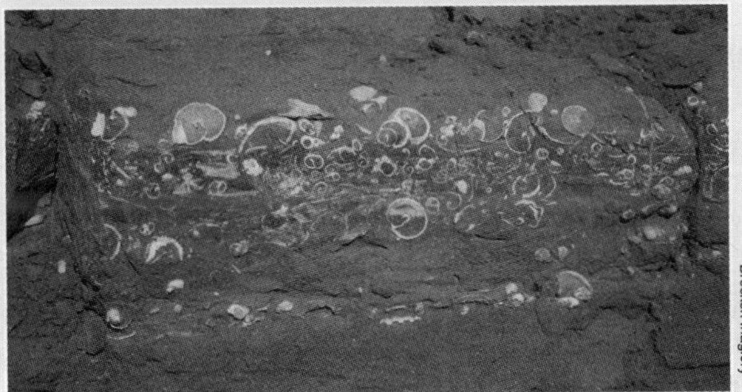

Brocken Inaglory

**Observation 2**: Different organisms are found in different rock layers. The oldest organisms are found in the deepest layers.

## Assumptions

Any investigation requires you to make **assumptions** about the system you are working with. Assumptions are features of the system you are studying that you assume to be true but that you do not (or cannot) test. Some assumptions about the geological systems described above include:

- Layers of sediment are always laid down horizontally.
- Layers of sediments are ordered youngest on top to oldest on the bottom, unless disturbed after formation.
- Organisms in the rock are there because they died at the same time as the sediments were being laid down (they weren't put there afterwards).

1. Read the two observations on sediments and fossils above and then answer the following questions:

   (a) Generate a hypothesis to explain the observation plant A is found above animal B in rock strata:

   Hypothesis: ______________________________

   (b) Describe one of the assumptions being made in your hypothesis: ______________________________

   (c) Generate a prediction about the relative age of fossils found in rock strata: ______________________________

   (d) How could you test your prediction? ______________________________

2. Form a group with three other classmates. Discuss what you would do if your test of the prediction did not support your hypothesis. Summarize your response and attach it to this page.


ISBN: 978-1-927309-37-7

## Occam's razor

▶ Occam's razor is a problem solving method which, at its simplest, states that among competing hypotheses, the one with the least number of assumptions should be used. Occam's razor helps rule out hypotheses or explanations that contain too many assumptions.

▶ For example there are two possible models explaining how the solar system is organized. The geocentric model states that the Sun and planets orbit the Earth. The heliocentric model states that all the planets, including Earth, orbit the Sun. Both can be used to calculate the position of the planets as we see them in the sky but the geocentric model makes many more assumptions, including that the laws of gravity don't apply to the Earth and Sun, and that the planets all have secondary "epicycles" along their orbits.

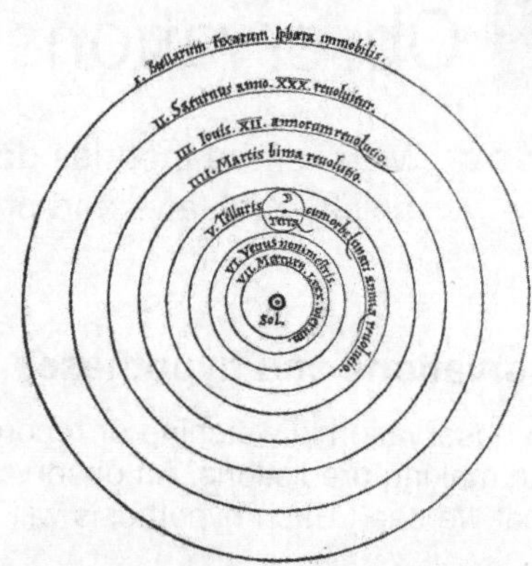

Copernicus' heliocentric model has just seven assumptions.

## Assumptions and the wider universe

For our universe to make sense, and for us to make sense of the universe, we have to assume certain ideas hold true everywhere in the universe and, if they don't, we have to be able to explain why not. There are essentially two rules that are assumed to be true in science.

**1** All of existence (i.e. the universe) is governed by rules (laws) that are the same everywhere, they are inviolable (they cannot be broken).

▶ This is essential for us to discover and understand the laws of the universe. There can't be one set of laws that apply in our universe today and a different set of laws tomorrow. Nor can there be different laws that apply to different parts of the universe. For example the movement of light in a vacuum must behave the same way today, as it did yesterday, as it does everywhere in the universe.

**2** These laws can be determined by observation of the universe around us.

▶ The second rule is equally important. What if the laws that govern the universe were not able to be understood? What if the number of laws was essentially infinite? No matter how carefully you observed something or how general your equations, you would never be able to write down a law that could be applied reliably to more than one situation.

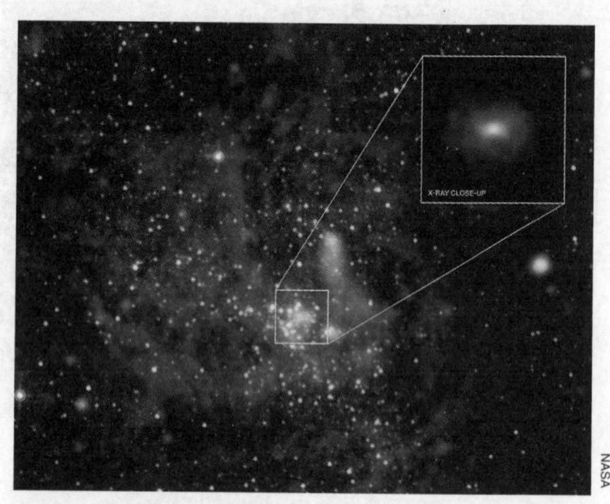


**Applying assumptions**

It has been observed that at the center of our galaxy about a dozen stars are orbiting a common point that appears empty. Some of the stars are moving through space at speeds of over 5000 km/s (the Sun moves at about 220 km/s). Assuming that the laws of gravity apply in the center of the galaxy the same as they apply here on Earth then it can be calculated that there must be an unseen object (called Sagittarius A*) with a mass of 4 million times the Sun holding the stars in their orbits.

3. (a) How does Occam's razor help to simplify explanations? ______________________

(b) Two students observed that two trees had toppled over in the night and that the grass around the trees was apparently undisturbed. Student A hypothesized that the wind had blown them over. Student B hypothesized that the trees had been knocked over by a large truck and the grass repaired after the truck had been removed.

Decide which of these two hypotheses is more likely and explain your choice: ______________________

4. Explain why it is important in science to assume that the laws of nature are universal, behaving in the same way everywhere and at every time:

______________________


**ISBN: 978-1-927309-37-7**

# 7 Tables and Graphs

**Key Idea**: Tables and graphs provide a way to organize and visualize data in a way that helps to identify trends.

- Tables and graphs are ways to present data and they have different purposes. Tables provide an accurate record of numerical values and allow you to organise your data so that relationships and trends are apparent.
- Graphs provide a visual image of trends in the data in a minimum of space. It is useful to plot your data as soon as possible, even during your experiment, as this will help you to evaluate your results as you proceed and make adjustments as necessary (e.g. to the sampling interval).
- The choice between graphing or tabulation in the final report depends on the type and complexity of the data and the information that you are wanting to convey. Sometimes, both are appropriate.

## Presenting data in tables

Table 1: Population, land area, and calculated population density in four US states.

| State | Population | Land area ($km^2$) | Population density (people $km^{-2}$) |
|---|---|---|---|
| Alabama | 4,871,547 | 135,754 | 35.9 |
| Florida | 20,636,975 | 170,307 | 121.2 |
| Montana | 1,032,949 | 380,847 | 2.7 |
| Texas | 27,469,114 | 695,662 | 39.5 |

- Tables provide a way to systematically record and condense a large amount of information. They provide an accurate record of numerical data and allow you to organise your data, making it easier to see patterns, trends, or anomalies.
- Table titles and row and column headings must be clear and accurate so the reader knows exactly what the table is about.
- Columns can be added for calculated values such as density, rate, and summary statistics (e.g. mean and standard deviation). For large data sets, it is often the summary statistic (e.g. mean temperature each year) that is plotted.
- Summary statistics make it easier to identify trends and compare different treatments. Rates are useful in making multiple data sets comparable, e.g. if recordings were made over different time periods.

## Presenting data in graphs

Fig. 1: Mean annual temperature in New Hampshire and Arizona

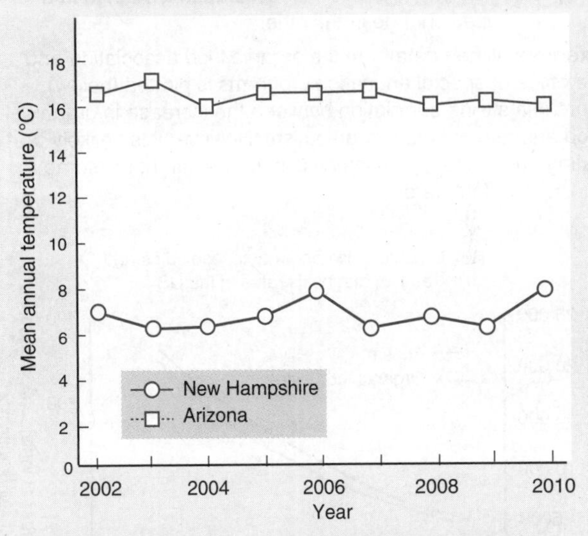


- Graphs are a good way of visually showing trends, patterns, and relationships without taking up too much space. Complex data sets tend to be presented as a graph rather than a table.
- Presenting graphs properly requires attention to a few basic details, including correct orientation and labeling of the axes, accurate plotting of points, and a descriptive, accurate title.

1. Describe the advantages of using a table to present information: __________

2. What is the benefit of including summary information (e.g. means or processed data) on a table? __________

3. What are the main advantages of presenting data in a graph? __________

4. Why might you include both graphs and tables in a final report: __________


**ISBN: 978-1-927309-37-7**


PRACTICES

**KNOW**

# 8 Correlation or Causation?

**Key Idea**: A correlation is a mutual relationship or association between two or more variables. A correlation between two variables does not imply that one causes change in the other.

- Researchers often want to know if two variables have any **correlation** (relationship) to each other. This can be achieved by plotting the data as a scatter graph and drawing a line of best fit through the data, or by testing for correlation using a statistical test.
- The strength of a correlation is indicated by the correlation coefficient (*r* or *R*), which varies between 1 and -1. A value of 1 indicates a perfect (1:1) relationship between the variables. A value of -1 indicates a 1:1 negative relationship, and 0 indicates no relationship between the variables.

## Correlation does not imply causation

You may come across the phrase "correlation does not necessarily imply causation". This means that even when there is a strong correlation between variables (they vary together in a predictable way), you cannot assume that change in one variable caused change in the other.

**Example**: When data from the organic food association and the office of special education programs is plotted (below), there is a strong correlation between the increase in organic food and rates of diagnosed autism. However it is unlikely that eating organic food causes autism, so we can not assume a causative effect here.

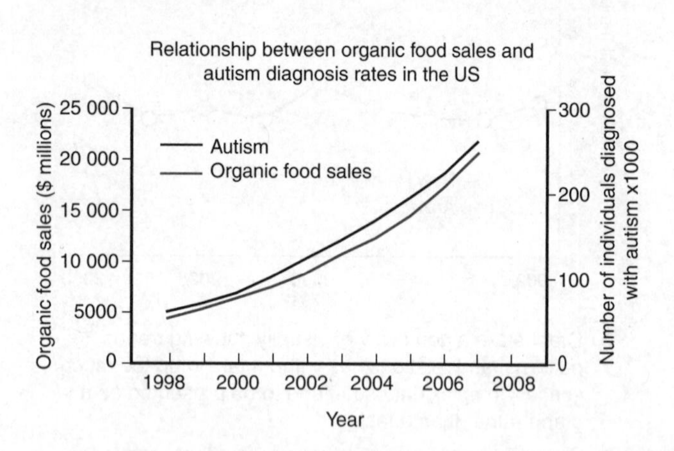


1. What does the phrase "correlation does not imply causation" mean?

2. A student measured the length of eruptions of Old Faithful and the time between the eruptions for a day and plotted a scatter graph of the results (right).

   (a) Draw a line of best fit through the data:

   (b) Using your line of best fit as a guide, comment on the correlation between eruption length and the time between eruptions:

## Drawing the line of best fit

Some simple guidelines need to be followed when drawing a line of best fit on your scatter plot.

- Your line should follow the trend of the data points.
- Roughly half of your data points should be above the line of best fit, and half below.
- The line of best fit does not necessarily pass through any particular point.
- The line of best fit should pivot around the point which represents the mean of the x and the mean of the y variables.

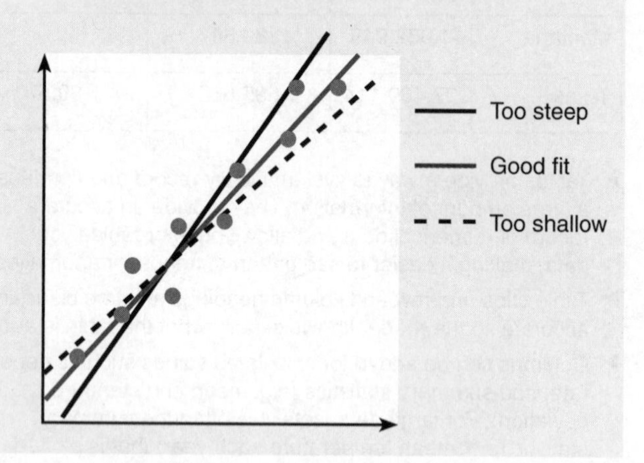


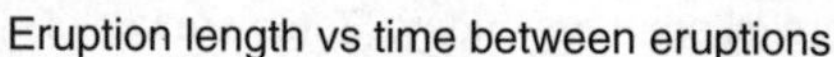

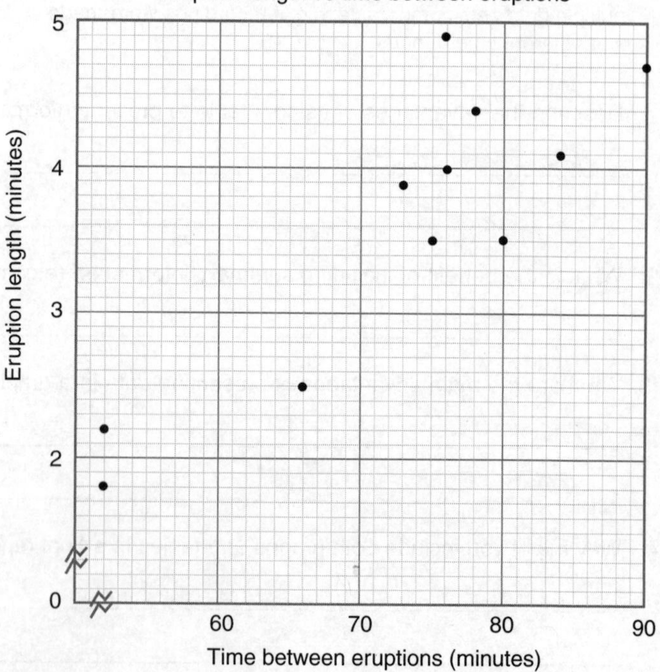

**ISBN: 978-1-927309-37-7**

# 9 Analyzing and Interpreting Data

**Key Idea**: Collected data needs to be processed interpreted to provide meaning information.

## Displaying processed data

Processed data is usually displayed in a table or graph. Tables and especially graphs can help make any trends in the data easier to see. Graphs can be used to predict values that do not appear in the original data. Sometimes many different data points may be plotted on the one graph so that a large amount of information is shown in a compact way.

### Pollen in sediments

- The data below shows the percentage of pollen in sediments from a region in northeastern United States, laid down over 15,000 years. By graphing the percentages of pollen from different plants beside each other and matching the pollen type to known trees and landscapes we can develop a picture of what the land looked like at any particular time in the last 15,000 years.
- We can see from the graph that about 5000 years ago the land was mostly hardwood forest including oak and beech trees. However 15,000 years ago we can see than the land was covered mostly by tundra. This is consistent with what is known about the climate conditions 15,000 years ago.

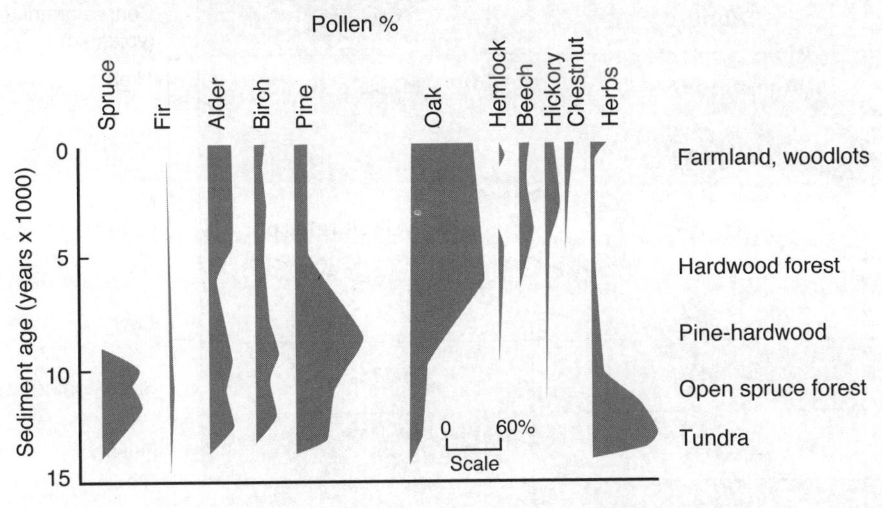


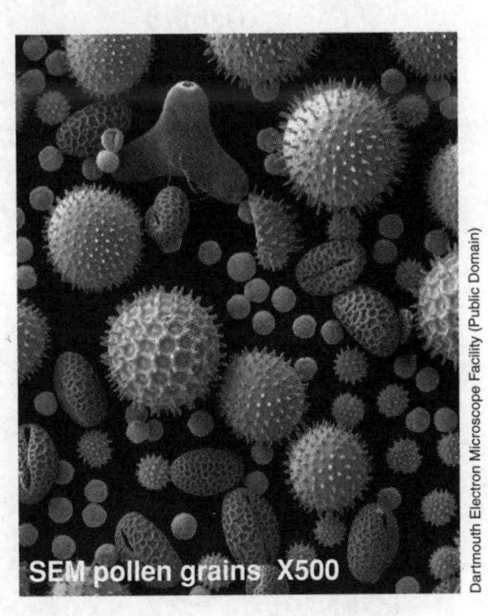


### Oxygen isotopes

- Oxygen comes in two common isotopes $^{16}O$ (light) and $^{18}O$ (heavy). In water, this causes water molecules with $^{18}O$ to be heavier than water containing $^{16}O$. Heavy water evaporates less than light water. Therefore there is less $^{18}O$ in the atmosphere and in rain than in seawater. Temperature affects the amount of $^{18}O$ in the atmosphere. The data below measures the difference in $^{18}O$ in rainwater compared to the ocean at different temperatures.

| Difference $^{18}O$ (%) in rainwater compared to seawater | Temp °C |
|---|---|
| -5.5 | -57 |
| -5.2 | -53 |
| -4.7 | -50 |
| -5.0 | -44 |
| -4.2 | -36 |
| -3.8 | -34 |
| -2.8 | -27 |
| -3.5 | -19 |
| -2.7 | -19 |
| -1.6 | -10 |
| -1.5 | 0 |
| -1.9 | 4 |
| -1.2 | 1 |
| -0.6 | 5 |
| -1.0 | 8 |
| -0.6 | 11 |
| -0.5 | 21 |


**ISBN: 978-1-927309-37-7**

1. (a) Draw a scatter graph of local temperature vs the difference in $^{18}O$ in rainwater compared to seawater.

   (b) How is the amount of $^{18}O$ in rainwater affected by temperature? ____________________

   (c) Near the poles, water in the atmosphere falls as snow and ice and freezes. Scientists can drill ice cores and measure the amount of $^{18}O$ in the ice at different levels. What could this information tell them?

2. The graph below displays information on different rock types. Study the graph and answer the questions below:

| | | | |
|---|---|---|---|
| Basalt | Andesite | Rhyolite | Fine grained (volcanic) |
| Gabbro | Diorite | Granite | Coarse grained (volcanic) |
| Pyroxene, Olivine (MAFIC MINERALS) | Amphibole, Plagioclase feldspar | Mica, Alkali feldspar, Quartz (FELSIC MINERALS) | Mineral content (High – Low) |
| 48-58% | 59-65% | 65-77% | Silica content |
| Fluid, high mobility 1160°C | | Viscous, low mobility 900°C | Fluidity |
| Basic | | Acidic | pH |

   (a) Identify the fine grained rock with a silica content of between 48-58%: ____________________

   (b) What is the pH of this rock type: ____________________

   (c) Is the rock fluid or viscous? ____________________

   (d) Identify the coarse grained highly acidic rock: ____________________

   (e) What is the most common mineral in the highest pH example of this rock? ____________________

3. Stalactites grow down from the top of cave roofs as rainwater deposits minerals on to them. The more rain there is, the faster the stalactites grow. They also tend to grow in annual rings as there is usually more rain in winter than in summer. Analyze the stalactite data below:

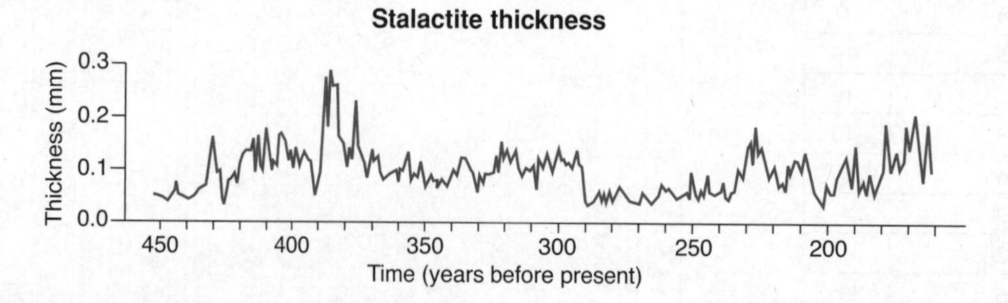


   (a) Identify the wettest period of time before present: ____________________

   (b) Identify the driest period of time before present: ____________________


**ISBN: 978-1-927309-37-7**

# 10 Descriptive Statistics and the Spread of Data

**Key Idea**: Descriptive statistics are used to summarize a data set and describe its basic features. The type of statistic calculated depends on the type of data and its distribution.

## Descriptive statistics

When we describe a set of data, it is usual to give a measure of central tendency. This is a single value identifying the central position within that set of data. Descriptive statistics, such as mean, median, and mode, are all valid measures of central tendency depending of the type of data and its distribution. They help to summarize features of the data, so are often called summary statistics. The appropriate statistic for different types of data variables and their distributions is described below.

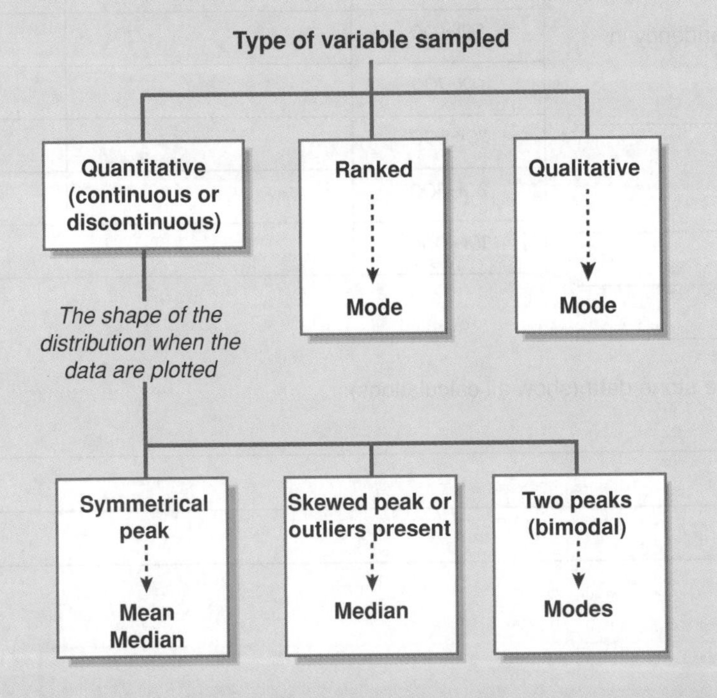


| Statistic | Definition and when to use it | How to calculate it |
|---|---|---|
| Mean | • The average of all data entries.<br>• Measure of central tendency for normally distributed data. | • Add up all the data entries.<br>• Divide by the total number of data entries. |
| Median | • The middle value when data entries are placed in rank order.<br>• A good measure of central tendency for skewed distributions. | • Arrange the data in increasing rank order.<br>• Identify the middle value.<br>• For an even number of entries, find the mid point of the two middle values. |
| Mode | • The most common data value.<br>• Suitable for bimodal distributions and qualitative data. | • Identify the category with the highest number of data entries using a tally chart or a bar graph. |

## Distribution of data

Variability in continuous data is often displayed as a frequency distribution. There are several types of distribution.

- Normal distribution (A): Data is spread symmetrically about the mean. It has a classical bell shape when plotted.
- Skewed data (B): Data is not centered around the middle but has a "tail" to the left or right.
- Bimodal data (C): Data which has two peaks.

The shape of the distribution will determine which statistic (mean, median, or mode) should be used to describe the central tendency of the sample data.

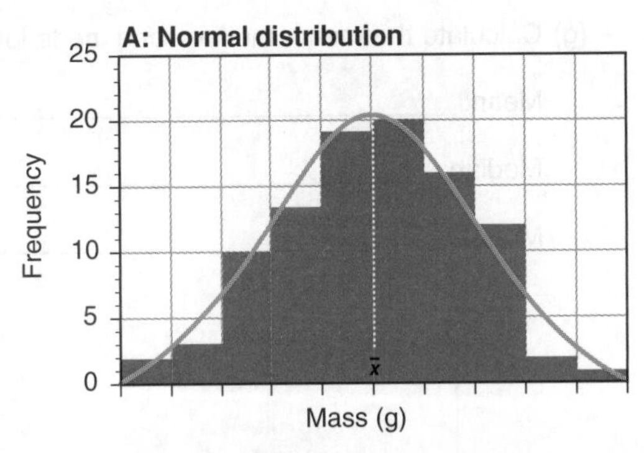


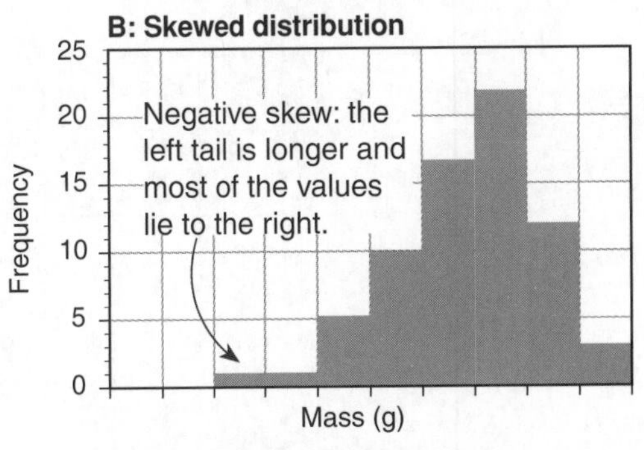


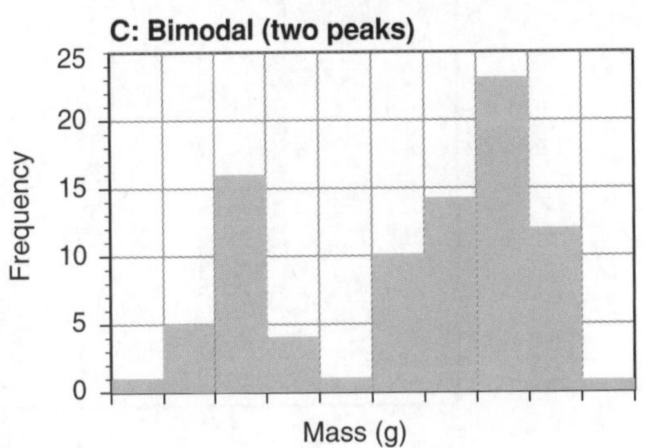



ISBN: 978-1-927309-37-7

1. A sample of stones on a beach was taken and their individual masses recorded as part of a study on beach stability.

| Stone mass (g) | | | | | | | | | |
|---|---|---|---|---|---|---|---|---|---|
| 881 | 335 | 909 | 632 | 706 | 359 | 881 | 284 | 607 | 290 |
| 795 | 439 | 229 | 182 | 383 | 719 | 509 | 322 | 578 | 686 |
| 488 | 375 | 143 | 204 | 161 | 160 | 419 | 147 | 68 | 91 |
| 167 | 459 | 151 | 135 | 197 | 180 | 115 | 314 | 414 | 83 |

(a) Draw up a tally chart in the space provided on the right for the stone masses.

(b) On the graph paper at the bottom of the page, draw a frequency histogram for the stone data.

(c) What type of distribution does the data have?

(d) Is this what you would expect?

(e) What would be the best measure of central tendency in the stone data set (mean, median, or mode)?

(f) Explain why you chose your answer in (e).

| Mass (g) | Tally | Total |
|---|---|---|
| 1-100 | | |
| 101-200 | | |
| 201-300 | | |
| 301-400 | | |
| 401-500 | | |
| 501-600 | | |
| 601-700 | | |
| 701-800 | | |
| 801-900 | | |
| 901-1000 | | |

(g) Calculate the mean, median, and mode for the stone data (show all calculations):

Mean:

Median:

Mode:


**ISBN: 978-1-927309-37-7**

# 11 Planning and Carrying Out an Investigation

**Key Idea**: Designing and carrying out an investigation carefully and recording the data in a methodical way makes it easier to analyze and interpret results.

The following activity will step you through a controlled investigation into an aspect of chemical weathering on rocks. You will analyze, tabulate, and graph data, and draw conclusions about chemical weathering of rocks.

### The aim

To investigate the effect of surface area on the reaction of acid with calcium carbonate (limestone).

### Hypothesis

If reaction rate is dependent on surface area, an increase in the surface area of limestone (calcium carbonate) will produce an increase in reaction rate.

### Background

Rainwater is very slightly acidic. When it comes in contact with limestone it reacts with it, forming carbon dioxide gas and calcium ions in solution. This causes the limestone to dissolve.

Limestone that is already partially eroded presents a greater surface area for rain to fall on and therefore a greater surface area for reaction.

### Experimental method

A roughly cubic, one gram piece of calcium carbonate ($CaCO_3$) was placed into excess 1 mol/L hydrochloric acid (HCl) at room temperature, 21°C (the reaction with rainwater is very slow due its weak acidity, so a stronger more concentrated acid was substituted). A stopwatch was started and the time taken for the $CaCO_3$ to completely dissolve recorded. One gram of $CaCO_3$ was then roughly crumbled and again placed into excess HCl and the time for it to completely dissolve recorded. Finally one gram of powered $CaCO_3$ was added to HCl and the time taken for it to completely dissolve recorded. The whole experiment was then repeated twice more.

| Surface area of sample | Time to dissolve (minutes) 1 | 2 | 3 |
|---|---|---|---|
| Low (Single chip) | 20 | 21 | 19.5 |
| Medium (Crumbled) | 5 | 5.5 | 4.8 |
| High (Powdered) | 1 | 1.5 | 1.2 |

### Controlling variables

To carry out an investigation fairly, all the **variables** (factors that could be changed) are kept the same, except the factor that is being investigated. These are called **controlled variables.** The variable that is being changed by the investigator is called the **independent variable**. The variable being measured is called the **dependent variable**.

1. Identify the independent variable for the experiment: ________________

2. Identify the dependent variable for the experiment: ________________

3. What is the sample size for each surface area of calcium carbonate? ________________

4. What assumption is this experiment based on? ________________


**ISBN: 978-1-927309-37-7**

5. Complete the table below:

| | Time to dissolve (minutes) | | | |
|---|---|---|---|---|
| **SA** | Sample 1 | Sample 2 | Sample 3 | Mean |
| **Low** | 20 | 21 | 19.5 | |
| **Medium** | 5 | 5.5 | 4.8 | |
| **High** | 1 | 1.5 | 1.2 | |

6. Use your calculations to plot a column graph on the grid below:

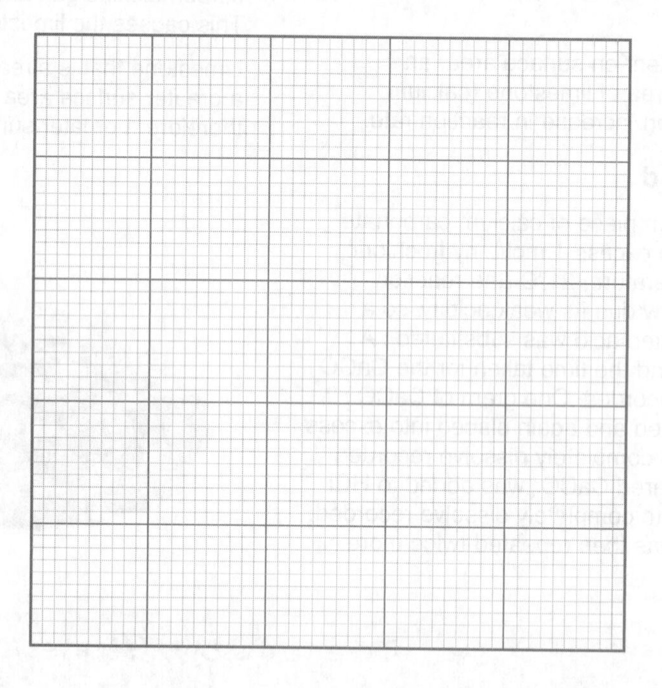

7. Write a conclusion based on the findings: ______

8. Discuss how this experiment relates to chemical erosion and how the experiment could be improved:

## Temperature and weathering

Some students decided to investigate the effect of temperature on erosion processes. They again used calcium carbonate chips and hydrochloric acid to simulate rainwater and rock but this time placed the $CaCO_3$ and acid into water baths to control the temperature of the solutions.

9. Write a hypothesis for this investigation: ______


**ISBN: 978-1-927309-37-7**

10. The students carried out the investigation using water bath temperatures of 15°C, 20°C, 25°C, and 30°C as their independent variable.

(a) Explain why the temperature range of 15°C to 30°C is appropriate for this investigation: ____________

(b) Identify some important variables that would be controlled in this investigation: ____________

11. Draw a diagram that outlines the method used by the students to carry out the investigation. Include labels for the equipment used:

12. For each temperature, the students carried out the procedure three times (three samples). Produce a table that would be appropriate for them to record and process their data in:

13. The students wanted to graph their results. What part of the results should the students graph and what would be the most appropriate type of graph to produce and why?


**ISBN: 978-1-927309-37-7**

# Earth's Place in the Universe

**Concepts and connections**
Use arrows to make your own connections between related concepts in this section of this book

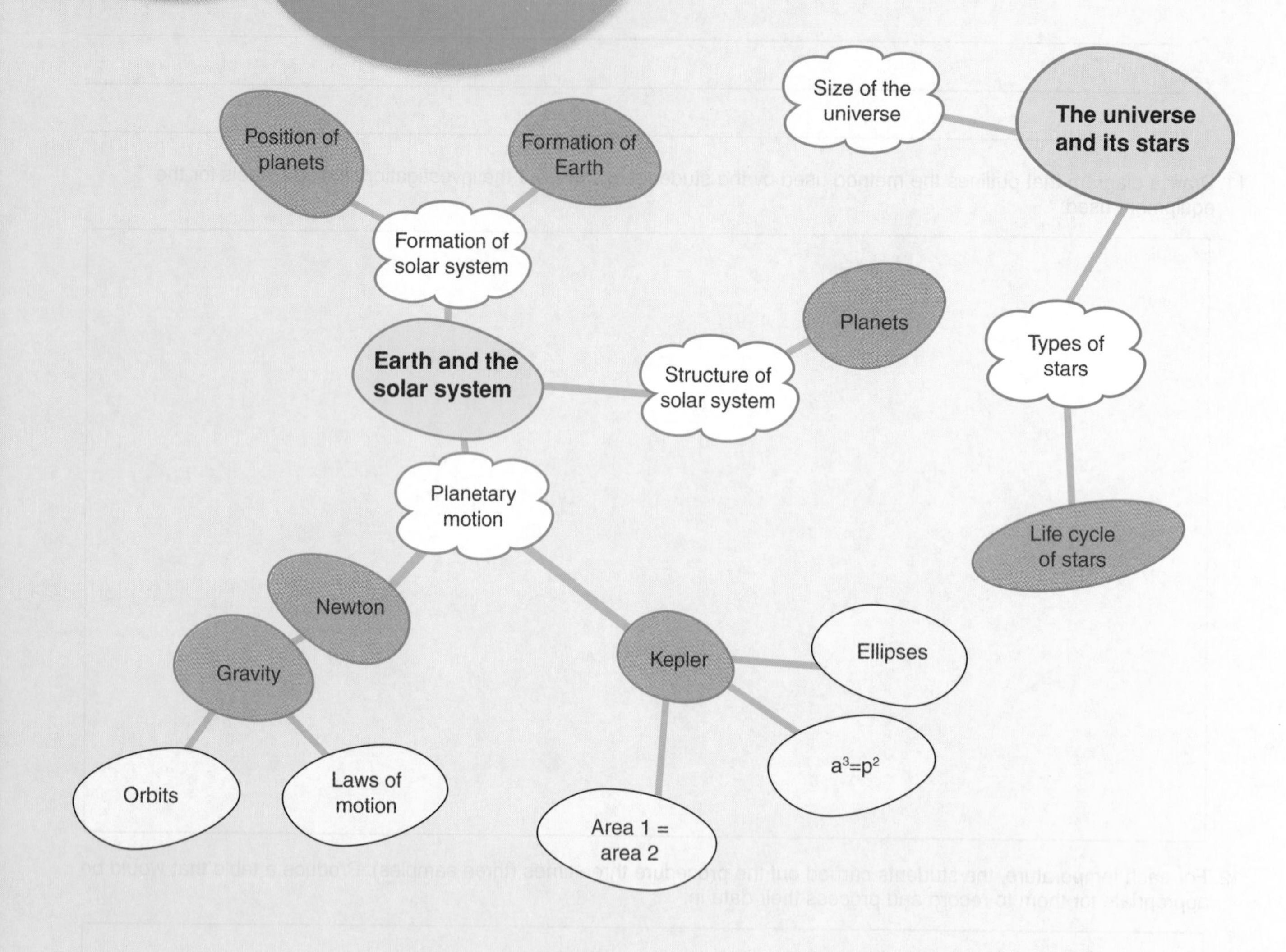


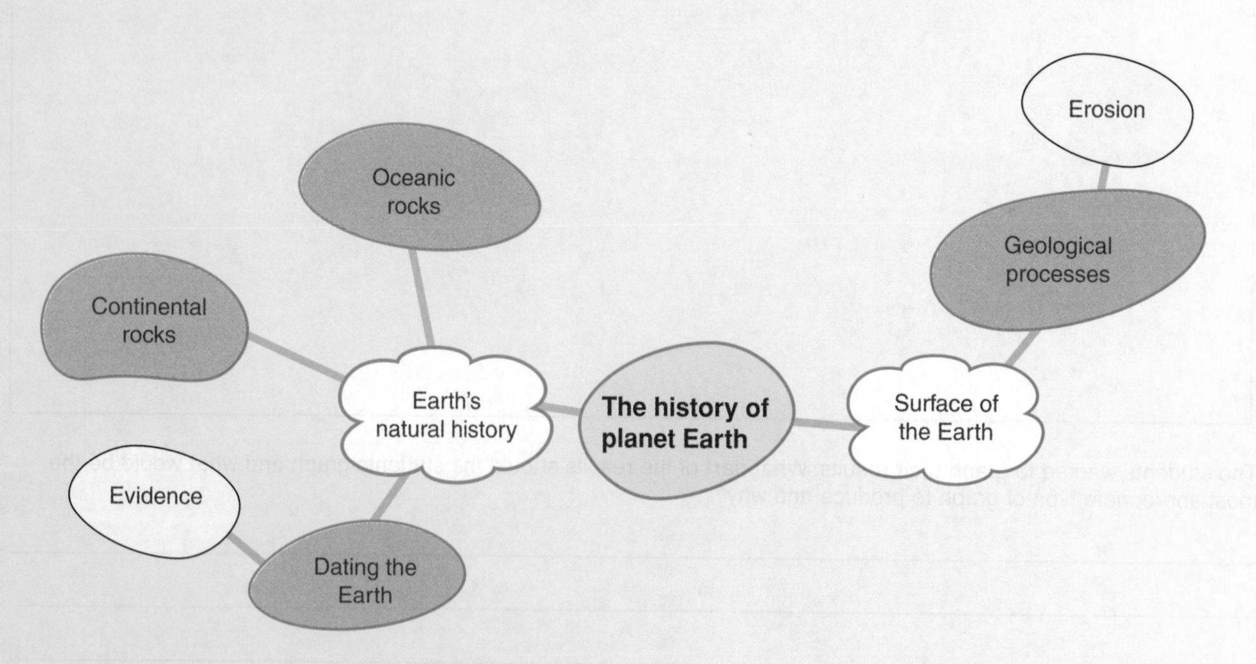

ESS1.A

# The Universe and its Stars

**Key terms**

atomic nuclei

Big Bang theory

cosmic microwave background

Doppler effect

electromagnetic radiation

element

event horizon

Hertzsprung–Russell diagram

light spectra

main sequence star

nebula

nucleon

nucleosynthesis

protostar

red shift

spectral analysis

star

stellar

Sun

sunspot cycle

## Disciplinary Core Ideas

***Show understanding of these core ideas***

| | | Activity number |
|---|---|---|
| | **The Sun is changing and will burn out** | |
| ☐ | 1 Recognize that our understanding of the universe continues to increase as the tools and technology available to us advance. | 12 13 |
| ☐ | 2 Describe the life cycle of the Sun, a main sequence star with a life span of ~10 billion years. Include reference to the changes during its existence, including variations in radiation, sunspot cycles, and non-cyclic variations. | 18 |
| ☐ | 3 Describe the evidence for changes in a star's luminosity during its lifetime and explain differences between the lifetimes of stars of different masses. | 17 18 19 |
| ☐ | 4 Describe how the study of stars' light spectra and brightness is used to identify their features, such as composition, movements, and distance from Earth. Include reference to O, B, A, F, G, K, and M standard stellar types. | 16 19 |
| | **The Big Bang theory is supported by many lines of evidence** | |
| ☐ | 5 Describe the Big Bang theory and the astronomical evidence supporting it. Include reference to the red shift, the cosmic microwave background, and the measured composition of stars and non-stellar gases. | 13-16 |
| | **Elements are produced by nucleosynthesis** | |
| ☐ | 6 Explain how nucleosynthesis creates new atomic nuclei from pre-existing nucleons (protons and neutrons) and releases electromagnetic energy. Distinguish between Big bang, stellar, and supernova nucleosynthesis. | 22 |
| ☐ | 7 Explain how nucleosynthesis and the different elements created varies as a function of the star's mass and the stage of its life cycle. | 20 21 22 |

DPC

DPC

NASA

## Crosscutting Concepts

***Understand how these fundamental concepts link different topics***

| | | Activity number |
|---|---|---|
| ☐ | 1 **SPQ** The significance of a phenomenon, such as nucleosynthesis, depends on the scale, proportion, and quantity at which it occurs. | 17 20 21 22 |
| ☐ | 2 **SPQ** The concept of orders of magnitude can be used to understand models of the universe at different scales. | 13 16-19 |
| ☐ | 3 **EM** In nuclear processes, atoms are not conserved, but the total number of protons plus neutrons is conserved. | 14 17-22 |

## Science and engineering practices

***Demonstrate competence in these science and engineering practices***

| | | Activity number |
|---|---|---|
| ☐ | 1 Develop and use an evidence-based model to illustrate the life span of the Sun and the role of nuclear fusion in the Sun's core. | 17 18 19 22 25 |
| ☐ | 2 Use an evidence-based model to describe the origin and nature of the universe. | 13 14 21 |
| ☐ | 3 Construct an explanation of the Big Bang theory based on evidence from a variety of sources and uniformitarianism. | 14 15 |
| ☐ | 4 Use multiple formats (e.g. oral presentation, text, diagrams, mathematics) to communicate scientific ideas about the way stars produce elements. | 20 21 22 |

# 12 Studying the Universe

**Key Idea**: Scientists and astronomers can study different aspects of the universe, its make up, and its history by using various devices for gathering data.

## Studying the universe

People have studied the universe for millennia. To begin with, people studied the sky visually, using only their eyes and noting how the stars, Moon, and Sun moved across the sky. Accurate study and mapping of the stars and Sun helped people keep track of the seasons, and important (usually religious) events. With the beginning of the Renaissance in Western Europe, studying stars became more academic and accurate as instrumentation became more advanced.

1

Simple devices such as Tycho Brahe's quadrant (left), in 1597, allowed the accurate measurement of the angle of the stars and planets above the horizon. Brahe produced an enormous amount of data on the movements and positions of stars and planets which was later used by Johannes Kepler to develop the laws of planetary motion.

2

In 1610, Galileo Galilei used a refracting telescope (above) to view the moons of Jupiter and became the first person to see objects orbiting another planet. He used his observations as part of his argument that the planets orbited the Sun instead of the Earth.

3

Once developed, optical telescopes became progressively sophisticated and larger. They included both refracting (using lenses) and reflecting (using mirrors) telescopes. However they only allowed observation of visible light. During the early 20th century, optical telescopes were used to confirm the universe extended beyond the Milky Way galaxy.

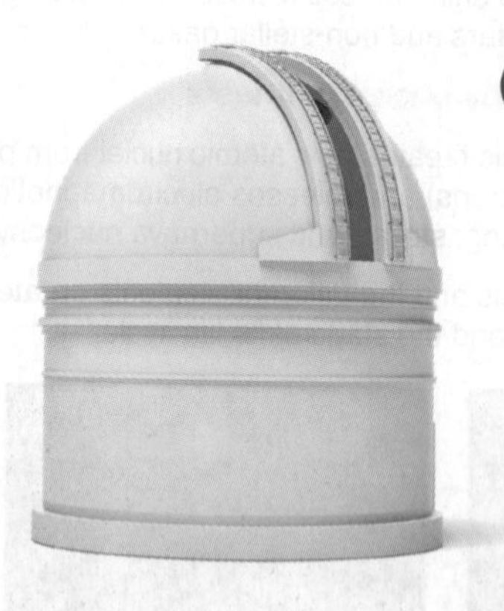

4

The use of radio telescopes to view the sky began in earnest after the Second World War. The use of radio astronomy allowed astronomers to see parts of universe that had been invisible to optical devices. Radio telescope observatories may have one huge radio telescope or many smaller ones joined together in an array.

5

With the advancement of the space age, astronomers have been able to launch observatories into space. This allows better detection of electromagnetic signals including gamma, x-ray, and infrared radiation. Space observatories include the Hubble Space Telescope (below) (optical), the Chandra Observatory (x-ray), and the Swift Gamma Ray Burst Explorer (gamma rays).

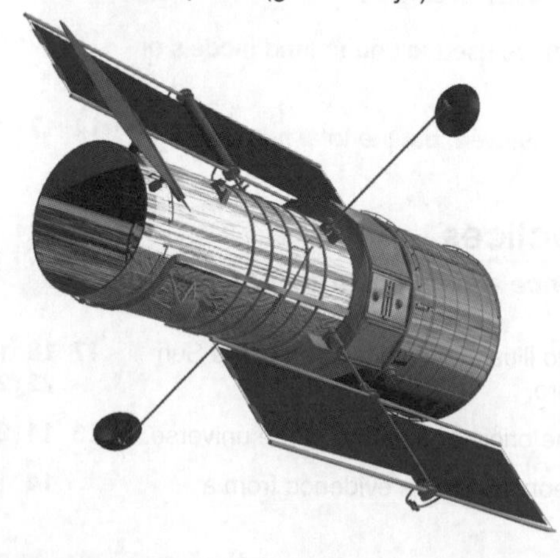

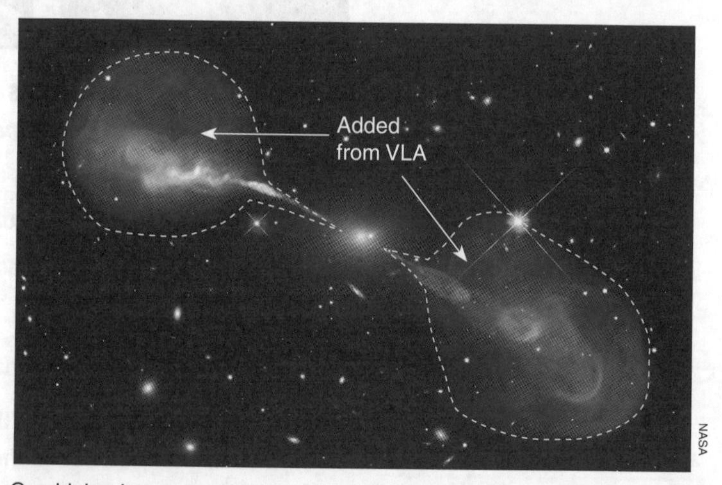


Combining images and information from different observatories allows us to gain a different view of celestial objects. The image above is a composite of an optical image taken from the Hubble Space Telescope and a radio image taken from the Very Large Array (VLA) in New Mexico.




ISBN: 978-1-927309-37-7

## Observing light waves

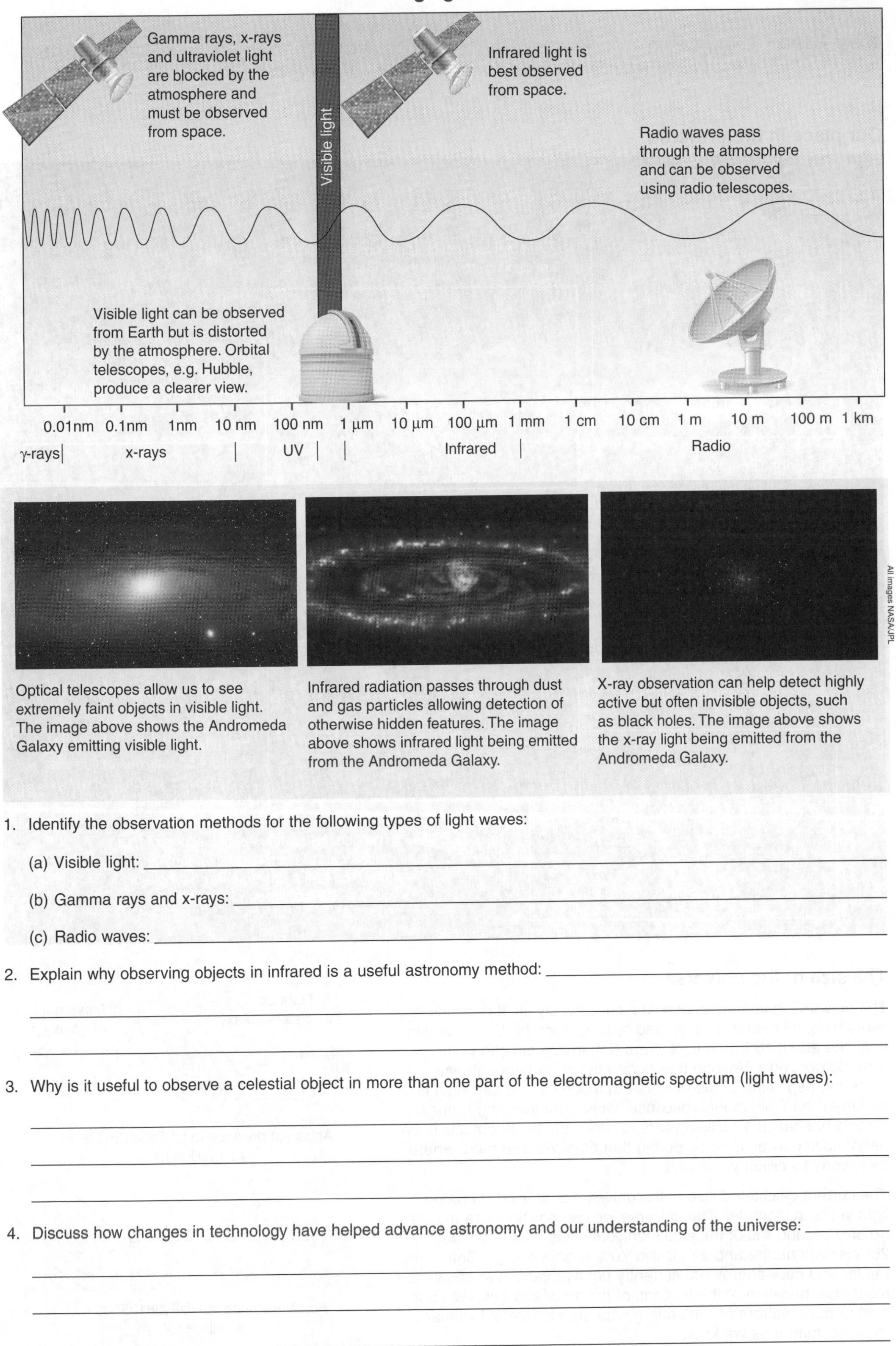


Optical telescopes allow us to see extremely faint objects in visible light. The image above shows the Andromeda Galaxy emitting visible light.

Infrared radiation passes through dust and gas particles allowing detection of otherwise hidden features. The image above shows infrared light being emitted from the Andromeda Galaxy.

X-ray observation can help detect highly active but often invisible objects, such as black holes. The image above shows the x-ray light being emitted from the Andromeda Galaxy.

1. Identify the observation methods for the following types of light waves:

   (a) Visible light: ____________________

   (b) Gamma rays and x-rays: ____________________

   (c) Radio waves: ____________________

2. Explain why observing objects in infrared is a useful astronomy method: ____________________

   ____________________

   ____________________

3. Why is it useful to observe a celestial object in more than one part of the electromagnetic spectrum (light waves):

   ____________________

   ____________________

   ____________________

4. Discuss how changes in technology have helped advance astronomy and our understanding of the universe: ____________________

   ____________________

   ____________________

   ____________________


**ISBN: 978-1-927309-37-7**

# 13 The Known Universe

**Key Idea**: The universe encompasses everything. All matter, all energy, and all space. The extent of the universe is uncertain as we are unable to measure its full size.

## Our place in the universe

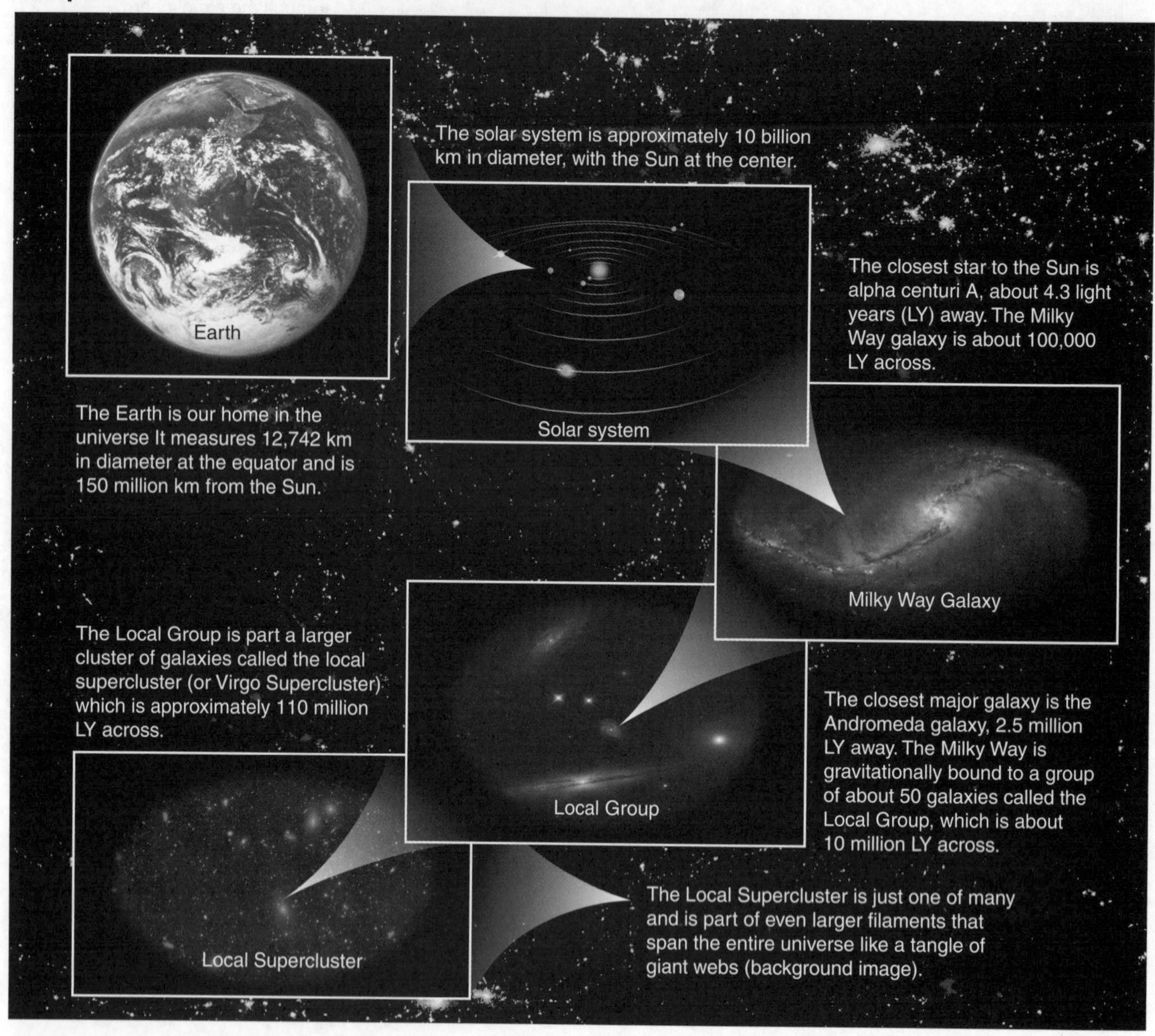


## The size of the universe

The universe is immense, almost inconceivably so. It contains everything, all matter, energy, and space. From the Earth we can only see about 13 billion light years (LY) in any direction (our "cosmic horizon"). Beyond this, light has not had enough time since the beginning of the universe to reach us. The universe is far larger than this though because, while light from the furthest objects has taken 13 billion years to reach us, those objects have been moving away from us during that time. We see them where they were 13 billion years ago.

The current calculated size of the universe is at least 93 billion light years in diameter. The universe we can possibly see makes up only around 4% of the actual universe. It is estimated that 26% is dark matter and a massive 70% is dark energy. Both dark matter and dark energy are currently unobservable. We know they must exist because of the motions of the stars and galaxies, but neither dark matter nor dark energy appears to interact with any "normal" matter as we know it.

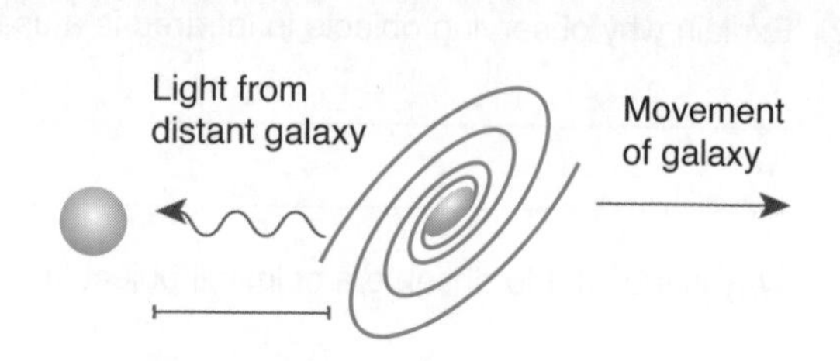

Apparent distance to UDFy38135539 = 13.1 billion LY

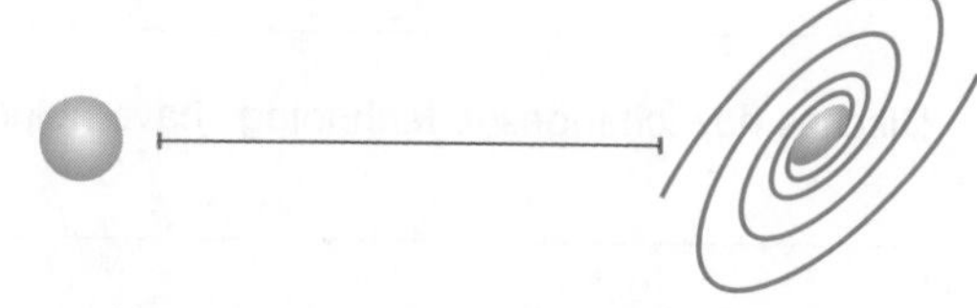
Actual distance to UDFy38135539 = 30 billion LY

**ISBN: 978-1-927309-37-7**

## The shape of the universe

If you draw a triangle on a flat piece of paper the internal angles of the triangle add up to 180°. But what happens if you draw a triangle on the surface of a ball? The internal angles add up to more than 180°. And what about the internal surface of a bowl? The internal angles of a triangle add up to less that 180°. Now imagine drawing a huge triangle on the "surface" of the universe. What would the internal angles of the triangle be? The answer depends on whether or not the universe is curved or flat. The answer has important consequences for the shape and fate of the universe.

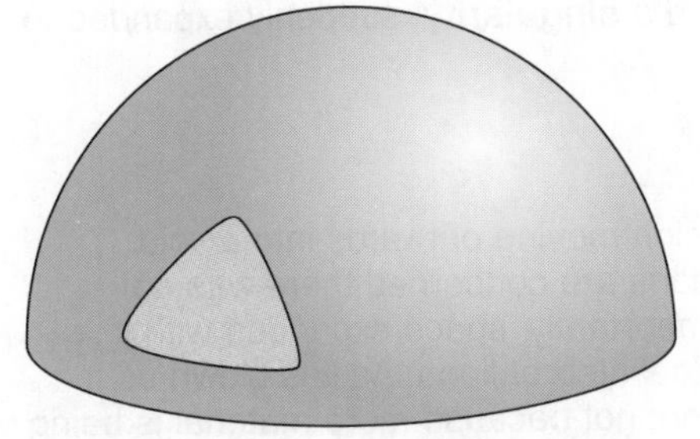

If the internal angles of your triangle add to be greater than 180°, then the universe is positively curved like a sphere. In this case, the universe is finite but has no edge. It will eventually slow its expansion and collapse back in on itself.

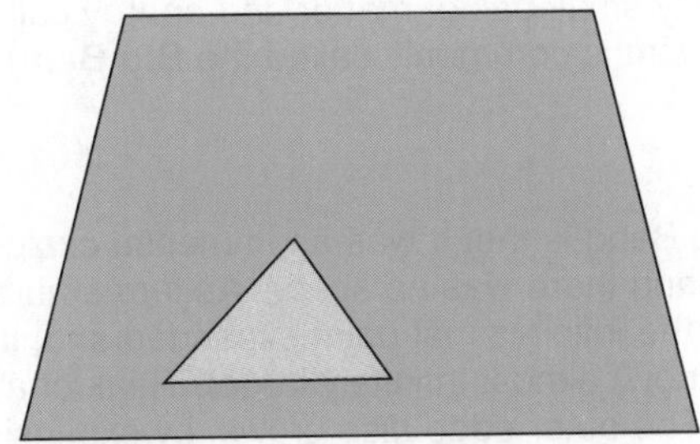

If the triangle's internal angles add to be equal to 180° then the universe is flat. In this case it will continue to expand forever, although the expansion will slow over an infinite amount of time. Current evidence suggests the universe is flat.

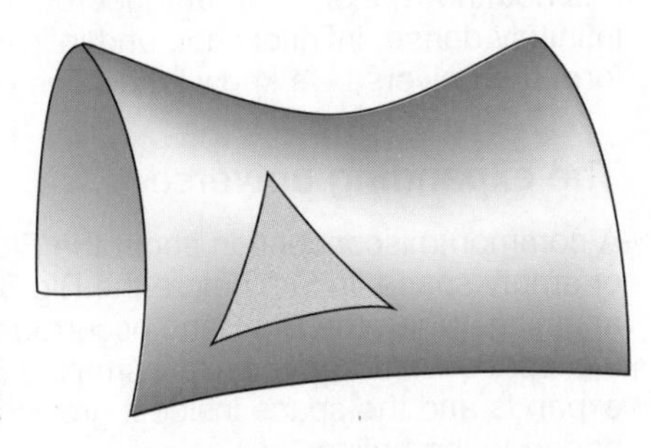

If your triangle has internal angles that add to less than 180° then the universe is negatively curved, like the shape of a saddle. It is infinite and will expand forever at an ever increasing rate.

1. What is the cosmic horizon? ______________________________

______________________________

______________________________

2. How big is the observable universe? ______________________________

3. Explain why the cosmic horizon is much smaller than the universe itself: ______________________________

______________________________

______________________________

______________________________

4. (a) What are the three components that make up the universe? ______________________________

   (b) What percentage of the universe comprises visible matter? ______________________________

5. (a) What is the distance to the Sun's nearest star? ______________________________

   (b) What percentage of the diameter of the Milky Way Galaxy is this? ______________________________

   (c) What is the distance to the nearest major galaxy? ______________________________

6. (a) Describe the path of two parallel light beams in a flat universe: ______________________________

______________________________

______________________________

   (b) Describe the path of two parallel light beams in a positively curved universe: ______________________________

______________________________

______________________________

7. How do we know dark matter and dark energy exist when they can not be observed directly? ______________________________

______________________________

______________________________

______________________________


**ISBN: 978-1-927309-37-7**

# 14 The Big Bang

**Key Idea**: The universe began in an event called the Big Bang, in which an infinitely small and dense object expanded, producing all the universe's matter, energy, and space.

It is not known exactly what triggered the expansion of the universe, but it is known that 13.8 billion years ago an infinitely dense, infinitely hot, and infinitely small ball of matter and energy called a **singularity**, suddenly expanded to form the universe we know today. The event is commonly called the **Big Bang**.

## The expanding universe

A common misconception about the Big Bang is that it was an immense explosion moving outwards into a void of empty space. In fact before the Big Bang there was no space. As far as humans are concerned there was no anything. When the Big Bang occurred, the infinitesimal point expanded and, importantly, space expanded with it (i.e. space went from infinitely small to (now) almost infinitely large). Think of a deflated balloon. As it is blown up it expands and the space inside it grows. The outer edge also grows, by expansion, not because more material is being added to the balloon.

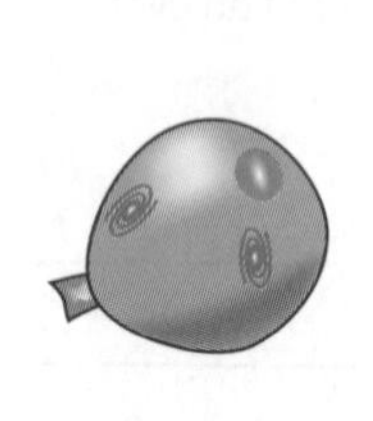

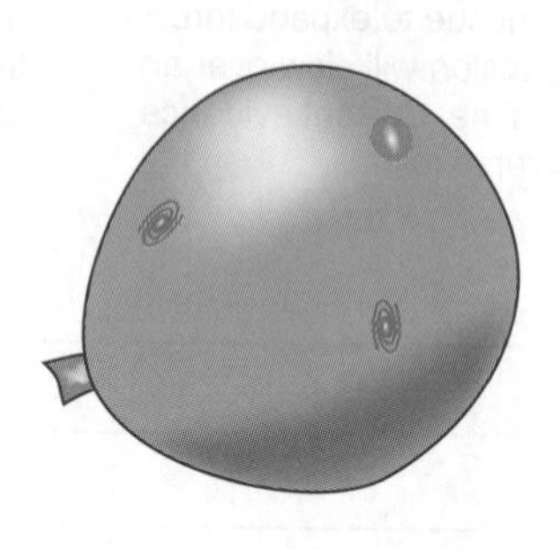

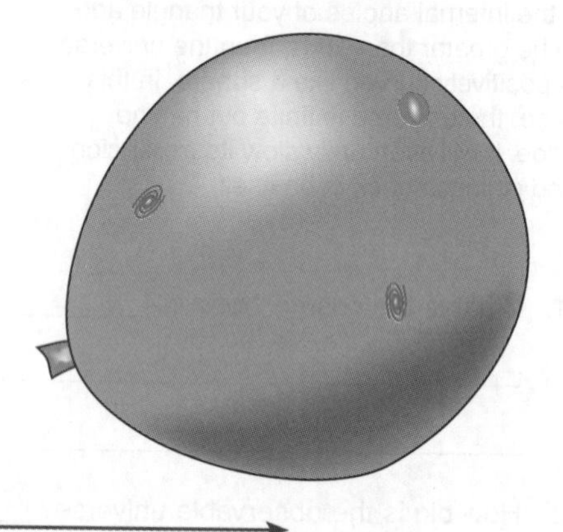

Time

The universe begins at time 0. Everything that currently exists in the universe was compressed to an infinitely small point called a singularity.

$10^{-43}$ seconds after the Big Bang (that's a decimal point (.) followed by 42 zeros then a 1) the temperature of the universe was $10^{32}$ °C. At this temperature, there is only one unified force. It is too hot for even elementary particles to form.

Between $10^{-38}$ and $10^{-35}$ seconds a process called inflation occurs. The universe grows by $10^{62}$ times. Inflation stops when the energy causing it is transformed into the matter and energy known today. The temperature drops to $10^{29}$ °C. Gravity separates from the other three fundamental forces.

**0** | **$10^{-43}$ seconds** | **$10^{-38}$ seconds**

The very early universe was filled with simple elementary particles and antiparticles including high energy photons (gamma rays), electrons, and positrons.

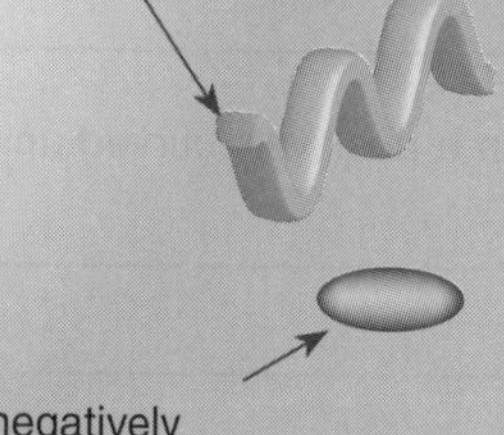

**Photon** - particles of light

**Graviton** - theoretical particles that transfer gravity between objects.

**Electron** - negatively charged particle. For an unknown reason, more electrons than positrons were produced in the Big Bang.

**Positron** - electron equivalent but with positive charge

**ISBN: 978-1-927309-37-7**

## An overview of the Big Bang and the evolution of the universe

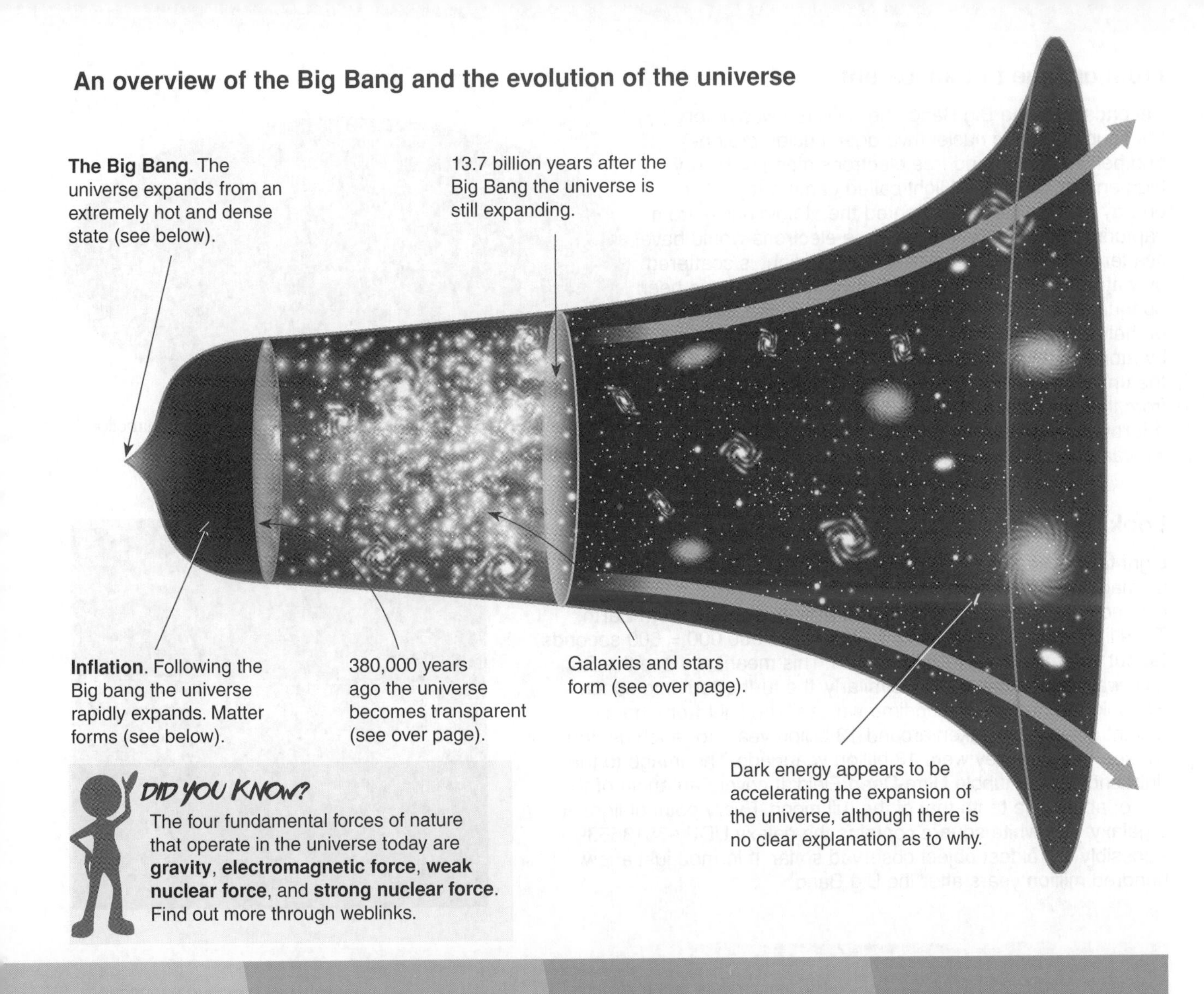


**DID YOU KNOW?**

The four fundamental forces of nature that operate in the universe today are **gravity**, **electromagnetic force**, **weak nuclear force**, and **strong nuclear force**. Find out more through weblinks.

**$10^{-12}$ seconds**: The temperature of the universe drops to $10^{15}$ °C. The four fundamental forces of the universe separate and particles such as gluons, and quarks form.

**$10^{-4}$ seconds**: The temperature of the universe drops to 1 trillion °C. Quarks combine to form protons and neutrons.

**5 seconds**: The universe continues to expand (but more slowly). The temperature drops to 5 billion °C. All positrons have been annihilated in collisions with electrons.

**3 minutes**: The universe's temperature reaches 1 billion °C, cool enough for protons and neutrons to join to together to form helium nuclei. Photons of light are still too energetic to let electrons join nuclei to form atoms.

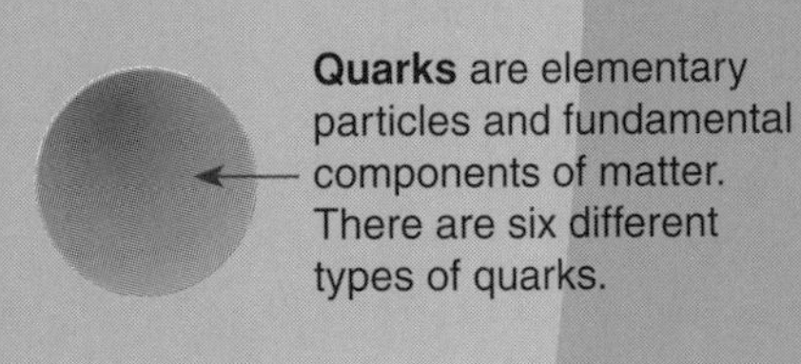


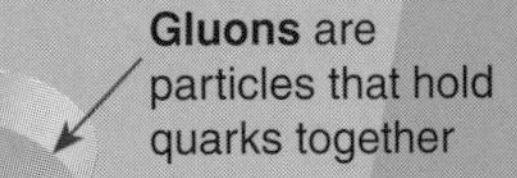


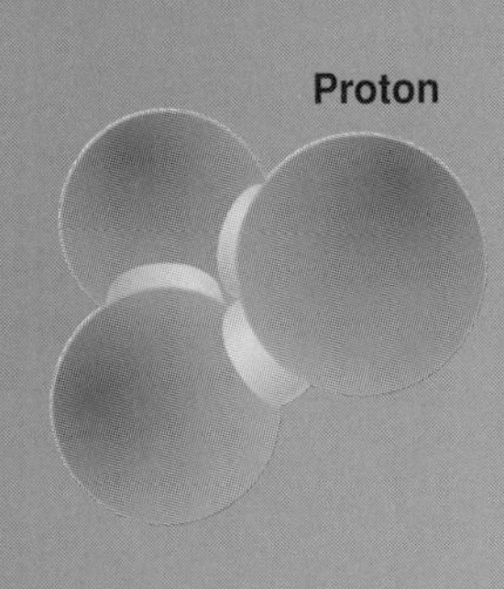


There are six different types of quarks. The combinations of three specific quarks produces protons and neutrons.

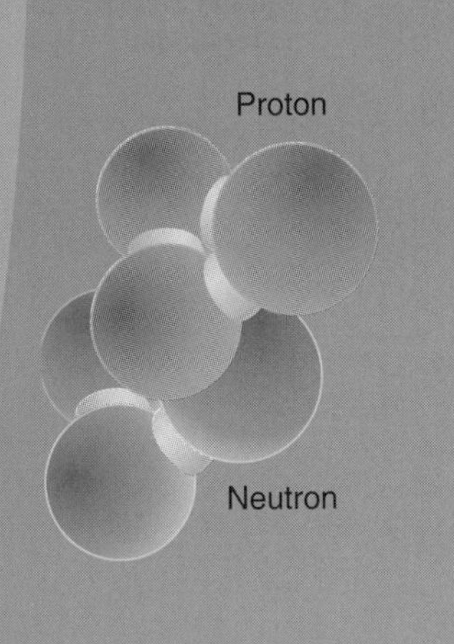



ISBN: 978-1-927309-37-7

## From opaque to transparent

Seconds after the Big Bang, the universe was a very hot soup of atomic nuclei (hydrogen nuclei (protons) and helium nuclei) and free electrons along with very high energy photons of light called gamma rays. The energy of the photons prevented the atomic nuclei from capturing the electrons. These free electrons would have scattered the photons much like visible light is scattered by water droplets in a fog. The universe would have been opaque. After 380,000 years the photons had lost most of their energy and electrons were able to be captured by atomic nuclei. Photons were no longer scattered and the universe became transparent. The photons of light from this time make up what is now called the **cosmic microwave background** (CMB). It is the oldest light that we can detect.

The CMB can be detected throughout the sky in every direction.

## Looking back in time

Light travels at about 300,000 km/s. But even at this almost unimaginable speed, light still takes time to travel the vast distances of space. The Sun is 150 million km away from Earth. Thus light from the Sun takes 150 million ÷ 300,000 = 500 seconds (about eight minutes) to reach Earth. This means we see the Sun as it was eight minutes ago. Similarly, the further out into space we look, the further back in time we see. The light from the most distant objects has taken around 13 billion years to reach us, thus we see them as they were 13 billion years ago. The image to the right shows the Hubble Ultra-Deep Field. It covers an angle of the sky of about one tenth that of the full moon. Every point of light is a galaxy. The white square contains the galaxy UDFy-38135539 - possibly the oldest object observed so far. It formed just a few hundred million years after the Big Bang.

The temperature of the universe reaches about 3000 °C. Photon energies are low enough for electrons to join atomic nuclei and form atoms. Photons are no longer scattered by these electrons and the universe becomes transparent.

Gravity has had long enough to pull clumps of matter together into huge structures called filaments. Galaxies will form along these filaments along with the first generation of stars. The explosion of first generation stars produces the elements heavier than carbon.

**380,000 years**

**300-500 million years**

The majority of the hydrogen and helium present in the universe today was formed during the early stages of the universe.

Hydrogen

Helium

Carbon


**ISBN: 978-1-927309-37-7**

1. How long ago did the Big Bang occur?
2. How long after the Big Bang did the following form:
   (a) Quarks:
   (b) Protons:
   (c) Atomic nuclei:
   (d) Atoms:
   (e) Galaxies:
3. Explain why the universe was opaque until around 380,000 years after the Big Bang:
4. What does the cosmic microwave background represent?
5. Read the following statement: "*The universe was formed when a dense ball of material exploded into space, forming the universe we see today*". Identify and comment on any errors in this statement:

The Sun forms as a second generation star. The heavy elements formed by the explosions of first generation stars form the planets orbiting it. The temperature of the universe is now -258°C. The Earth forms about 500,000 years after the Sun.

## 9 billion years

The planets of our solar system formed from the dust around the proto-Sun.

Presently the temperature of the universe is -270°C, just a few degrees above absolute zero. The universe is still expanding and evidence suggests the expansion is accelerating.

## 13.7 billion years

Andrew Pontzen and Fabio Governato

At the largest scale, the galaxies are collected together along huge filaments. It is suggested these filaments formed as a result of ordinary matter interacting with dark matter.


**ISBN: 978-1-927309-37-7**

# 15 Evidence for the Big Bang

**Key Idea**: Numerous lines of evidence for the Big Bang include the CMB, current movement of the galaxies, and the composition of very old stars and galaxies.

When Albert Einstein published his equations for the General Theory of Relativity they allowed for several solutions to how the universe could behave. In 1924, Alexander Friedmann solved the equations to show that the universe could be expanding (it also could be contracting, or static). Before the end of the decade, observations by Edwin Hubble of the movement of distant galaxies confirmed that the universe was expanding. Since then, the discovery of the cosmic microwave background and the hydrogen and helium composition of early stars has added weight to the theory that the universe was once much smaller, hotter, and denser than it is now.

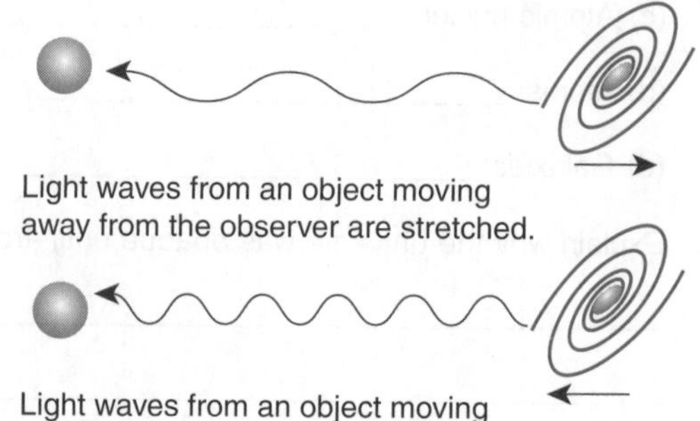

Light waves from an object moving away from the observer are stretched.

Light waves from an object moving towards the observer are compressed.

## 1: Red-shifted galaxies

- Light travels at a constant speed no matter whether an object is moving towards or away from you. What changes is the wavelength of the light. Light waves emitted from an object moving towards you are compressed into the blue part of the spectrum. Light waves emitted from an object moving away from you are expanded into the red part of the spectrum. This is called the **Doppler effect**. It also occurs in sound waves and is the reason why a car moving towards you sounds higher pitched than a car moving away from you. This property of light can be exploited to work out whether a distant galaxy is moving towards or away from us and how fast it is moving.
- Light from a galaxy can be separated into its spectrum. Elements in the galaxy absorb some of the wavelengths of light in this spectrum and produce dark bands. The spectrum of a galaxy moving towards us will have these characteristic bands shifted towards the blue end of its spectrum, while a galaxy moving away from us will have these bands shifted towards the red end of its spectrum and will be **red-shifted**.
- In 1929, Edwin Hubble used the 100 inch Hooker telescope to measure the red-shift of various galaxies. Hubble found that all but the very nearest galaxies were red-shifted, meaning they were moving away from us. Importantly, he also found that the further away they were, the more red-shifted they were and the faster they were moving away.

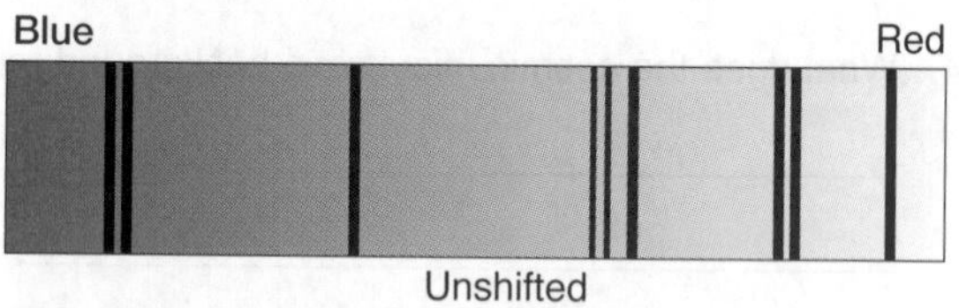


Unshifted

The spectral lines of a stationary galaxy are represented above.

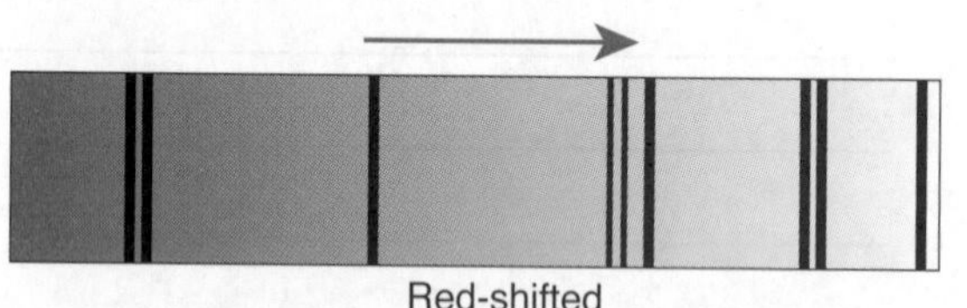

Red-shifted

The spectral lines of a galaxy moving away from us will beshifted towards the red end of the spectrum.

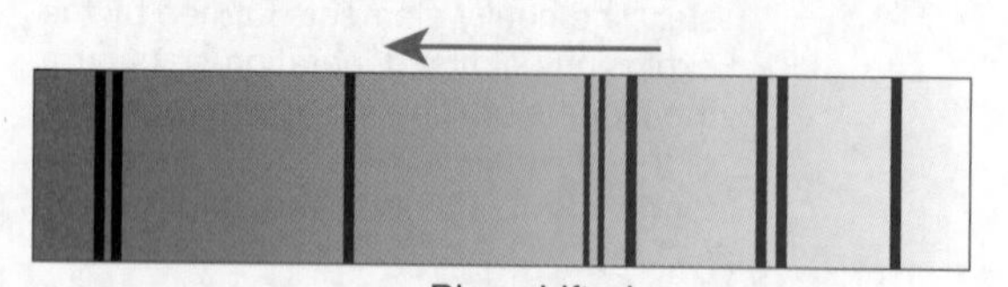

Blue-shifted

If the galaxy is moving towards us the spectral lines will be shifted into the blue end of the spectrum.

## 2: Cosmic microwave background

- It was realized as far back as the late 1940s that the heat and energy left over from the Big Bang should still be present and able to be detected. Astrophysicists Ralph Alpher and Robert Herman reasoned that the expansion of the universe since the Big Bang should have stretched the wavelength of the high energy radiation to somewhere in the microwave region of the electromagnetic spectrum, with a temperature of about 5 K (5°C above absolute zero).
- About 15 years later, Arno Penzias and Robert Wilson, working for Bell Telephone Laboratories found that the communications equipment they were working with produced a steady background radio noise no matter how much they cleaned it or what direction they pointed it. Inspection of the background noise showed it to have a "noise temperature" of 4.2 K. The wavelength of this radio noise was measured at 7.35 cm, within the microwave region of the spectrum. Refinements of measurements have now placed the average temperature at 2.728 K. Fluctuations in the temperature indicate precursors to the large scale structures we see in the universe today.

Darker regions in the image of the CMB above show cooler temperatures, brighter regions show higher temperatures. The contrast of the image is 30,000 times (the temperature fluctuations are *very* small. Between the "hot" and "cold" regions there is a temperature difference of 0.0002 K).




**ISBN: 978-1-927309-37-7**

## 3: Composition of early galaxies and stars

- Big Bang theory states that during the few seconds after the Big Bang a small amount of protons and neutrons fused to produce helium and a trace amount of lithium. Heavy elements were not formed because the conditions for the formation of heavier elements require a much greater time (tens of thousands of years) than the Big Bang lasted for. Therefore, when we measure the elements in distant young galaxies we should expect to see large amounts of hydrogen (about 75% of all elements), smaller amount of helium (about 24%), and trace amounts of lithium and other elements.

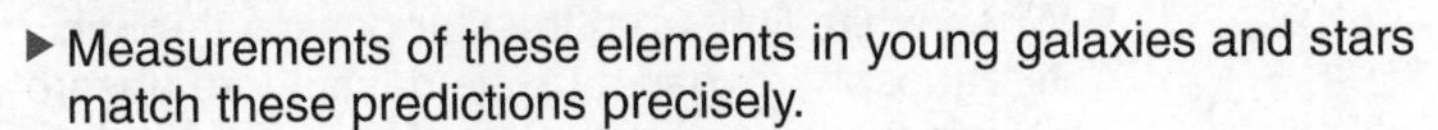

- Measurements of these elements in young galaxies and stars match these predictions precisely.

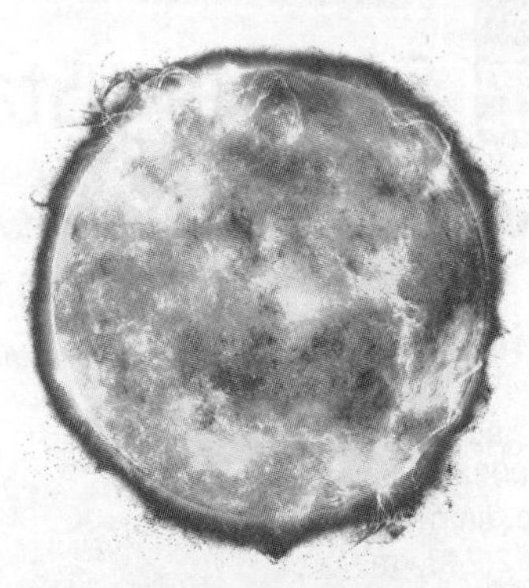

Young stars comprise 75% hydrogen and 25% helium by mass, exactly what would be expected if hydrogen and helium were formed during the Big Bang.

1. Explain what is meant by red-shifted: ______

2. Study the spectral diagrams below. Compare them to the unshifted spectral diagram opposite. What can you say about galaxy A and galaxy B?

A

B

______

3. Explain how the red-shifting of galaxies provided evidence for the Big Bang: ______

4. (a) What is the cosmic microwave background? ______

   (b) Explain how the discovery of the CMB provided evidence for the Big Bang: ______

5. Why could heavy elements not form during the Big Bang? ______

6. Study the three pieces of evidence for the Big Bang in this activity carefully. What is common to all these pieces of observable evidence and why does this make them much more powerful than if they did not have this common feature?

______


**ISBN: 978-1-927309-37-7**

# 16 Studying Stars

**Key Idea:** By measuring aspects such as absorption spectra, motion, and luminosity, a detailed picture of star types can be developed.

The Sun has a diameter of 1,392,000 km. Compare this to the diameter of the Earth at 12,742 km.

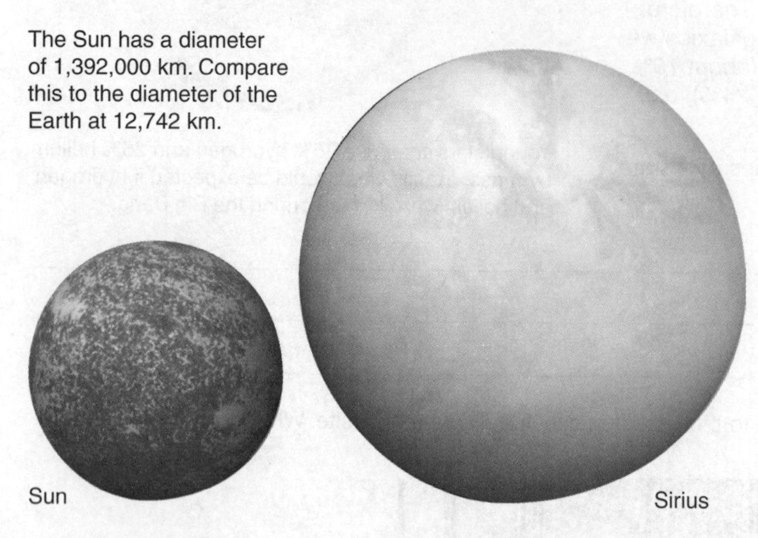


The sizes of the stars are almost incomprehensible. Each pair of the stars shown in this activity are to scale, yet even using this scale it is difficult to imagine the size of a star such as Canis Majoris (opposite), one of the largest stars spotted so far.

## What's in a star?

- The composition of a star can be determined by its absorption spectrum.
- When white light from the star passes through its atmosphere, gaseous elements there absorb certain wavelengths, leaving dark bands in the electromagnetic spectrum. These dark lines can be compared to known element absorption lines (after taking into account the effect of the temperature of the star on the absorption lines).

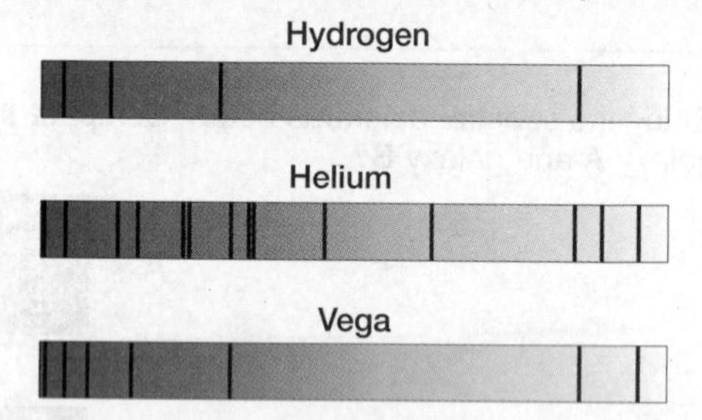


The absorption spectrums of hydrogen and helium and the star Vega. Vega shows hydrogen absorption lines and weaker helium lines.

## Measuring the mass and size of a star

- The mass and size of a star can be measured by observing their effect on other objects, their absolute luminosity (the total amount of energy emitted, which is related to brightness), and their temperature.
- A binary system (two stars) is needed for an accurate assessment of mass. In a binary system, Kepler's laws of motion can be used to relate the mass of the stars to their period of orbit and the size of the orbit (both from observation).
- The size of the star is related proportionally to its luminosity (large stars are usually brighter than small stars). Luminosity can be calculated based on a star's distance from Earth (measured by parallax) and the use of the inverse square law (luminosity decreases in proportion to distance).

Aldebaran (right) is an orange giant star, 61 million km in diameter. It is about 65 light years away and 425 times more luminous than the Sun. At this scale the Sun is about the size of this dot ▪

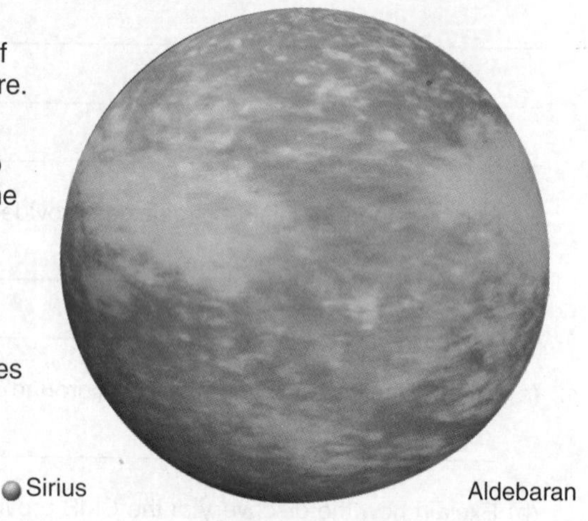


1. What are the two main elements in stars? ______________________________

2. Give a brief explantation of how the composition of a star can be determined: ______________________________

______________________________

______________________________

______________________________

3. The five stars shown in this activity are in pairs. Each pair is to scale. Using the diameter of the Sun as a starting point calculate the diameter of the other stars:

(a) Sirius: ______________________________

(b) Aldebaran: ______________________________

(c) Betelgeuse: ______________________________

(d) Canis Majoris: ______________________________




ISBN: 978-1-927309-37-7

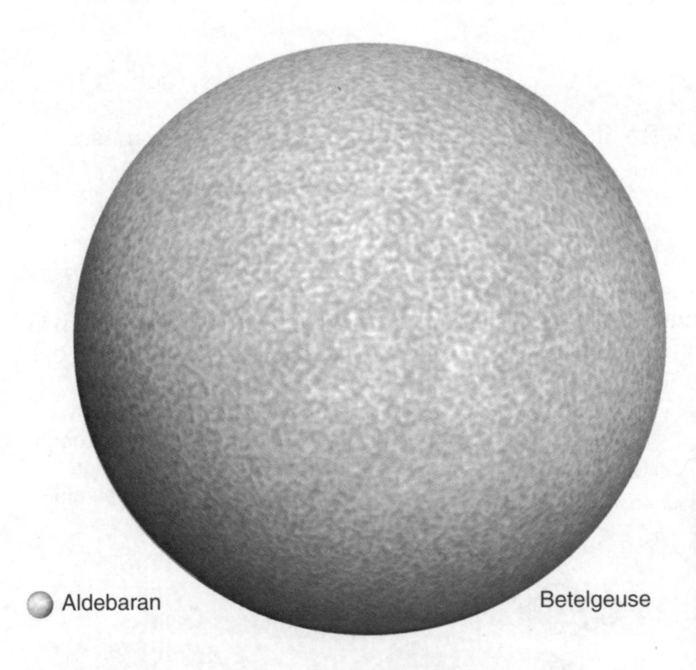


## Brightness

- A star's apparent magnitude is how bright the star is to the naked eye. The scale is logarithmic: a magnitude 1 star is 2.5 times brighter than a magnitude 2 star. The absolute magnitude is the apparent magnitude of the star from a distance of 10 parsecs (32.6 light years). If the distance to the star is known (from parallax) then the absolute magnitude can be calculated from apparent magnitude.
- The Sun has an absolute magnitude of 4.83. The star Sirius has an apparent magnitude of -1.44 (this seems counter-intuitive but the original scale started at 0 and missed some of the brightest stars in the sky). It has an absolute magnitude of 1.45.

| Star | Apparent magnitude | Absolute magnitude | Distance (light years) |
|---|---|---|---|
| Sun | -26.8 | 4.83 | 0 |
| Aldebaran | 0.75 | -2.1 | 65 |
| Betelgeuse | 0.42 | -2.9 | 640 |
| Canis Majoris | 7.9 | -9.4 | 3900 |
| Vega | 0.03 | 0.58 | 25 |

## Star color and temperature

- The color of a star is related to its temperature. The hottest stars are blue, the coolest are a dull red.
- The color (spectral class) of stars and their characteristics are shown below:

| Spectral class | Surface temperature (K) | Color | Examples |
|---|---|---|---|
| O | 30,000 | Blue | Mintaka |
| B | 20,000 | Blue - white | Rigel |
| A | 10,000 | White | Vega, Sirius |
| F | 7000 | Yellow - white | Canopus |
| G | 6000 | Yellow | Sun |
| K | 4000 | Orange | Aldebaran |
| M | 3000 | Red | Betelgeuse |

Canis Majoris (right) is a red hypergiant and one of the largest stars known.

4. Identify two ways of calculating a star's mass: ____________________

5. Explain the difference between apparent magnitude and absolute magnitude: ____________________

6. Which of the stars in the table top right:

   (a) Is the brightest to an observer on Earth using the naked eye? ____________________

   (b) Is the absolute brightest star: ____________________

   (c) How many times brighter is Sirius than the Sun? ____________________

NEED HELP? See Activity 3

7. Use the table above left to answer the following questions:

   (a) What is the color and surface temperature of Sirius? ____________________

   (b) What is the spectral class of the Sun? ____________________

   (c) What is the color of the hottest star in the table? ____________________


**ISBN: 978-1-927309-37-7**

# 17 The Sun

**Key Idea**: The Sun is the center of the solar system. It contains most of the solar system's mass and provides light and heat by nuclear fusion of hydrogen into helium.

The Sun contains 99.8% of all the mass in the solar system. It has a diameter of 1,392,000 km and is more than 330,000 times more massive than the Earth. The Sun formed about 4.5 billion years ago and will continue to shine with little change for at least another 4 billion years. When the Sun reaches about 10 billion years old, the hydrogen in its core will be exhausted. The core will shrink and the Sun will swell to form a red giant, with a diameter reaching out to the orbit of the Earth.

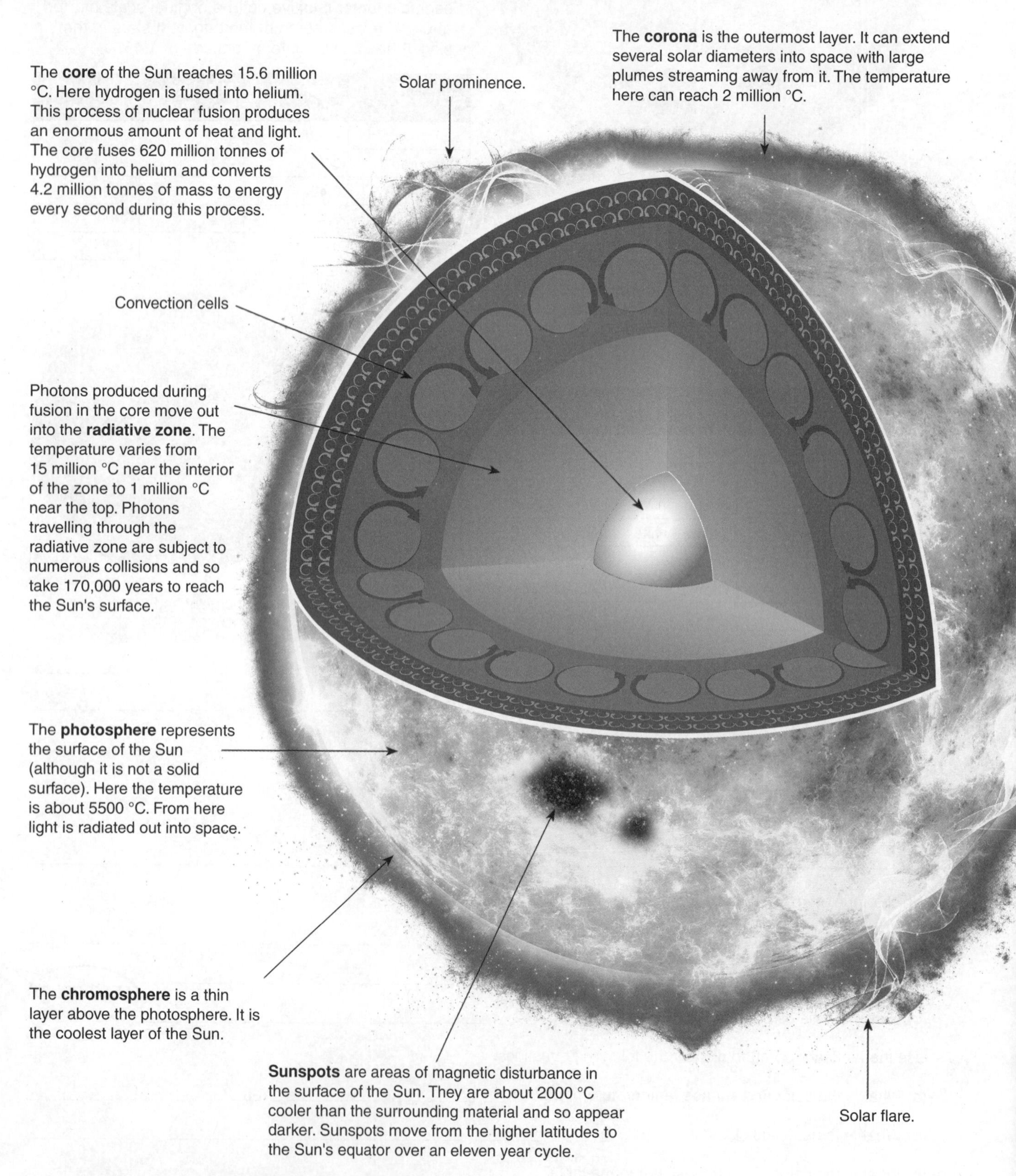

**ISBN: 978-1-927309-37-7**

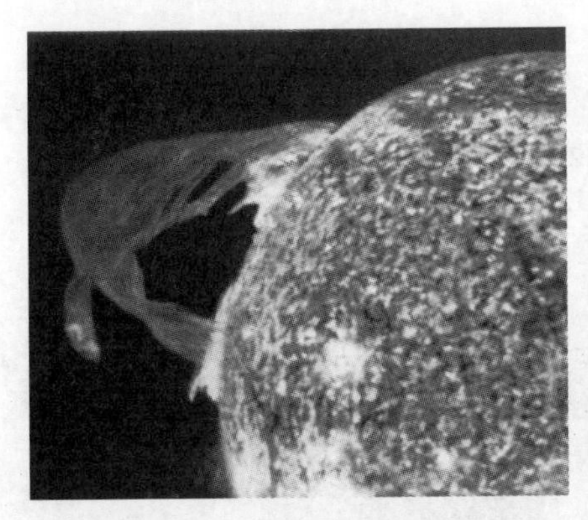

Solar prominences are loop shaped outbursts. They follow along the magnetic lines of the Sun.

Solar flares are sudden releases of energy that can extend thousands of kilometers into space.

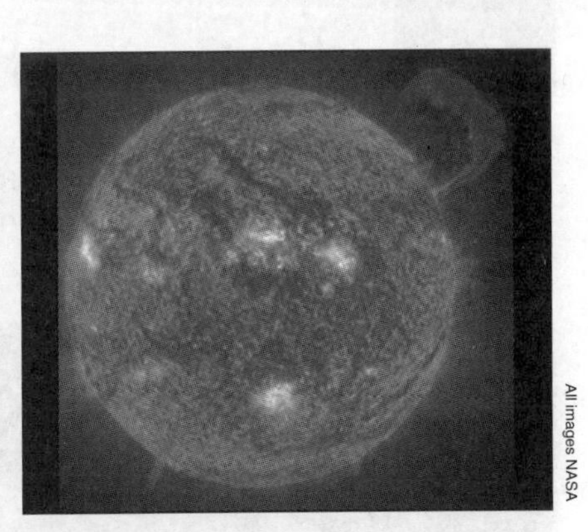

The Sun has a mass of about 1.98 x $10^{30}$ kg. Some of this mass is lost during nuclear fusion and as the solar wind and mass ejections.

All images NASA

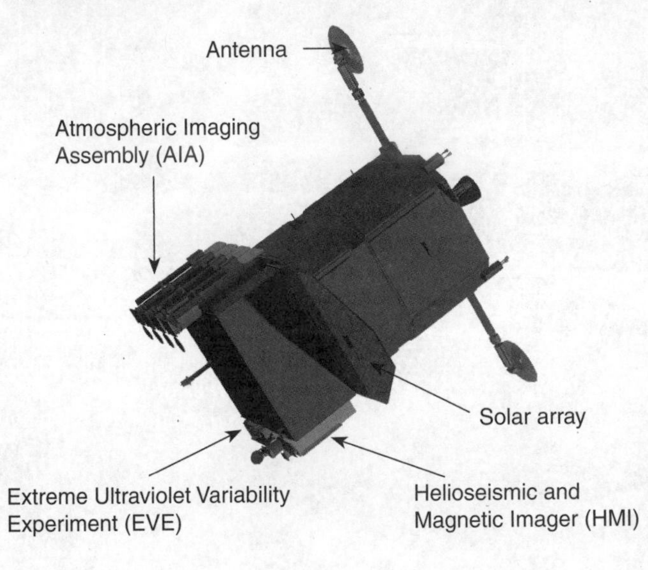


There have been and still are numerous space probes studying the Sun. One of the latest is the Solar Dynamics Observatory (SDO), (left). The SDO's mission is to:

- ▶ Measure extreme ultraviolet light from the Sun
- ▶ Make high-resolution measurements of the magnetic field over the entire visible disk
- ▶ Make images of the chromosphere and inner corona at several temperatures.

These observations will help us understand how and why the Sun's magnetic field changes and the effect the magnetic energy has when it is released into the Sun's atmosphere.

1. (a) What type of reaction produces the Sun's energy? ___

   (b) Which two elements are involved in this process? ___

2. Why do Sunspots appear darker than the rest of the Sun's surface? ___

3. (a) How do the photons emitted by the Sun originate? ___

   (b) Why do the photons take so long to reach the surface of the Sun? ___

4. What is a solar prominence? ___

5. (a) In one day how much mass is converted into energy in the Sun: ___

   (b) The Sun produces 3.8 x $10^{26}$ joules of energy per second. What is the total energy output of the Sun in one day?

6. Why is the photosphere considered the "surface" of the Sun? ___


**ISBN: 978-1-927309-37-7**

# 18 Life Cycle of Stars

**Key Idea**: Stars form from gigantic clouds of gas (nebulae). The size and life cycle of the star depends on the mass of gas from which the star formed.

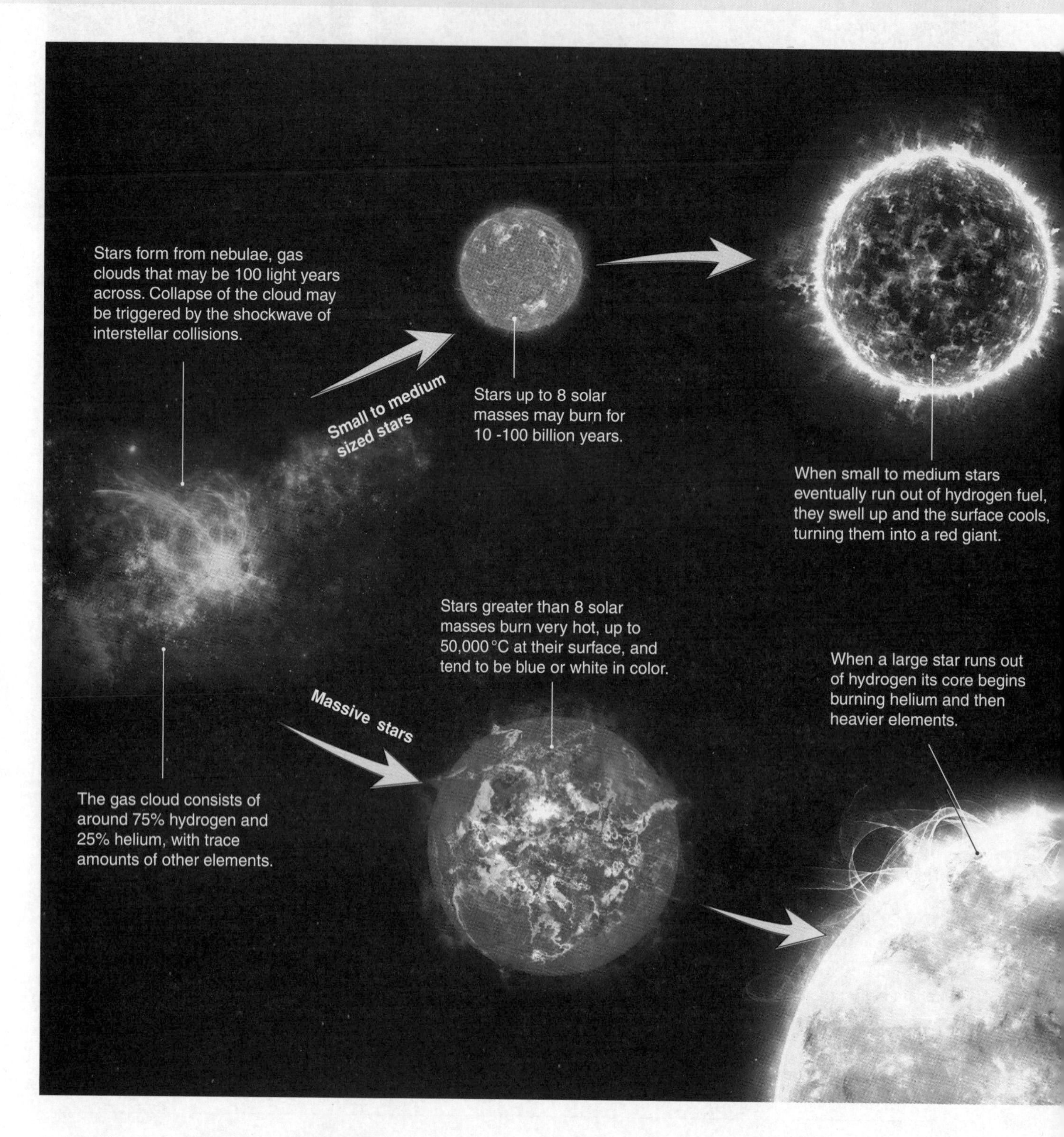


1. (a) What is the composition of a nebula? ______________________

   ______________________

   (b) What causes the collapse of a nebula? ______________________

   ______________________

2. What is the fate of a star less than 8 solar masses? ______________________

   ______________________

KNOW

WEB

CCC

CCC
EM

PRACTICES

**ISBN: 978-1-927309-37-7**

- A star forms from a cloud of dust and gas called a **nebula** (*pl.* nebulae) which may be many light years across. The nebula may begin to collapse due to a nearby shockwave and continue to shrink under gravity, releasing heat as it gets smaller. The heat eventually reaches temperatures hot enough to start nuclear reactions and ignite the star.
- The path of a star's life cycle depends on its mass. A very large star burns its fuel very quickly and may eventually end its life in a **supernova**. Stars similar to the Sun burn their fuel far more slowly and end as **white dwarf** stars.

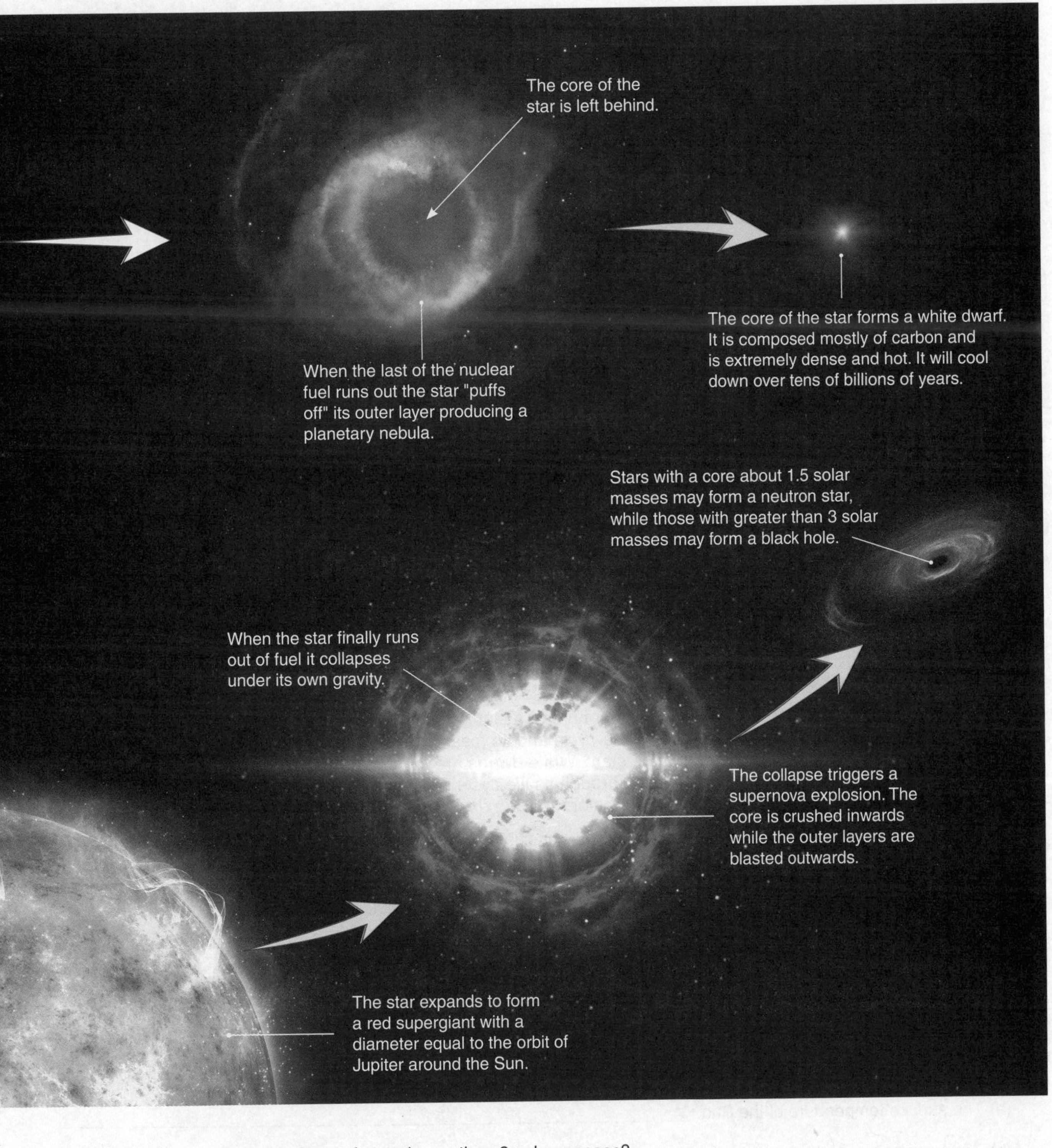


3. What are the two possible fates of stars larger than 8 solar masses? ______________________

______________________

4. What is a planetary nebula? ______________________

______________________

______________________

5. What is the element that forms at the core of a red giant star? ______________________


**ISBN: 978-1-927309-37-7**

# 19 Interpreting Hertzsprung-Russell Diagrams

**Key Idea**: The Hertzsprung-Russell diagram is a scatter graph showing luminosity vs spectral class. It can be used in determining the life stage of a star.

## Hertzsprung-Russell diagram

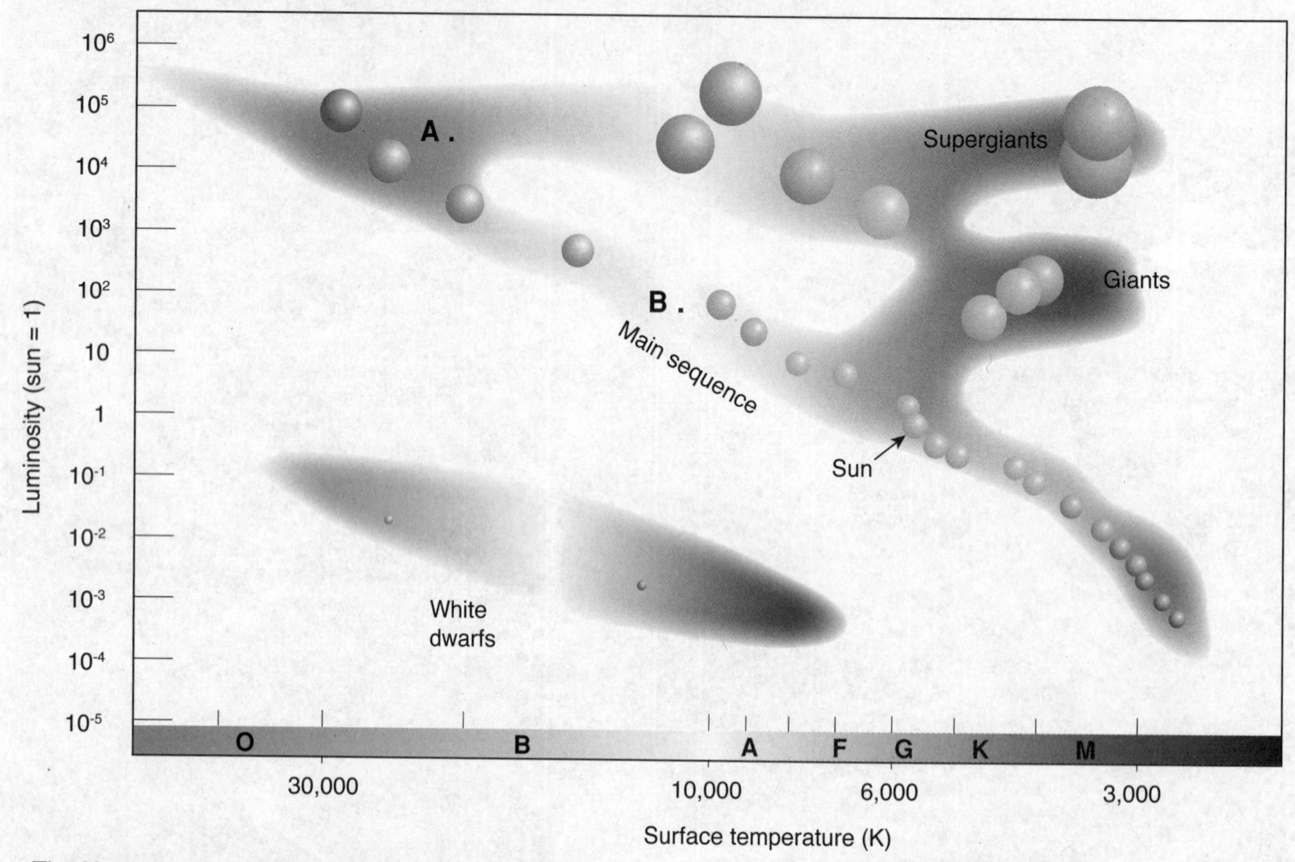


- The Hertzsprung-Russell diagram plots the temperature of a star against its luminosity. The position of the star on the graph tells us about its present stage in its life cycle and its mass. For most of their life, stars are found in the main sequence and are called main sequence stars. White dwarfs are found near the bottom left while giant stars are found in the top right. As a star progresses through its life cycle it will move off the main sequence. A star like the Sun will move to the upper right (red giants) before moving down to the lower left (white dwarfs).
- The diagram may list surface temperature or spectral classification along the horizontal axis. The diagram can also be used to estimate the distance to a star using the star's apparent magnitude and the absolute magnitude of a similar star with a known distance. The observed star is shifted vertically on the graph to overlap the reference star. The difference in magnitude shift to match the stars relates directly to the distance of the observed star.

## Star spectral classes

- Originally stars were classified on the strength of the hydrogen lines in their absorption spectrum. These classifications were eventually rearranged to reflect the temperature of the star. Type O stars are the hottest, with surfaces of tens of thousands of degrees and shine in the blue part of the spectrum. M type stars are cool and a dull red. Each class has ten divisions (0-9). The Sun is a G2 class star (a yellow dwarf).

1. Use the Hertzsprung-Russell diagram to determine the following:

   (a) The surface temperature of the Sun: ______

   (b) The temperature and luminosity of the star at the point labeled A: ______

   (c) The temperature and luminosity of the star at the point labeled B: ______

2. Why are most stars found in the main sequence part of the diagram? ______

______

______

______

KNOW

WEB

CCC

CCC

PRACTICES


**ISBN: 978-1-927309-37-7**

# 20 Red Giants

**Key Idea**: Red giant stars form when sun-like stars use up their supply of hydrogen and enter the final stages of their life cycle.

## Running out of hydrogen

- When the core of a star similar to the Sun (up to eight solar masses) begins to run out of hydrogen (after about 10 billion years) the core contracts due to gravity, increasing the pressure and temperature. This causes hydrogen fusion to begin in a shell around the core, while helium fuses to carbon in the core. The extra heat and light produced expands the star's outer layers. Because of the extra surface area, the surface temperature drops, but luminosity increases up to 1000 times because of the extra size.
- Stars with masses 0-8 times that of the Sun do not produce enough gravitational pressure in the core to progress fusion beyond carbon and oxygen. As the last of the helium is converted to carbon, fusion stops and the outer layers are lost as a planetary nebula (nothing to do with planets). The core will shine as a white hot, extremely dense white dwarf star.

Red giant stars have a core of helium fusing to carbon and oxygen surrounded by a core of hydrogen fusing to helium.

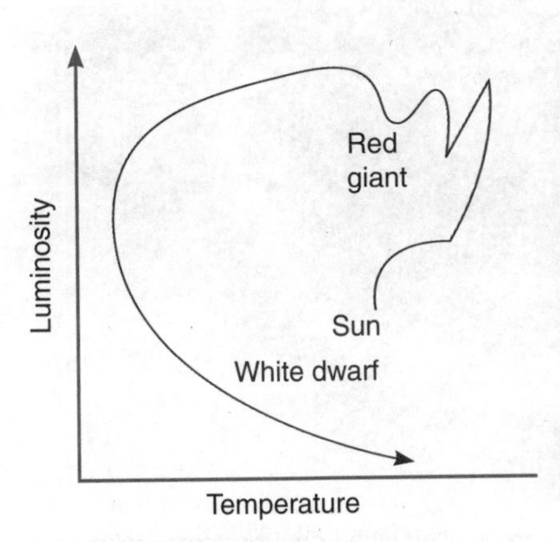


As a Sun-like star nears the end of its life, it moves off the main sequence as a red giant. Over the final two billion years, the star may move back and forth about the Hertzsprung-Russell diagram as its core progressively burns hydrogen and helium in different layers.

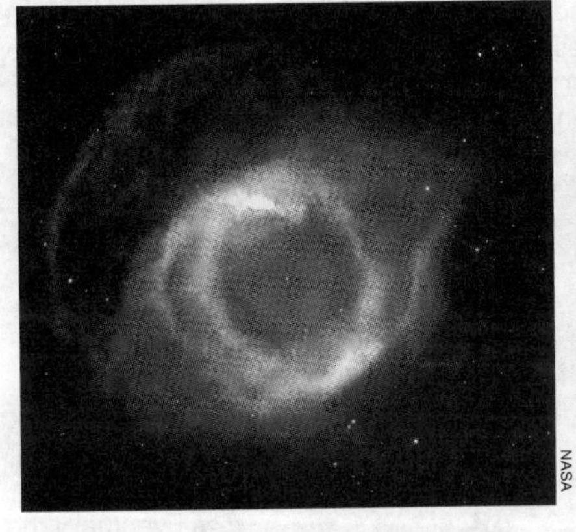


Eventually the star runs out of hydrogen and helium. The core temperature and pressure are too low to continue fusion and a core of carbon and oxygen forms. The hot outer layers are shed into space as a planetary nebula, leaving behind a dense white hot core.

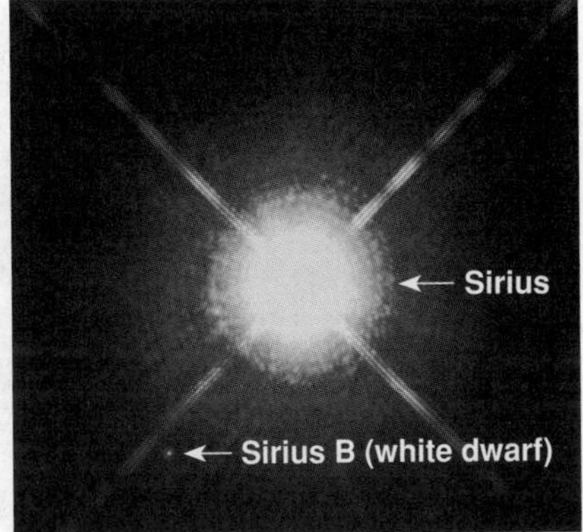


White dwarfs are the left over cores of a Sun-like star. They can be a million times more dense than the original star's average density, with a density of $10^6$ g/cm$^3$. No longer supporting nuclear fusion, the white dwarf slowly cools to a black dwarf over about a quadrillion years.

1. Give a brief description of why a sun-like star swells to a red giant near the end of its life:

2. Explain why red giants have very high luminosity even though they have very low surface temperatures:

3. What is the cause of a planetary nebula?

4. What is the composition of a white dwarf?

5. Why is it highly unlikely that there are any black dwarfs in the universe?


**ISBN: 978-1-927309-37-7**


PRACTICES  CCC EM CCC SPQ WEB 20 KNOW

# 21 Supernovae

**Key Idea**: A supernova is the explosive end of a massive star. It may result in a neutron star or a black hole. It is also responsible for forming the heavy elements.

Supernovae are some of the most violent events in the universe. They produce so much energy in one monumental explosion that they can outshine a whole galaxy. The luminosity of such an explosion may be five trillion times more than the Sun. Depending on the size of the original star a neutron star or black hole may form immediately after the explosion.

1. Stars greater than eight times the mass of the Sun produce enormously high temperatures in their core. The heat and light from their nuclear reactions pushes the gases out away from the core. However their immense gravity pulls the gases back in. For the majority of its life the giant star is at equilibrium between these opposing forces.

2. In the core, hydrogen fuses into helium. Eventually the hydrogen in the core begins to be used up. The core contracts under gravity forming a core of helium surrounded by a shell of hydrogen. As the core contracts and temperature and pressure rise, the outer layers are pushed further outwards and the star forms a red supergiant. The helium core ignites and begins to produce heavier elements such as carbon and oxygen.

The heavy elements sink to the core where they ignite and fuse to form heavier elements. The star may form onion like layers of heavier and heavier elements undergoing fusion.

3. Eventually the core is filled with iron. Iron atoms take more energy to fuse than the fusion produces, and nuclear fusion stops. When the core of inert iron reaches about 1.4 solar masses the core no longer produces the force to hold the rest of the star up. The force of the star's gravity takes over and the core and the star collapses.

1. (a) What is a supernova? ____________________

   (b) What size stars can undergo a supernova? ____________________

2. During the life of a massive star, what keeps the star from collapsing under its own gravity? ____________________

3. Why does a massive star form onion-like layers of elements near the end of its life? ____________________

4. What event triggers the collapse of a massive star? ____________________

**ISBN: 978-1-927309-37-7**

4 The implosion is cataclysmic and takes just a few seconds. It is calculated that the speed of the collapsing layers may reach 23% of the speed of light and the core reaches 100 billion K. Protons and electrons are crushed together to form neutrons. Neutrons are forced together and the core becomes so dense that the falling layers rebound outwards in a titanic explosion. The pressures produced are high enough to cause the iron and other elements in the core to fuse into even heavier elements such as gold.

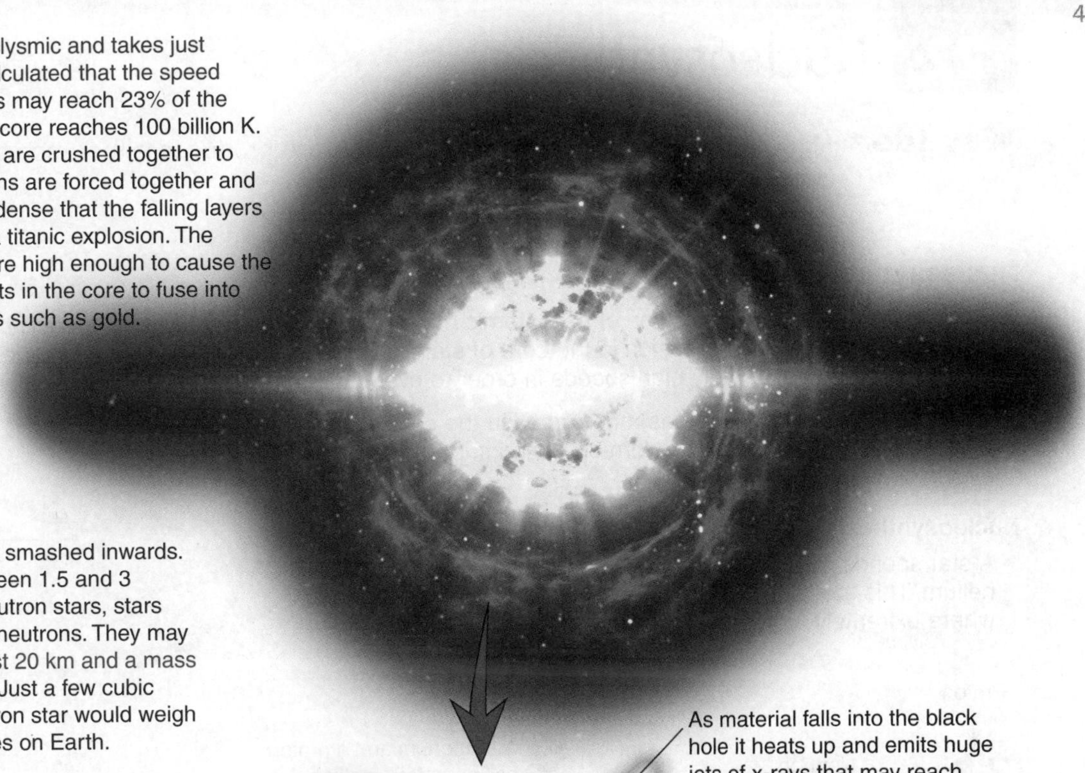

5 The core of the star is smashed inwards. Stars with cores between 1.5 and 3 solar masses form neutron stars, stars composed entirely of neutrons. They may have a diameter of just 20 km and a mass twice that of the Sun. Just a few cubic centimeters of a neutron star would weigh at least 5 trillion tonnes on Earth.

6 In stars with cores greater than 3 solar masses, nothing is strong enough to withstand the force of gravity pulling the core inwards and it collapses to a black hole, a point with no dimensions and gravity so great that not even light can escape.

As material falls into the black hole it heats up and emits huge jets of x-rays that may reach millions of kilometers into space.

**Event horizon** - the point at which nothing, not even light can escape from the black hole's gravity.

**Accretion disk** - material swirling around the black hole forms a disk.

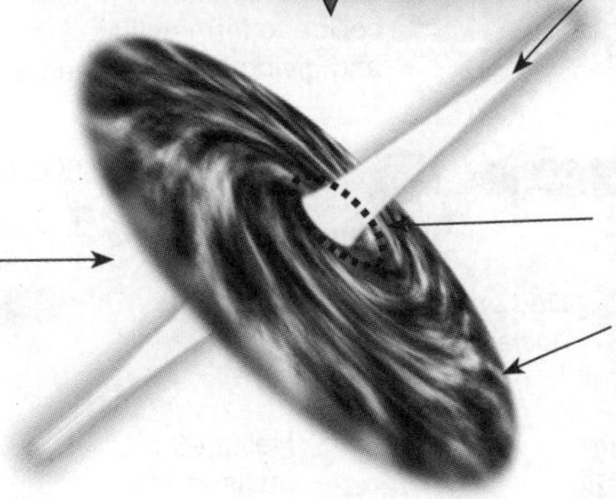

NASA/ESA

Supernova 1994D (arrowed left) in galaxy NGC 4526 taken by the Hubble Space Telescope. Notice how the bright supernova easily matches the brightness of the entire galaxy beside it.

Cygnus X-1 (right) was the first black hole discovered. It has a mass of about 15 times that of the Sun and an event horizon of just 44 km. As the black hole can not be seen, the image is of the x-rays emitted by superheated gas surrounding it.

NASA

5. How does a neutron star form? ______________________

6. (a) What is a black hole? ______________________

(b) How does a black hole form? ______________________

(c) What is the event horizon of a black hole? ______________________


**ISBN: 978-1-927309-37-7**

# 22 Nucleosynthesis

**Key Idea**: Most of the elements we know of today were produced either moments after the Big Bang or during the life or death of a star.

## Nucleosynthesis

- Nucleosynthesis is the production of new atomic nuclei from pre-existing ones. Nucleosynthesis most commonly occurs in nature by nuclear fusion in the core of stars. Nuclear fusion requires enormous energy as the nuclei must be accelerated to extremely high speeds in order to overcome the repulsive forces that normally keep them apart.
- Elements up to iron (26 protons) are formed in the core of stars. Elements heavier than iron are formed during supernovae, the explosion of giant stars.

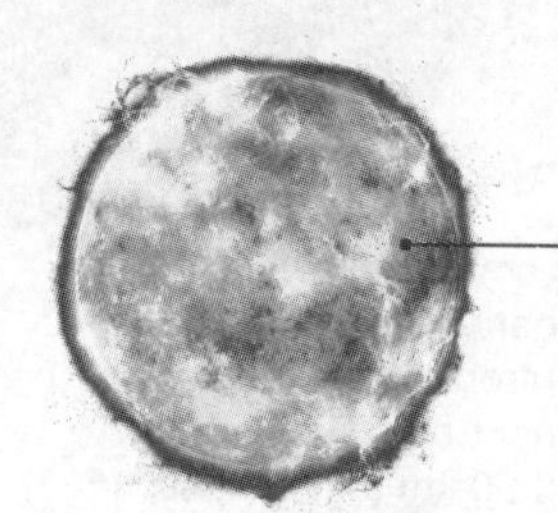

A star is a huge nuclear fusion reactor. The fusion reaction shown below is called the proton-proton chain reaction.

### Nucleosynthesis of helium

- A star spends most of its existence converting hydrogen into helium. This **stellar nucleosynthesis** occurs in the core of a star where extremely high temperatures and pressures are found.

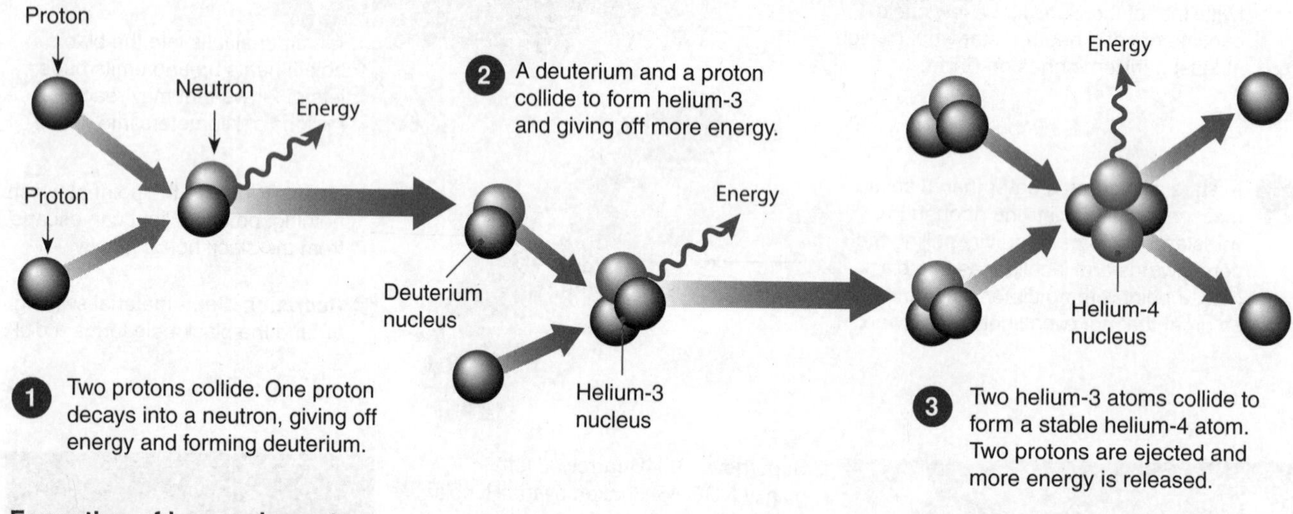


### Formation of heavy elements

- Fusion of iron atoms with other atoms to form heavy atoms requires energy. The energy comes from a supernova. **Supernova nucleosynthesis** occurs during the explosion of a massive star. Atoms, protons, and neutrons are smashed together and produce heavy atoms such as uranium, lead, and gold.

The nucleus of an iron atom is extraordinarily stable (right). More energy is required to fuse iron atoms with other atoms than is produced from their fusion. Thus when iron forms inside a star the process of nucleosynthesis stops and the star collapses.

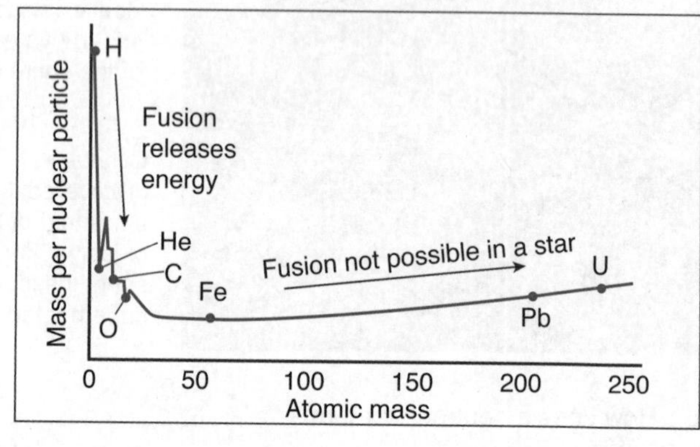


1. Define nucleosynthesis: ______________________________

2. Why does the formation of iron lead to the collapse of a star? ______________________________

3. Study the nucleosynthesis of helium diagram. How many protons go to form helium-4? ______________________________

4. How do elements heavier than iron form? ______________________________

PRACTICES


ISBN: 978-1-927309-37-7

# 23 Chapter Review

Summarize what you know about this topic so far under the headings provided. You can draw diagrams or mind maps, or write short notes to organize your thoughts. Use the introduction and the images and hints included to help you:

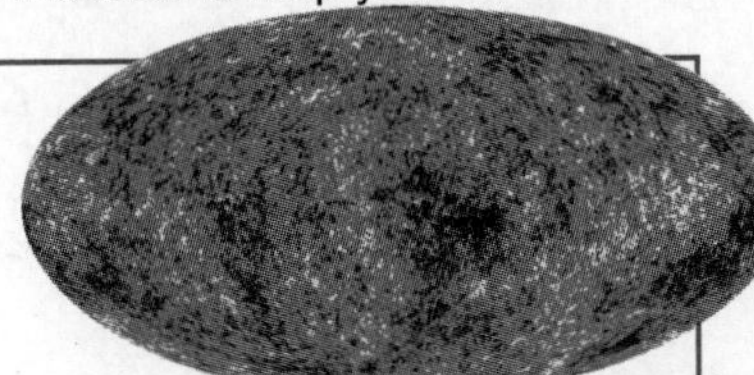

The universe and the Big Bang

HINT: The size of the universe and the evidence for the Big Bang.

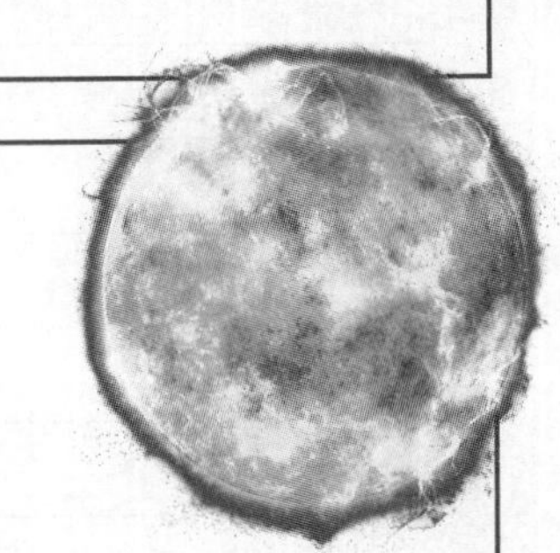

Stars and stellar evolution

HINT: The composition of stars and how their size affects their evolution.

# 24 KEY TERMS AND IDEAS: Did You Get It?

1. Test your vocabulary by matching each term to its definition, as identified by its preceding letter code.

|  | |  |
|---|---|---|
| Big Bang theory | A | Scatter plot in which the luminosity of the stars are plotted against their temperature. |
| cosmic microwave background | B | The production of heavy elements from the nuclear fusion of lighter ones, e.g. four hydrogen atoms fusing to form a helium atom. |
| element | C | The star in the center of our solar system. |
| Hertzsprung–Russell diagram | D | Microwave radiation that permeates the universe and is a remnant of the Big Bang itself. |
| main sequence star | E | A phenomenon in which the light waves from a galaxy moving away from the observer are stretched. This has the effect of moving absorption lines towards the red end of the spectrum. |
| nebula | F | A large interstellar cloud of dust and gas. |
| nucleosynthesis | G | A giant luminous sphere of plasma undergoing nuclear fusion held together by its own gravity. Its composition is primarily hydrogen and helium. |
| red shift | H | Chemical substance consisting only of atoms with the same number of protons. |
| star | I | Star that appears in the band of stars than runs from the top left to bottom right of a Hertzsprung-Russell diagram when its luminosity is plotted against its temperature. |
| Sun | J | The prevailing cosmological model for the evolution of the universe, stating that the universe expanded form a infinitely dense point 13.7 billion years ago to the universe we see today. |

2. The Sun is medium sized yellow star. Explain how its life cycle will differ from a giant blue star:

3. Briefly describe three pieces of evidence for the Big Bang:

(a)

(b)

(c)




ISBN: 978-1-927309-37-7

# 25 Summative Assessment

1. Use the list below to draw and label a star's potential life cycle. Include a brief description of each stage. Identify the pathway that represents our Sun.

   *protostar, small-medium star, large-star, white dwarf, black hole, neutron star, supernova, red giant, red super-giant, planetary nebula, nebula:*

2. The Big Bang occurred 13.7 billion years ago. During this time hydrogen and some helium formed (about 75% and 25% of all matter respectively). Explain why other heavier elements were not formed and where those heavier elements came from instead:


**ISBN: 978-1-927309-37-7**


PRACTICES

TEST

3. A group of astronomers studying a cluster of 20 stars produced data on the color, temperature, and luminosity. The data are tabulated below:

| Star | Temperature (K) | Color | Luminosity (solar units) |
|---|---|---|---|
| 1 | 3000 | M | 100,000 |
| 2 | 3500 | M | 1000 |
| 3 | 4000 | M | 4500 |
| 4 | 6500 | G | 55,000 |
| 5 | 3200 | M | 3000 |
| 6 | 3500 | M | 90,000 |
| 7 | 3100 | M | 30,000 |
| 8 | 5900 | G | 50,000 |
| 9 | 4000 | M | 90,000 |
| 10 | 3200 | M | 800 |

| Star | Temperature (K) | Color | Luminosity (solar units) |
|---|---|---|---|
| 11 | 4500 | M | 85,000 |
| 12 | 5900 | G | 50,000 |
| 13 | 4900 | K | 52,000 |
| 14 | 3900 | M | 7000 |
| 15 | 4000 | M | 2300 |
| 16 | 5000 | K | 46,000 |
| 17 | 4200 | M | 67,000 |
| 18 | 3900 | M | 34,000 |
| 19 | 5700 | K | 58,000 |
| 20 | 3800 | M | 2000 |

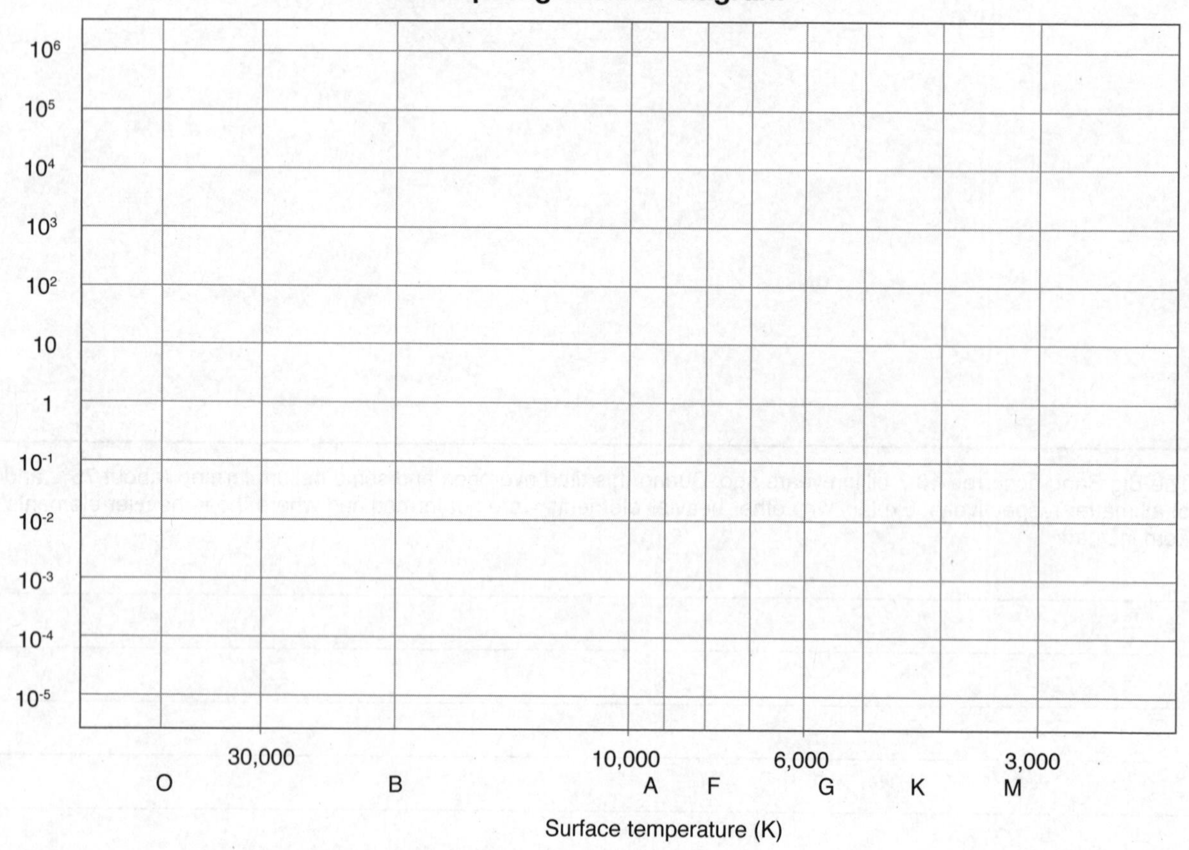


(a) Plot the data on the Hertzsprung-Russell diagram above:

(b) Describe the spread of the data on the graph: __________

(c) What does this tell you about the stars in the cluster and therefore about the cluster itself? __________


**ISBN: 978-1-927309-37-7**

ESS1.B

# Earth and the solar system

## Key terms

apparent retrograde motion

eccentricity

exoplanet

gravity

Kepler's laws

nebula

newton (N)

Newton's law

orbit

planet

retrograde motion

satellite

solar system

## Disciplinary core ideas

Activity number

***Show understanding of these core ideas***

**The motions of orbiting objects show common features**

| | | Activity number |
|---|---|---|
| ☐ | 1 The solar system began as a nebula, which collapsed, forming the Sun and a flattened disk from which the planets, moons, asteroids, and other bodies formed. Features of the planetary orbits and the composition of the gas planets provide evidence of how the solar system formed. | 26 |
| ☐ | 2 The solar system consists of the Sun, eight planets, a number of dwarf planets and other smaller bodies. The Sun is the overwhelming source of gravity in our Solar system. Its mass causes the planets to describe elliptical orbits around it. | 27 38 |
| ☐ | 3 The four inner planets are rocky. Information about these planets and their orbits comes from space probes, satellites, and planetary rovers. | 28-31 |
| ☐ | 4 The four outer planets are mainly gas. Information about these planets and their orbits comes from unmanned spacecraft. The presence of Neptune was predicted from the orbit of Uranus and discovered using Newton's laws of gravity. | 32-35 |
| ☐ | 5 The solar system includes many dwarf planets. These bodies do not meet the gravitational and orbital criteria to be classed as planets. Planets outside the solar system can be found using empirical evidence from stars other than our Sun. | 36 37 |
| ☐ | 6 The apparent retrograde motion of planets viewed from a point on Earth is the result of the motions of the planets relative to each other. | 42 |
| ☐ | 7 The orbits of planets follow Kepler's laws, which are mathematical rules describing common features of the motions of orbiting objects. | 38 39 40 |
| ☐ | 8 Newton's Law of Universal Gravitation can be applied to Kepler's laws to derive a mathematical model of how all orbiting objects in the universe behave. | 40 |
| ☐ | 9 Orbits can change due to the gravitational effects from, or collisions with, other objects in the solar system. | 32 43 |

**Cyclical changes in the Earth's orbit and tilt cause climate cycles**

| | | Activity number |
|---|---|---|
| ☐ | 10 Changes in Earth's orbit, tilt, and rotation cause cyclical changes on Earth. | 43 44 |

NASA

## Crosscutting concepts

***Understand how these fundamental concepts link different topics***

| | | Activity number |
|---|---|---|
| ☐ | 1 SPQ Algebraic thinking is used to examine data and predict the motion of orbiting objects in the solar system. | 38-43 |
| ☐ | 2 CE Empirical evidence enables us to support claims about the causes of planetary orbits and environmental cycles. | 31 32 34 35 42 43 |

## Science and engineering practices

***Demonstrate competence in these science and engineering practices***

| | | Activity number |
|---|---|---|
| ☐ | 1 Use mathematical representations to describe and predict the motion of orbiting objects in the solar system. | 35-43 47 |
| ☐ | 2 Use a model to describe the behavior of planetary bodies and their satellites. | 27 42-44 |

# 26 Formation of the Solar System

**Key Idea**: The solar system formed 4.5 billion years ago out of a huge nebula of dust and gas.

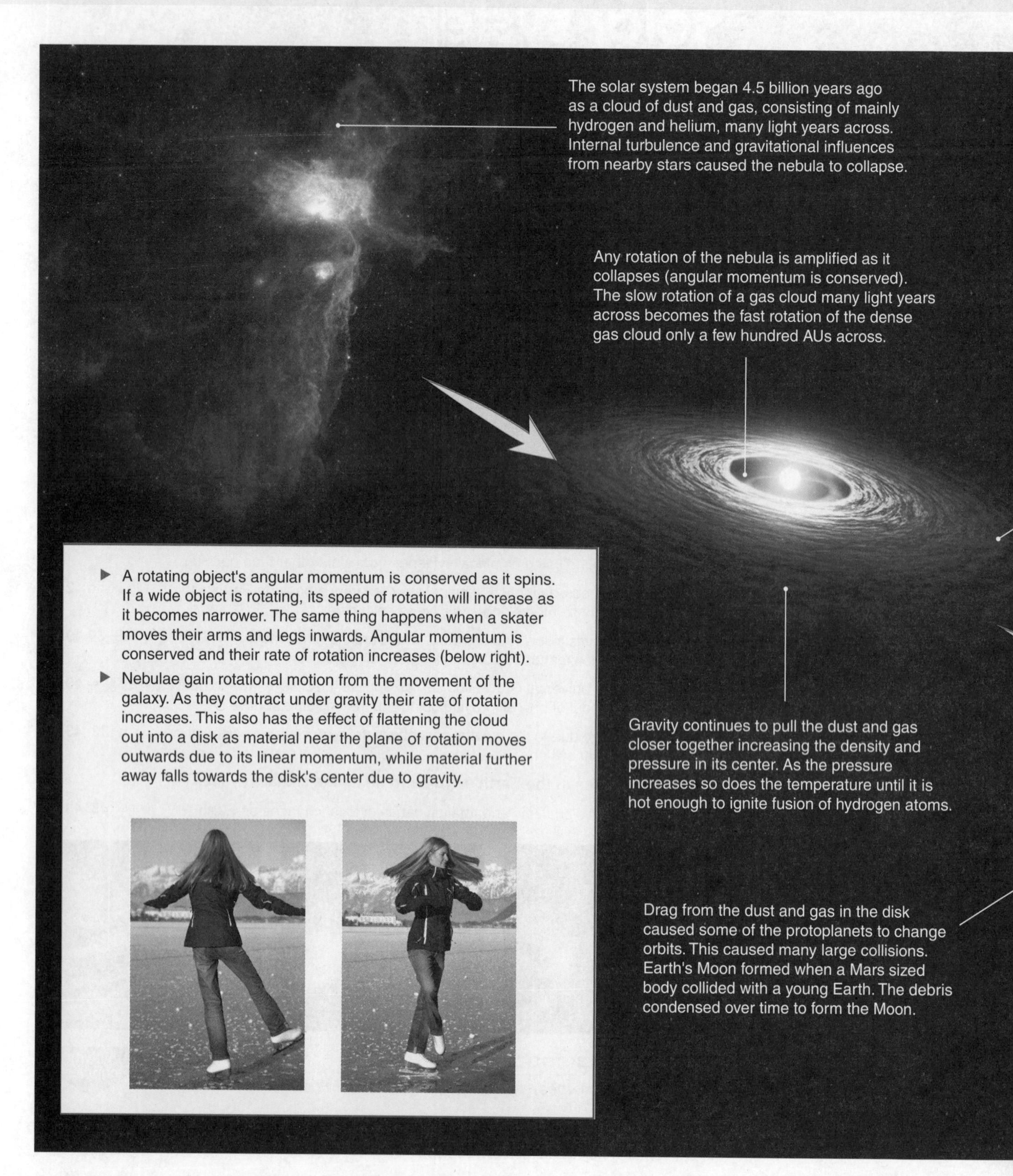


- A rotating object's angular momentum is conserved as it spins. If a wide object is rotating, its speed of rotation will increase as it becomes narrower. The same thing happens when a skater moves their arms and legs inwards. Angular momentum is conserved and their rate of rotation increases (below right).
- Nebulae gain rotational motion from the movement of the galaxy. As they contract under gravity their rate of rotation increases. This also has the effect of flattening the cloud out into a disk as material near the plane of rotation moves outwards due to its linear momentum, while material further away falls towards the disk's center due to gravity.

1. How does a large slowly spinning cloud of gas become a fast spinning disk? ______________________




ISBN: 978-1-927309-37-7

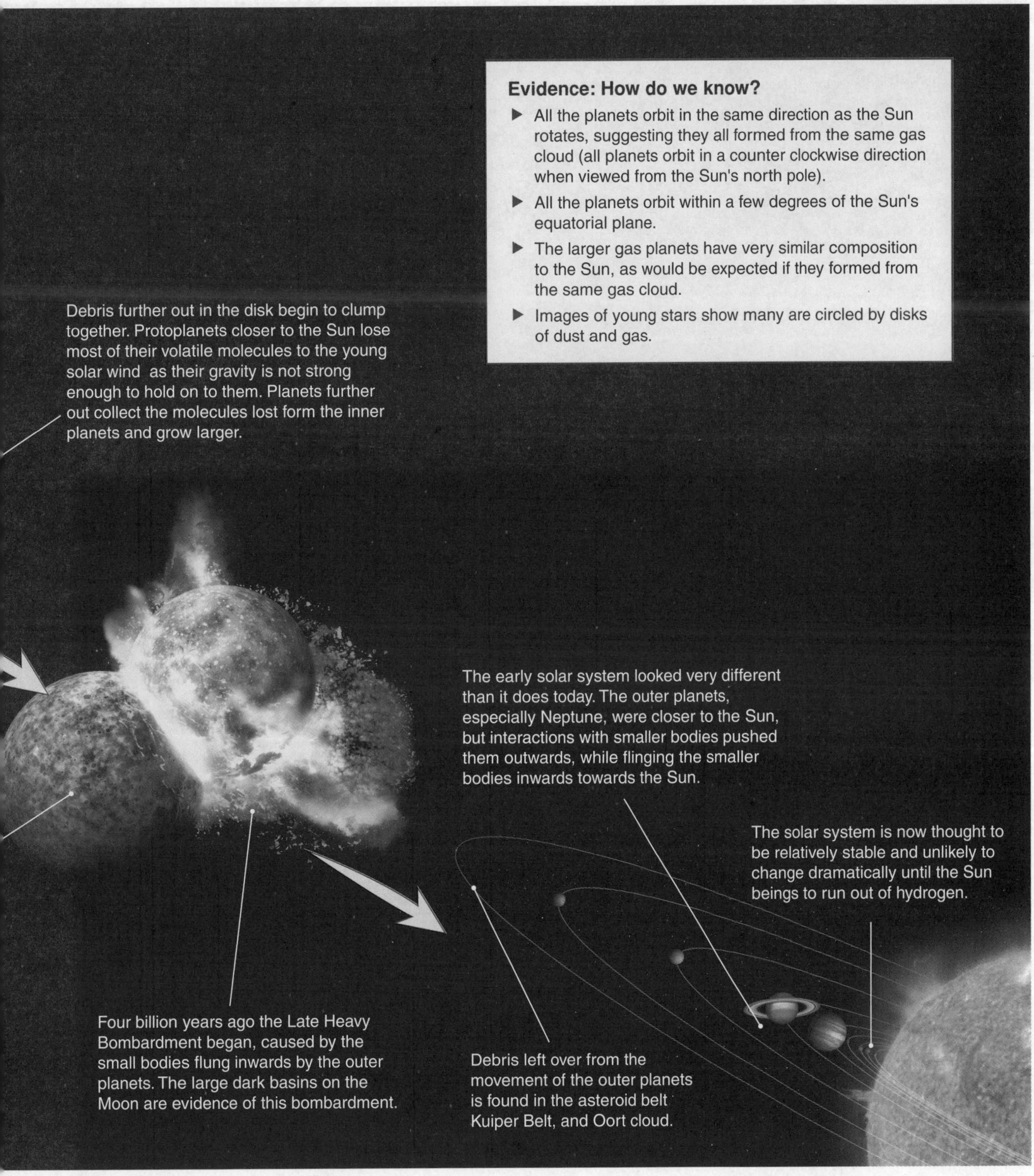

**Evidence: How do we know?**

- All the planets orbit in the same direction as the Sun rotates, suggesting they all formed from the same gas cloud (all planets orbit in a counter clockwise direction when viewed from the Sun's north pole).
- All the planets orbit within a few degrees of the Sun's equatorial plane.
- The larger gas planets have very similar composition to the Sun, as would be expected if they formed from the same gas cloud.
- Images of young stars show many are circled by disks of dust and gas.

Debris further out in the disk begin to clump together. Protoplanets closer to the Sun lose most of their volatile molecules to the young solar wind as their gravity is not strong enough to hold on to them. Planets further out collect the molecules lost form the inner planets and grow larger.

The early solar system looked very different than it does today. The outer planets, especially Neptune, were closer to the Sun, but interactions with smaller bodies pushed them outwards, while flinging the smaller bodies inwards towards the Sun.

The solar system is now thought to be relatively stable and unlikely to change dramatically until the Sun beings to run out of hydrogen.

Four billion years ago the Late Heavy Bombardment began, caused by the small bodies flung inwards by the outer planets. The large dark basins on the Moon are evidence of this bombardment.

Debris left over from the movement of the outer planets is found in the asteroid belt Kuiper Belt, and Oort cloud.

2. Describe three pieces of evidence for the current theory of the formation of the solar system: ______________________

_______________________________________________

_______________________________________________

_______________________________________________


**ISBN: 978-1-927309-37-7**

# 27 The Solar System

**Key Idea**: The solar system has eight planets, a number of dwarf planets, and countless numbers of asteroids and comets.

1. How many planets are there in the solar system? ______________________

2. What is the main difference between the inner planets and the outer planets? ______________________

______________________

3. What feature separates the inner planets from the outer planets: ______________________

______________________




ISBN: 978-1-927309-37-7

- The solar system consists of the Sun, eight planets, a number of dwarf planets, and other smaller object such as comets and asteroids. Planets orbit the Sun, have enough gravity to form a sphere, and sweep their orbit clear of debris. The dwarf planets are not large enough to sweep their orbits clear of debris.
- The four inner planets all have rocky surfaces, while the four outer planets are mainly gas. The inner planets are separated from the outer planets by the asteroid belt.

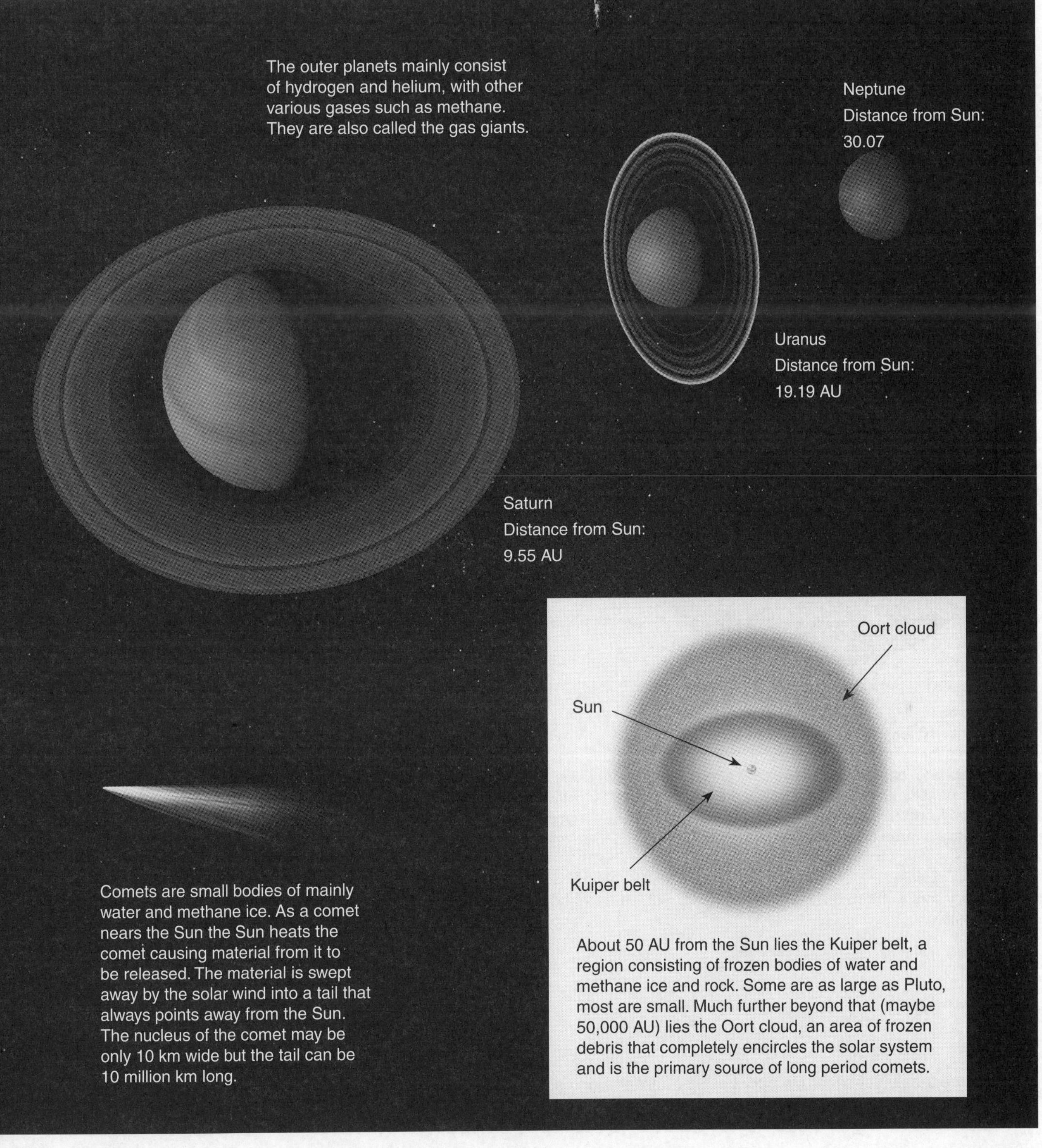


4. Name the three areas in the solar system where small frozen, rocky debris is found: ______________________

______________________________________________

5. (a) Which planet is nearest the Sun? ______________________

(b) Which planet is furthest from the Sun? ______________________

(c) Recall the definition of one AU. Calculate the distance of Uranus from the sun in km: ______________________


**ISBN: 978-1-927309-37-7**

# 28 Mercury

**Key Idea**: Mercury is the planet closest to the Sun and the smallest. Missions to Mercury have shown it to be a harsh, hot, and cratered planet.

| Statistics | |
|---|---|
| Radius (km) | 2440 |
| Year (Earth days) | 87.97 |
| Rotation (Earth days) | 58.7 |
| Mass (Earth =1 ) | 0.055 |
| Satellites | 0 |
| Mean temperature °C | 167 |

Mercury's surface is covered with deep craters. Near the north pole, ice has been detected in permanently shadowed parts of these craters. Mercury has a very weak magnetic field which implies it must still have a molten iron core. It rotates on its axis very slowly compared to its orbital period (a Mercurian day is twice as long as a year). Because of the relationship between Mercury's rotation and its orbit, the Sun in Mercury's sky rises from the east to its zenith, then moves back to the east, before tracking to the west and sunset. Mercury has a very thin atmosphere of oxygen, sodium, and hydrogen, but it is continuously lashed by high energy solar winds that stripped away most of the atmosphere long ago.

Core largely made of iron and about 3700 km in diameter

Crust of silicate rock

Mantle (fluid)

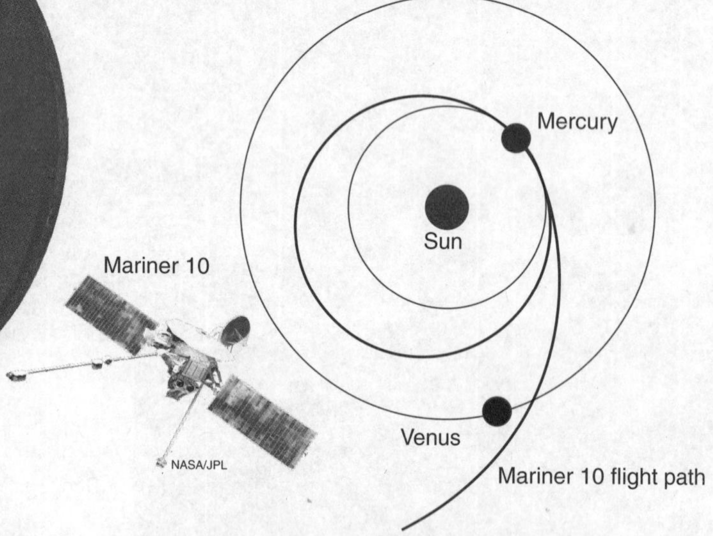

The Mariner 10 and Messenger probes included instruments for measuring the atmosphere, magnetosphere, and surface of Mercury. Calculations of Mercury's density (5.4 g/cm$^3$) show it is only slightly less dense than Earth (the most dense planet), so that it is likely to have a large iron core.

### Missions to Mercury

Although Mercury is relatively close to the Earth, it is one of the least visited planets with only two space probes visiting it. Mariner 10 flew by in 1974 and it was not until 2013 that the Messenger mission finished mapping the entire surface of the planet.

Mercury

Sun

Mariner 10

Venus

NASA/JPL

Mariner 10 flight path

The Mariner 10 mission was the first time a space probe used the gravity from another planet (Venus) to bend its trajectory (a gravitational assist). It then orbited the Sun once for every two times Mercury orbited, meeting the planet two more times, the first probe to return to a planet.

1. Mercury has a mean temperature of 167°C yet ice has been detected in deep craters. Explain how this presence of ice is possible.

2. Describe the evidence for Mercury having a molten iron core:

3. Describe how space probes use a gravitational assist to change course:




ISBN: 978-1-927309-37-7

# 29 Venus

**Key Idea**: Venus is similar in size and density to the Earth but its surface is the hottest of any planet in the solar system.

| Statistics | |
|---|---|
| Radius (km) | 6052 |
| Year (Earth days) | 224.7 |
| Rotation (Earth days) | 243 (retrograde) |
| Mass (Earth =1) | 0.82 |
| Satellites | 0 |
| Mean temperature °C | 464 |

Compared to Earth, Venus' atmosphere is 50 times as dense and its atmospheric pressure is 92 times greater. It is composed of carbon dioxide with clouds of sulfuric acid. Its surface temperature is 464°C, the hottest of any planet. Space probes orbiting Venus have been equipped with radar in order to scan the surface as the thick atmosphere makes photography of the surface impossible. The Russian Venera landers have been the only probes to reach and photograph Venus' surface, each operating only a few hours before being crushed by the atmospheric pressure. Unlike Earth, Venus carries most of its carbon in the atmosphere as carbon dioxide. This has produced a run-away greenhouse effect that has led to what are often described as Hellish conditions.

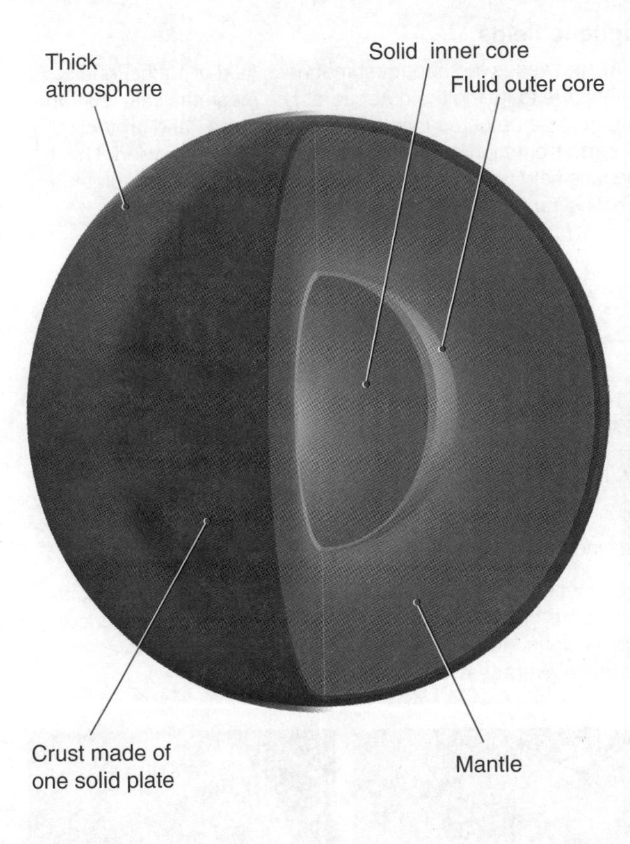


### Venus and the greenhouse effect

Why are the conditions on Venus so nightmarish? Part of the answer is that Venus is closer to the Sun than Earth. Any water it once had was boiled off and held in the atmosphere, beginning a run-away greenhouse effect. Sulfur dioxide from volcanoes and carbon dioxide stayed in the atmosphere instead of being rained out, accelerating the greenhouse effect even further. Radar imaging of Venus by the Magellan space probe has shown the crust is one huge plate, unlike Earth. It is also covered with at least 50,000 volcanoes that simply punch through the thin crust. Strangely Venus has a very weak magnetic field, meaning it does not contain a large iron core.

Venus' surface is covered with volcanoes (above). With no tectonic activity, the volcanic material piles up on itself.

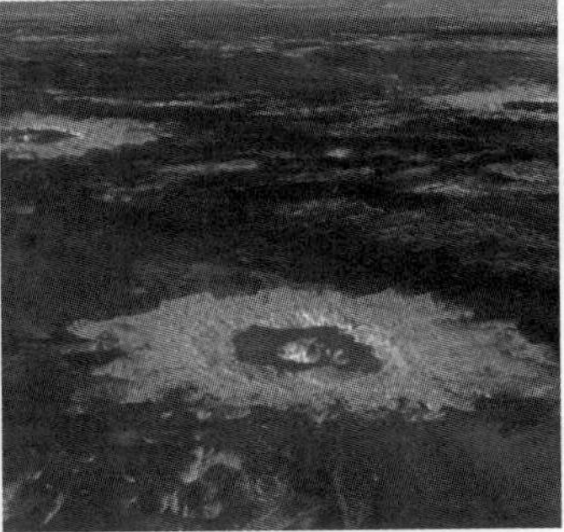

NASA/JPL

Radar on probes such Magellan were able to see through Venus' clouds and produce the high resolution topology maps.

1. (a) The density of Earth's atmosphere is 1.217 kg/m$^3$. Calculate the density of Venus' atmosphere: ______________________

   (b) For every 10 m descended in water, the pressure rises by the equivalent of 1 Earth atmosphere. Calculate the depth of water equivalent to the pressure of Venus' atmosphere and use this to suggest why this makes landing a probe on the surface so difficult:

   ______________________

2. What does the weak magnetic field tell you about the internal structure of Venus? ______________________

3. Explain why the study of Venus' atmosphere is useful when studying climate change here on Earth: ______________________


ISBN: 978-1-927309-37-7

# 30 Earth

**Key Idea**: Earth is the third planet from the Sun. It has a large magnetic field produced by a solid metal core surrounded by a liquid outer core.

| Statistics | |
|---|---|
| Radius (km) | 6371 |
| Year (Earth days) | 365.25 |
| Rotation (hours) | 23 h 56 min* |
| Mass | $6.0 \times 10^{24}$ kg |
| Satellites | 1 |
| Mean temperature °C | 15 |

* Rotation with respect to the stars (sidereal rotation).

Earth is our home in space and the only planet, as far as we know, to support life. It is a geologically active planet, with a thin crust composed of many plates that constantly shift over its surface. There is a strong magnetic field caused by the movement of a molten outer core. Earth's tilt with respect to the Sun results in seasons. Its atmosphere is composed of 21% oxygen, 78% nitrogen, and 1% trace gases. Unlike Venus, Earth carries only a small amount of carbon dioxide gas in the atmosphere adding to a limited greenhouse effect. The atmosphere keeps the planet warm but also allows excess heat to radiate into space. It is the only planet with large amounts of liquid water on the surface.

Outer core
Inner core
Crust
Outer mantle
Inner mantle

### Magnetic fields

Earth possesses the strongest magnetic field of all the inner planets due to its molten outer core. The magnetic field is strong enough to be detected using a simple compass and protects the Earth from high energy particles from the solar wind. The magnetic field produces the Van Allen belts, areas of high energy electrons captured from the solar wind.

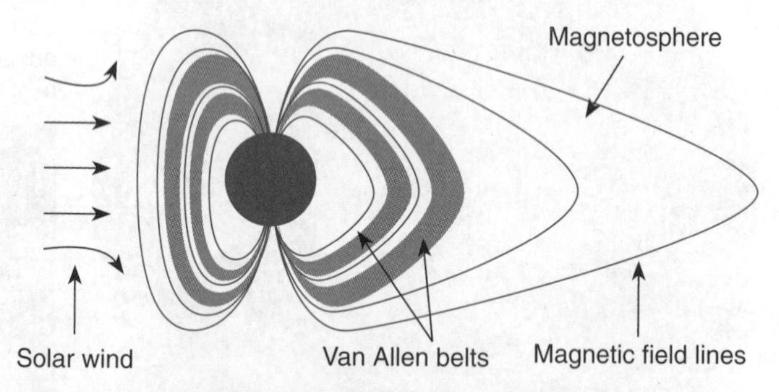


The magnetic field is not aligned with the Earth's rotational poles, there is a 10° difference. The field wanders over time, evidence of which is found in the magnetic lines in rocks. It also reverses direction on average every million years.

Earth's orbit is almost circular. Even so there is a 5 million km difference between when it is closest to the Sun and furthest away.

Artificial satellites orbiting the Earth do everything from providing communications to studying its surface. Many satellites are in Low Earth Orbit.

To maintain LEO, a satellite must be travelling at about 7.8 km/s. At this speed it will complete an orbit of the Earth every 90 minutes.

The Moon orbits the Earth every 27 days. The Moon has a synchronous rotation with the Earth so only one side ever faces the planet.

All images NASA

1. What is unusual (compared to other planets) about how water is found on Earth? ______

2. How is Earth's magnetic field formed and why is it important to life on Earth? ______

3. What causes the seasons on Earth? ______




**ISBN: 978-1-927309-37-7**

# 31 Mars

**Key Idea**: Mars is the most explored of all the extraterrestrial planets. Evidence shows it once had liquid water on its surface. However it is now dry and frozen.

| Statistics | |
|---|---|
| Radius (km) | 3386 |
| Year (Earth days) | 686.97 |
| Rotation (Earth days) | 1.03 |
| Mass (Earth = 1) | 0.11 |
| Satellites | 2 |
| Mean temperature °C | -63 |

There have been many orbiter and lander missions to Mars, including the spectacularly successful Spirit and Opportunity twin rovers. Evidence shows that Mars once had liquid water on its surface (and may still have beneath the surface). Despite its small size and distance from the Sun, Mars has similarities to Earth. Its axial tilt is almost the same as Earth's so it experiences seasons (although each is twice as long as those on Earth). Its day length is also very similar to Earth's. However it has lost most of its water and atmosphere to space due to its weak gravity, making it frigidly cold, and there is little protection from high energy particles from space.

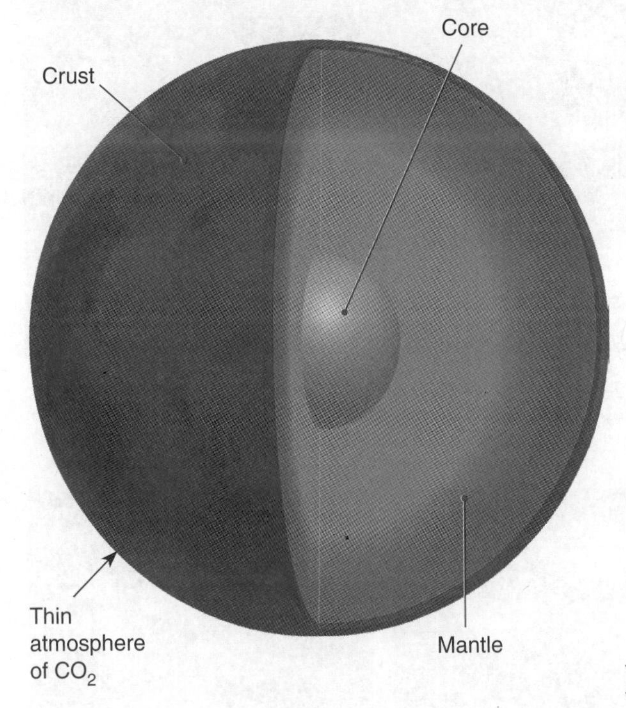


## Gravity on Mars

Because of its small size, the force of gravity on Mars is just 38% of Earth's. For every 1 kg here on Earth, the force of gravity (weight) is about 10 Newtons (N). Therefore a person with a mass of 100 kg has a weight of 1000 N. However on Mars, a 100 kg person has a weight of just 380 N (their mass remains the same).

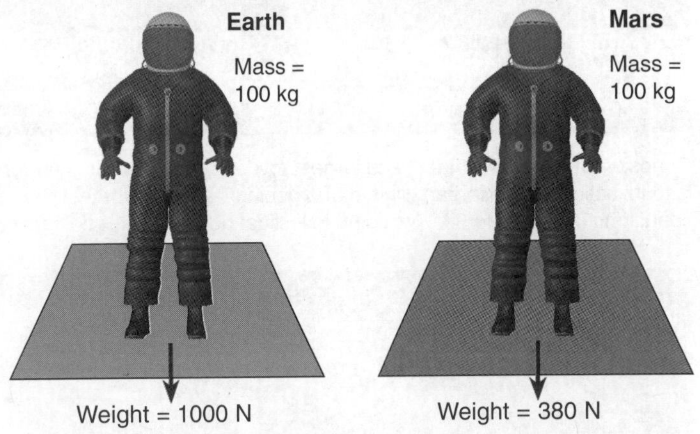


Your mass, the amount of matter in you, remains the same everywhere. Your weight (force due to gravity) changes depending on where you are.

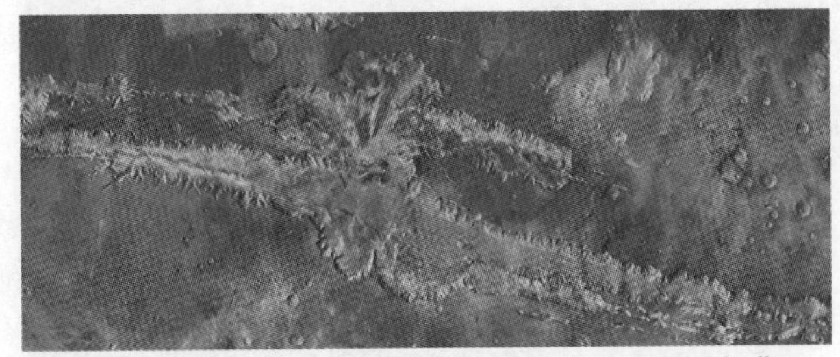

Mars is covered in interesting geological features including ancient dry river beds, huge volcanoes, and deep canyons. Mariner Valley is 3000 km long and formed by rifting in Mars' crust.

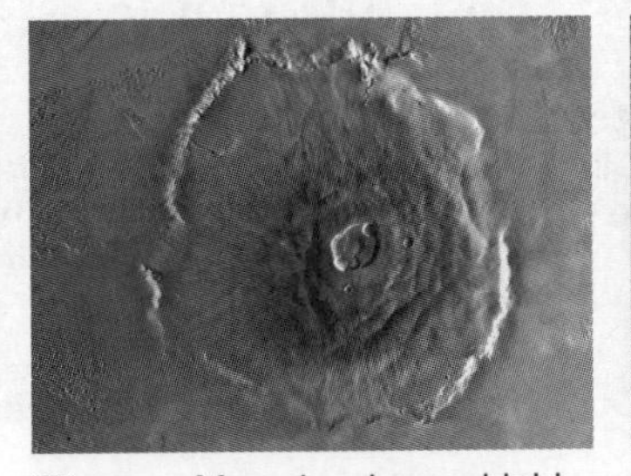

Olympus Mons is a huge shield volcano 25 km high, the largest in the solar system

Sedimentary rock like this is evidence of Mars having surface lakes and liquid water millions of years ago.

All images NASA

1. Identify a piece of evidence for Mars having surface water in the past: ______________________

2. Give a reason why Mars no longer has surface water: ______________________

3. The maximum a person on Earth can lift is about 270 kg. On Mars, a person could lift about 440 kg. Explain why?

______________________


**ISBN: 978-1-927309-37-7**

## Exploring Mars

There have been over forty missions to Mars, more than any other planet. An enormous amount of data has been gathered about the planet as a result. This data is useful for comparing geological processes here on Earth and in paving the way for a future manned mission to the red planet.

### Curiosity

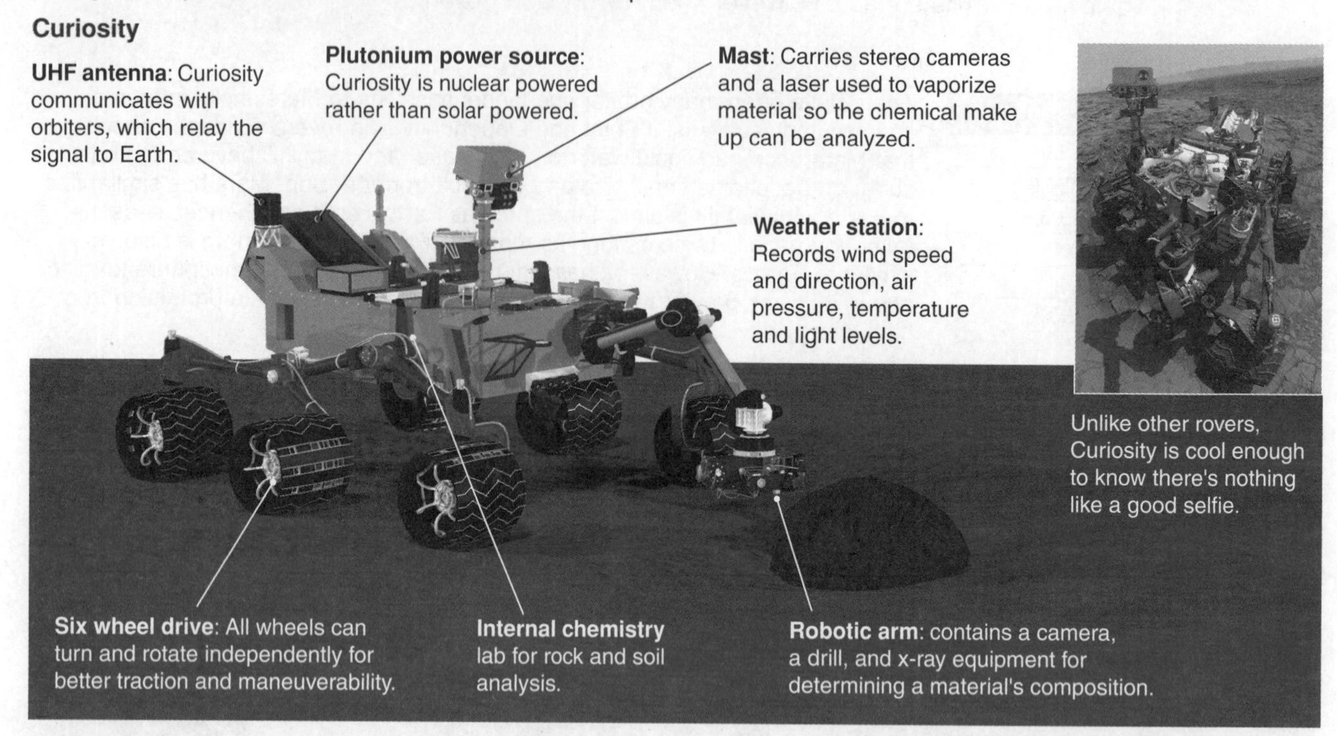


Unlike other rovers, Curiosity is cool enough to know there's nothing like a good selfie.

Curiosity (above) is the latest and largest rover sent to Mars. It is about the size of a small car and has a mass of 174 kg. Its mission is to investigate the Martian climate and geology, assess whether the target site inside Gale Crater has ever offered environmental conditions favorable for life, including investigating for water, and carry out studies relevant for any future human exploration.

Curiosity is so large that parachutes would not be enough to slow its descent. Instead, a "sky-crane" was used to slow the descent and lower Curiosity to the ground.

Curiosity is the fourth rover to land on Mars. The original Sojourner rover is tiny in comparison. The rover Opportunity has been on Mars since 2004 and travelled more than 40 km.

The data and experience gathered and the technology developed from the rover programs will be used in future manned missions to Mars.

All images NASA

4. Why were parachutes not used to slow Curiosity on its final descent? ______________________________

______________________________________________________________________

5. How does Curiosity communicate with Earth? ______________________________

______________________________________________________________________

6. List some reasons why landers and rovers are sent to explore Mars rather than manned missions: ______________

______________________________________________________________________

______________________________________________________________________

______________________________________________________________________

______________________________________________________________________

______________________________________________________________________


**ISBN: 978-1-927309-37-7**

# 32 Jupiter

**Key Idea**: Jupiter is the largest of all the solar system's planets. Its influence on the solar system is second only to the Sun.

| Statistics | |
|---|---|
| Radius (km) | 69,911 |
| Year (Earth years) | 11.86 |
| Rotation (hours) | 9.93 |
| Mass (Earth =1 ) | 318 |
| Satellites | 67 |
| Mean temperature °C | -120 |

Jupiter is massive. It has more mass and more volume than all the other planets put together. Like the Sun, it is made mostly of hydrogen and helium. Its atmosphere is divided into bands caused by its rapid rotation and high speed winds. Its lower hemisphere is dominated by the Great Red Spot, a storm three times the diameter of Earth that has been raging for at least 400 years. Jupiter has at least 67 moons, the four largest were first spotted by Galileo and are of particular interest due to their varied environments. Jupiter has a massively strong magnetic field that extends past Saturn.

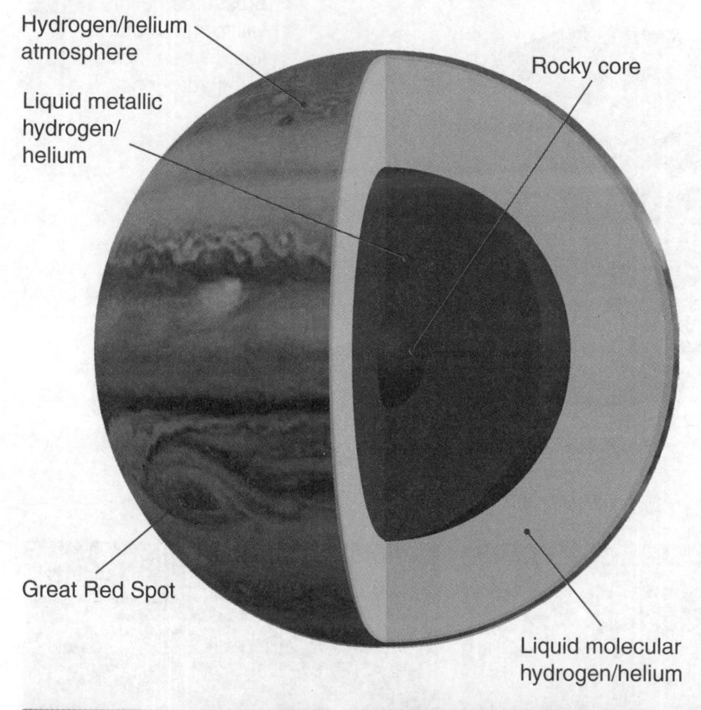


### The king of the asteroids

Jupiter's gravity has a large effect on asteroids. It prevents asteroids from having orbits whose orbital period would be in resonance with its own, e.g. 2:1, 3:1, and 4:1 orbits for asteroids compared to Jupiter do not exist.

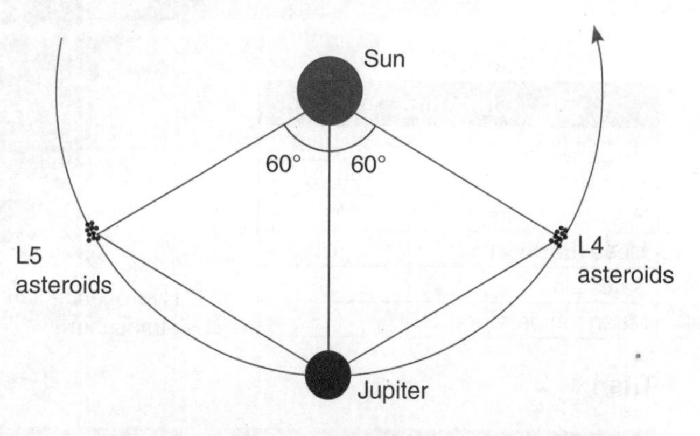


Trojan asteroids are caught in gravitationally stable areas ahead or behind Jupiter called Lagrangian points (L4 and L5). Once in these areas the asteroids are doomed to run before and trail after Jupiter forever.

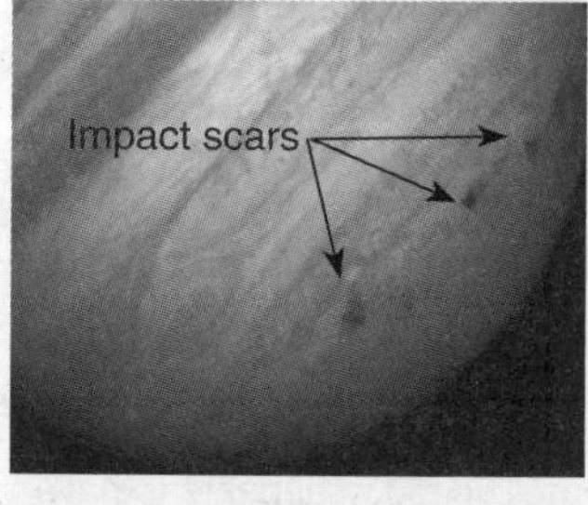


Jupiter's enormous gravity was demonstrated in 1994. The comet Shoemaker-Levy passed too close to Jupiter and was torn apart, slamming into it.

Jupiter's moon Io is the most geologically active body in the solar system. Jupiter's gravity flexes Io, heating its interior creating many volcanoes.

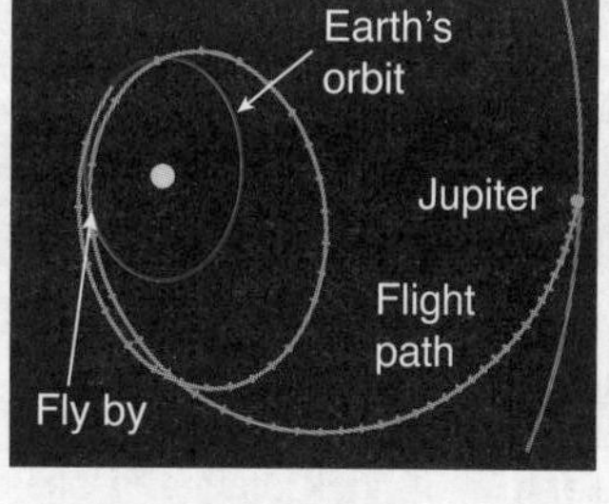


Juno is the latest NASA space probe to reach Jupiter, entering orbit in July 2016. During its five year journey, Earth provided a gravity assist speed boost.

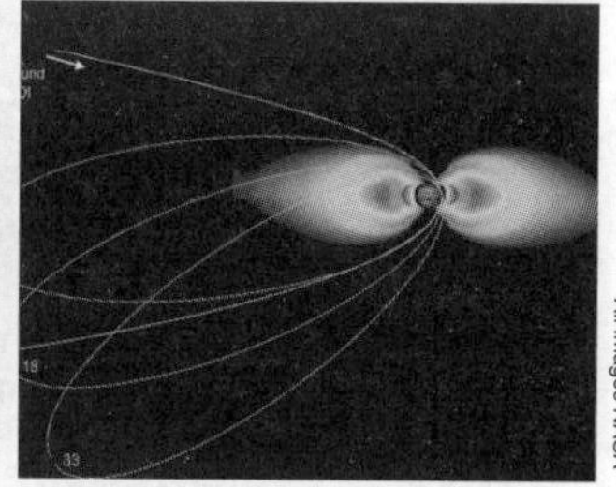

Juno will fly closer to Jupiter than any other space probe. Juno has had to be specially built to withstand the enormous radiation given off by Jupiter.

All images NASA

1. Use the information in this activity and the other activities on planets to calculate how many times more massive (in Earth masses) Jupiter is than the other planets combined.

_______________________________________________

2. Describe three examples of Jupiter's gravitational effect on objects in the solar system: ____________________

_______________________________________________

_______________________________________________

3. Juno's launch trajectory would not reach Jupiter. How then, did Juno reach Jupiter? ____________________

_______________________________________________


ISBN: 978-1-927309-37-7

# 33 Saturn

**Key Idea**: Saturn is encircled by a huge ring system made of chunks of ice and dust. It is the second largest planet but the least dense.

Saturn's most striking feature is, of course, its rings, which were first spotted (but not correctly identified) by Galileo in 1610. Although its mass is much greater than Earth's, its density is less than water and it would float if a body of water big enough to support it was ever found. Like Jupiter, Saturn spins very quickly and so bulges out at the equator making it look somewhat flattened. Saturn's atmosphere is made mostly of hydrogen and helium. Perhaps of more interest is Saturn's moon Titan which is often described as a frozen primordial Earth.

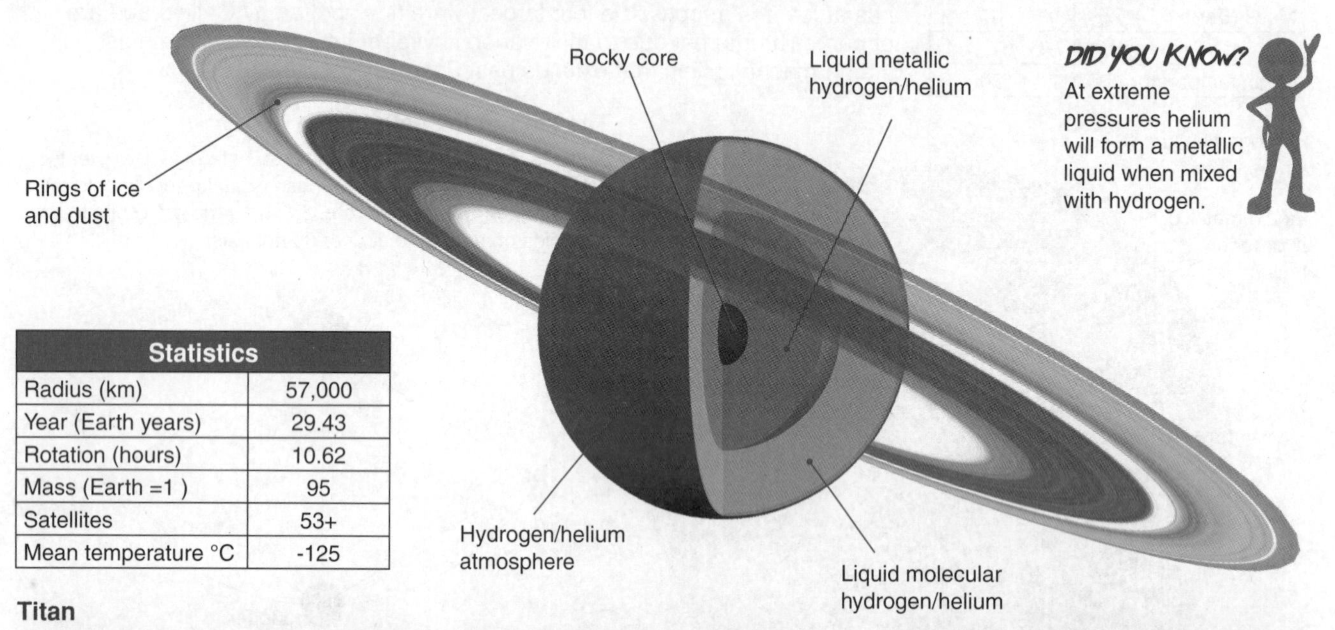


**DID YOU KNOW?** At extreme pressures helium will form a metallic liquid when mixed with hydrogen.

| Statistics | |
|---|---|
| Radius (km) | 57,000 |
| Year (Earth years) | 29.43 |
| Rotation (hours) | 10.62 |
| Mass (Earth =1 ) | 95 |
| Satellites | 53+ |
| Mean temperature °C | -125 |

### Titan

Titan is the second largest moon in the solar system. Its atmosphere is mostly nitrogen, similar to Earth. Its surface level atmospheric pressure is also similar to Earth's. At higher altitude is a thick smog of hydrocarbons. Titan has many of the ingredients for life but at -180°C its unlikely we will find any there.

In 2005 the Cassini space probe dropped the Huygens Lander into the clouds of Titan. As the lander parachuted down, it took images of lakes and rivers of liquid methane. Huygens landed on a river delta. Rocks rubbed smooth by flowing liquid methane indicate erosion happens on other planets the same way as on Earth.

All images NASA

The Cassini space probe has scanned part of Titan's surface with radar, mapping huge oceans of methane and hydrocarbons. The moon appears to have a hydrocarbon cycle much like Earth's water cycle, the only other place known to have such a complex system. Interestingly radar indicates the oceans have no waves.

1. Saturn is 95 times more massive than Earth, yet it would float on water. Why is this? ______________________________

______________________________

______________________________

2. Saturn is mainly composed of which two gases: ______________________________

3. Identify three features of Titan that are very much like those found here on Earth: ______________________________

______________________________

______________________________

______________________________




ISBN: 978-1-927309-37-7

# 34 Uranus

**Key Idea**: Uranus is the third largest planet. It took Voyager 2 nine years to reach it, the only visit by a interplanetary probe from Earth.

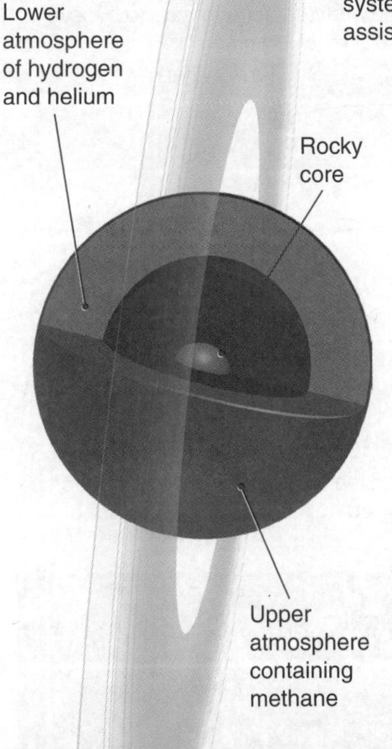


| Statistics | |
|---|---|
| Radius (km) | 25,362 |
| Year (Earth years) | 84.01 |
| Rotation (hours) | 17.19 |
| Mass (Earth =1 ) | 14.5 |
| Satellites | 27 |
| Mean temperature °C | -210 |

Uranus was discovered in 1757, the first planet discovered that was unknown to the ancients. Uranus is unusual in that instead of spinning like a top around the Sun, it rolls like a ball. Its axial tilt is 97.9°. This may have been caused by an ever increasing wobble while it was forming which eventually tipped it over. This is partially evidenced by the fact that its satellites orbit around its axial tilt.

### The Grand Tour

August and September 1977 saw the launches of what are arguably the greatest interplanetary missions so far, the launches of Voyagers 1 and 2 on the Grand Tour. The planets of the solar system were aligned in such a way that mission controllers could take advantage of the gravitational assist technique to send a probe to visit all the outer planets.

Voyager 1 travelled below and around Saturn, then up and out of the plane of the solar system.

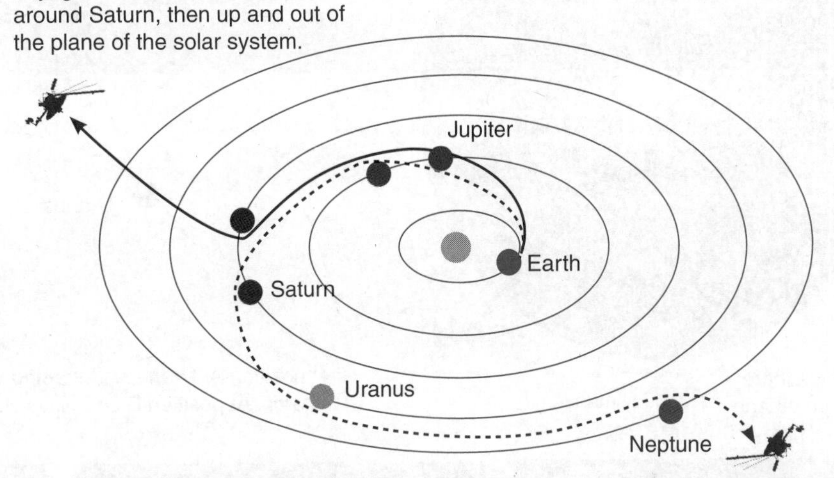


As a probe approaches (effectively falling towards) a planet it picks up speed. The probe swings around the planet, but instead of continuing in an orbit around the planet, the probe is moving fast enough to break free, effectively being flung out away like a rock from a slingshot.

Voyager 2 passed by Neptune then down below the plane of the solar system. It is the only space probe to visit Uranus or Neptune.

Uranus' magnetic field is tilted 60° relative to its rotation. It would be very difficult to use a compass there.

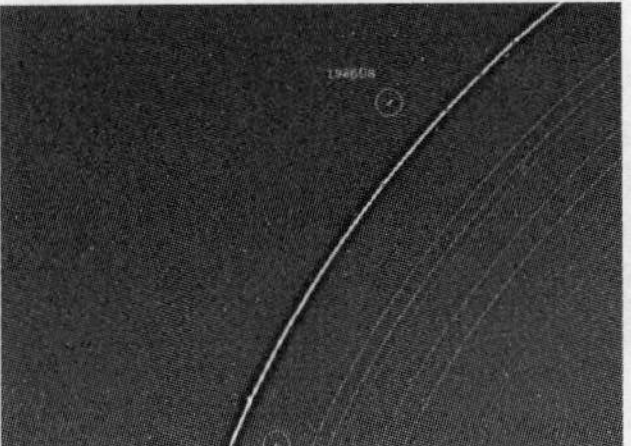

Uranus has very thin dark planetary rings, photographed by Voyager 2.

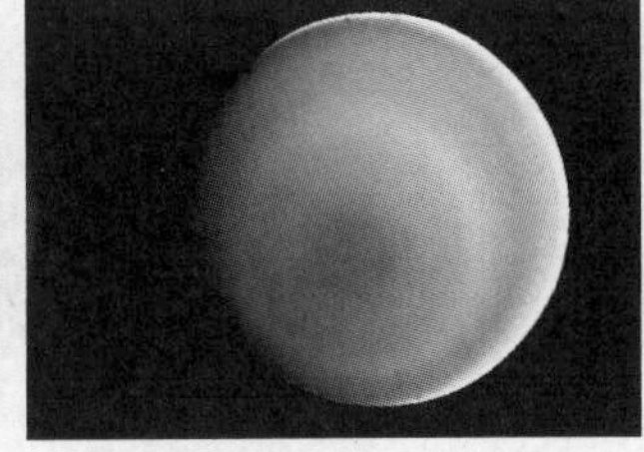

Because Uranus is on its side, it experiences extreme seasonal changes with darkness lasting decades.

Uranus' moon Miranda has a strange fractured surface possibly from stretching caused by Uranus' gravity.

All images NASA

1. (a) What is unusual about Uranus' angle of rotation? ______________________

   (b) Comment on its alignment with Uranus' magnetic field: ______________________

   ______________________

2. In 2012 Voyager 2 reached 100 AU from the Sun. Calculate this distance in kilometers: ______________________

   ______________________


**ISBN: 978-1-927309-37-7**

# 35 Neptune

**Key Idea**: Neptune is the last planet in the solar system. It was discovered using Newton's laws of gravity to calculate its position in the sky.

| Statistics | |
|---|---|
| Radius (km) | 24,622 |
| Year (Earth years) | 165 |
| Rotation (hours) | 16.1 |
| Mass (Earth =1 ) | 17 |
| Satellites | 14 |
| Mean temperature °C | -200 |

Neptune had been observed many times before its official discovery. There is evidence to show that Galileo observed it in 1613. However no one recognized it as a planet. In 1845, in a triumph of Newton's laws of motion, Neptune's position was calculated a year before it was observed. Neptune's blue color comes from methane in its upper atmosphere. One of its moon's, Triton, is particularly interesting as it orbits in the opposite direction to the way the planet spins. Neptune was reached by Voyager 2 in 1989, the only time it has been visited by a probe from Earth.

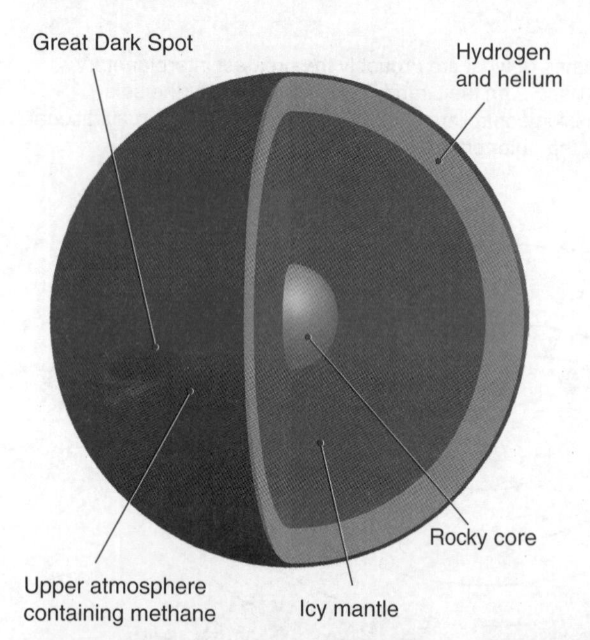


## The pull of gravity shows the way

In 1781 the orbit of Uranus was calculated and tables of its orbit and future position were published in 1821. Differences between these calculations and Uranus' observed orbit led astronomers to believe that another planet lay beyond the orbit of Uranus.

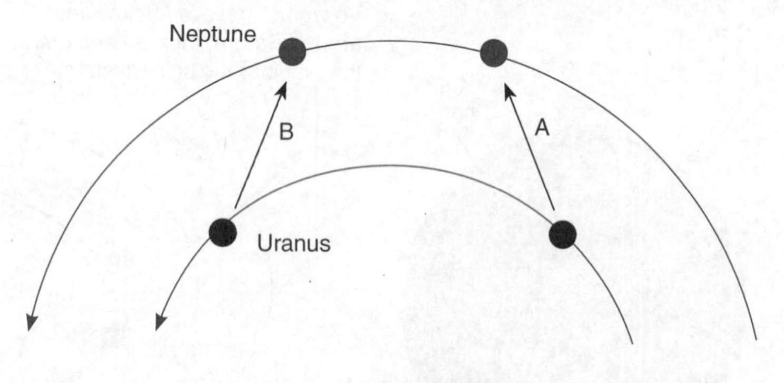


At position A, Uranus is attracted to Neptune and is accelerated along its orbit. At position B Uranus is slowed by its attraction to Neptune.

Neptune's atmosphere is very active with winds up to 2000 km/h. Voyager 2 viewed the Great Dark Spot (a large storm)in 1989, but pictures taken by Hubble in 1994 showed it had disappeared.

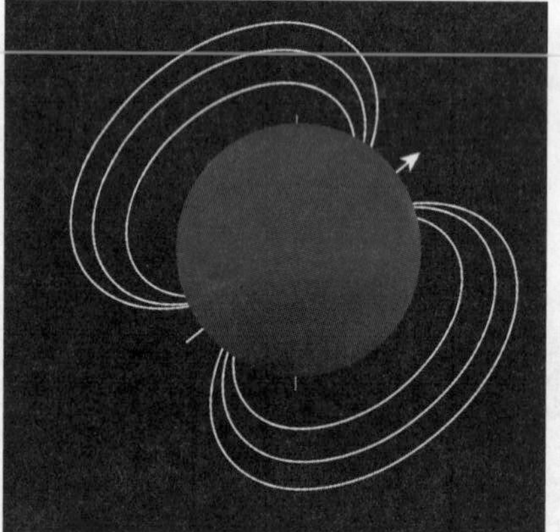

Like Uranus, Neptune's magnetic field is not in line with its rotation, being at about 47° relative to the planet's axis of rotation. Like the other gas planets Neptune radiates more heat than it receives.

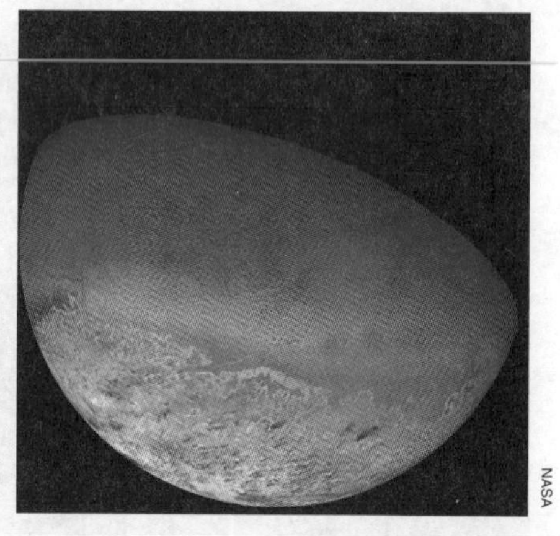


Neptune's moon Triton orbits in the opposite direction to Neptune's rotation. It must therefore have been captured by Neptune rather than forming with it. At -235 °C, it is one of the coldest places in the solar system.

1. Explain the importance of observation and calculations for predictions in finding Neptune: ______________________

______________________

______________________

2. Why would it be difficult to navigate on Neptune using a compass? ______________________

______________________




**ISBN: 978-1-927309-37-7**

# 36 Dwarf Planets

**Key Idea**: There were once nine planets in the solar system. In 2006 Pluto was demoted to a dwarf planet. There are currently five recognized dwarf planets.

When Pluto was discovered in 1930 it was classified as the ninth planet. However by 2006, nearly a dozen Pluto sized objects had been discovered orbiting the sun. This caused a problem. If Pluto was a planet then surely all these near Pluto-sized object should also be planets. The issue was resolved in 2006 when the International Astronomical Union adopted three rules for defining a planet. Pluto and all objects like it were now called dwarf planets as they did not follow all three rules.

Ironically the New Horizons spacecraft, which took this picture of Pluto, was launched the same year Pluto was demoted from planetary status.

### How to be a planet

1. The object must orbit a star (e.g. the Sun) (but not anything else or it is a moon).
2. The object must have enough gravity to form itself into a sphere (i.e. it's round, like the Earth).
3. The object must clear its neighborhood of other objects (i.e. other objects along its orbit are captured or flung out its orbit).

### Dwarf planets

There are five recognized dwarf planets (below) but there are many other dwarf planet-like objects discovered but yet to be recognized or named officially. Many of the dwarf planets are part of what are collectively called Kuiper belt objects. The Kuiper belt is an area beyond the orbit of Neptune that contains the icy-rocky remnants from the formation of the solar system. Pluto, once a planet, is now recognized as one of these.

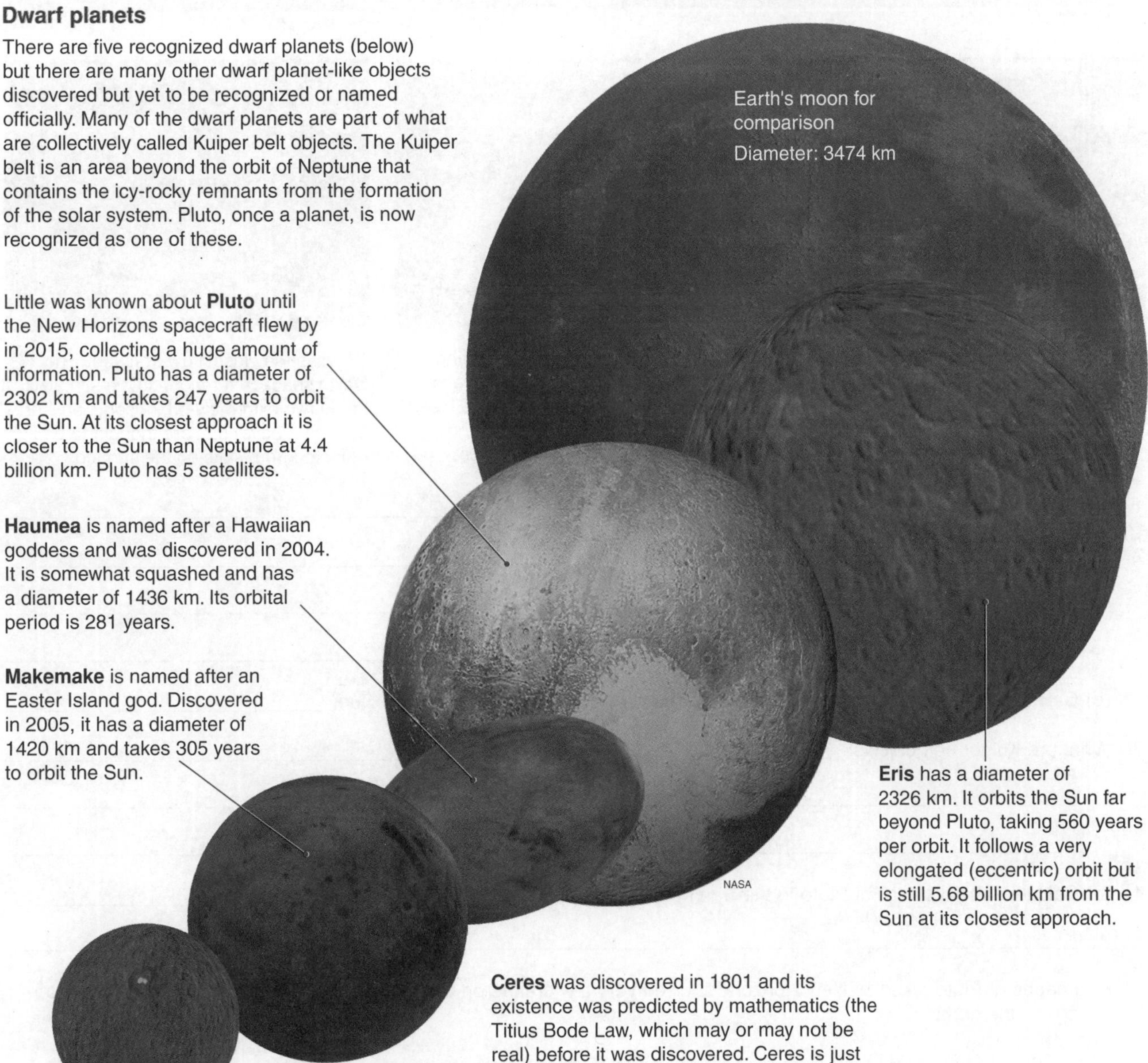


Little was known about **Pluto** until the New Horizons spacecraft flew by in 2015, collecting a huge amount of information. Pluto has a diameter of 2302 km and takes 247 years to orbit the Sun. At its closest approach it is closer to the Sun than Neptune at 4.4 billion km. Pluto has 5 satellites.

**Haumea** is named after a Hawaiian goddess and was discovered in 2004. It is somewhat squashed and has a diameter of 1436 km. Its orbital period is 281 years.

**Makemake** is named after an Easter Island god. Discovered in 2005, it has a diameter of 1420 km and takes 305 years to orbit the Sun.

**Eris** has a diameter of 2326 km. It orbits the Sun far beyond Pluto, taking 560 years per orbit. It follows a very elongated (eccentric) orbit but is still 5.68 billion km from the Sun at its closest approach.

**Ceres** was discovered in 1801 and its existence was predicted by mathematics (the Titius Bode Law, which may or may not be real) before it was discovered. Ceres is just 952 km across and resides in the asteroid belt, taking 4.6 years to orbit the Sun.


**ISBN: 978-1-927309-37-7**

## New Horizons

Until 2015, Pluto and the Kuiper belt objects (KBOs) were unexplored. Even the best Hubble images showed Pluto as little more than a fuzzy disc. It was thought Pluto was little more than a ball of frozen nitrogen and methane. The July 2015 flyby by the New Horizons spacecraft showed it as an apparently active world with high mountains and mutlicolored plains.

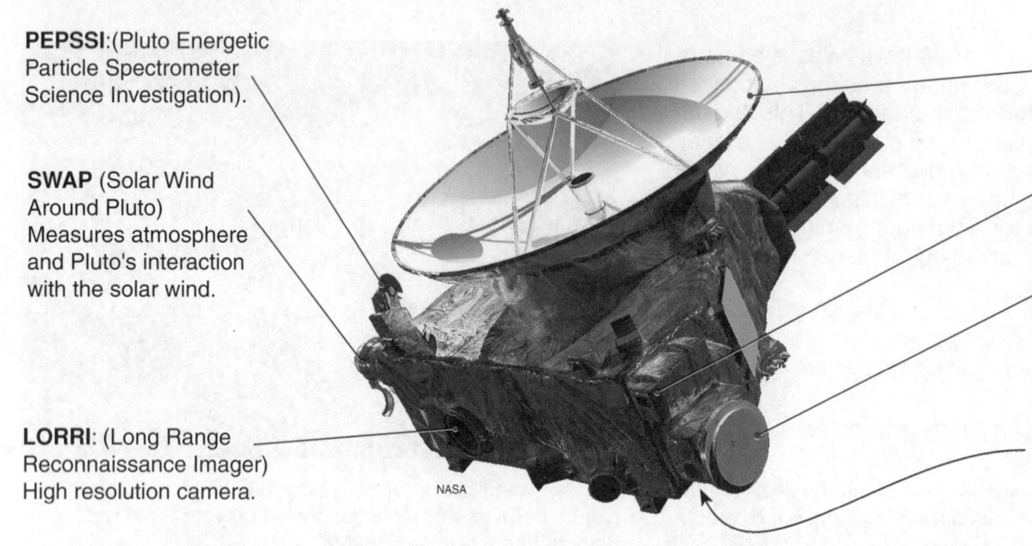


**REX**: (Radio Science Experiment) Measures atmospheric composition and temperature.

**Alice**: Ultraviolet imaging spectrophotometer. Looks at the structure of Pluto's atmosphere.

**Ralph**: Visible and infrared imager/spectrophotometer. Provide color composition and thermal maps.

**Student Dust Counter** experiment. Measures the size number and density of interplanetary dust (rear).

Close up images of Pluto showed young high ice mountains and deep canyons that suggest recent geological activity. Some of the mountains are 3400 m meters high.

Pluto has one of the most contrasting surfaces in the solar system. Above is Tombaugh Regio, nicknamed "The Heart" because of it shape. It is a thousand kilometer wide plain of nitrogen ice.

Ceres is the closest dwarf planet. The Dawn spacecraft reached Ceres in 2015. It found two extremely bright spots on Ceres' surface (above) that are thought to be light reflected from salts on the surface.

1. Which rule applying to planets do the dwarf planets not follow? ____________________

2. (a) Explain what is meant by saying Eris had a particularly eccentric orbit: ____________________

   ____________________

   ____________________

   (b) Calculate the distance of Eris to the Sun in AU at its closest approach to the Sun: ____________________

3. What are Kuiper belt objects? ____________________

   ____________________

   ____________________

4. (a) Calculate the distance of Pluto to the Sun in AU at its closest approach to the Sun: ____________________

   ____________________

   (b) Images of Pluto taken by New Horizons showed very few craters on the surface. What might this suggest about the age of the surface?

   ____________________

   ____________________


**ISBN: 978-1-927309-37-7**

## A dusty experiment

The New Horizons spacecraft was not quiet while flying the billions of kilometers to Pluto. In 2007 it flew past Jupiter, using the planet for a gravitational assist and took images of the planet and its moons using LORRI. New Horizons has also been counting dust. Attached to the spacecraft, facing the direction of travel, is the Student Dust Counter (below), built by students at the University of Colorado Boulder. This counts the number and density of the dust particles in space, continually sending the information back to Earth.

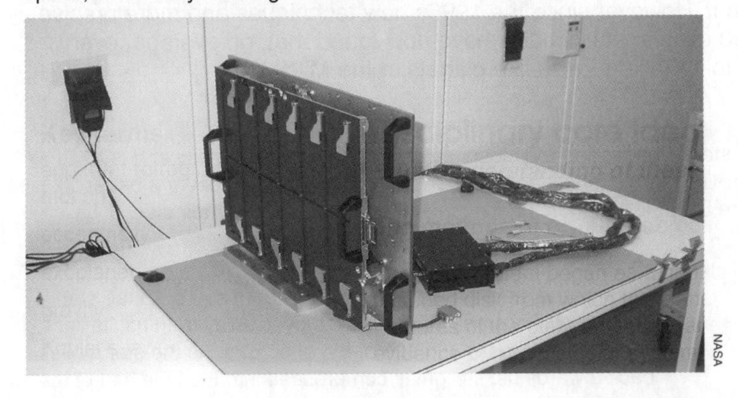
NASA

The table below shows the number of dust particle "hits" per AU travelled from Earth by the New Horizons.

| Distance (AU) | No. of hits* | Particles/$km^3$** |
|---|---|---|
| 1-2 | 404 | 21.5 |
| 2-3 | 227 | |
| 3-4 | 662 | |
| 4-5 | 1143 | |
| 5-6 | 291 | |
| 6-7 | 1270 | |
| 7-8 | 708 | |
| 8-9 | 274 | |
| 9-10 | 1687 | |

* These are minimum numbers as the counter was periodically switched on and off.

** This is a minimum amount as the spacecraft is actually travelling in a curve rather than a straight line.

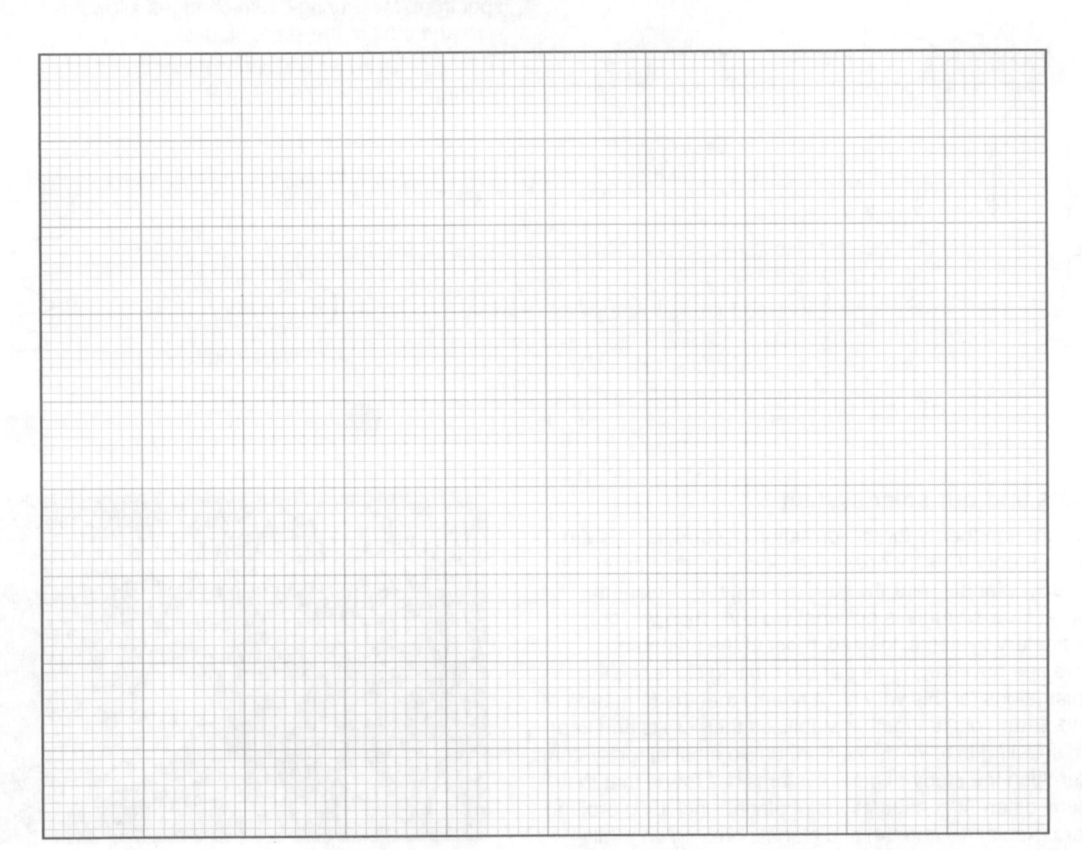

5. (a) The Student Dust Counter (SDC) has an active area of 0.125 $m^2$. Use this and the number of hits data to calculate the number of particles per cubic kilometer of space (the particle density of space) and fill in the table above. The following equations will help you:
Volume of space travelled = active area of SDC (in $km^2$) x distance travelled in km.
Particle density of space = number of hits ÷ volume of space travelled (the first one is done for you).

(b) Draw a histogram of the particle density of space from 1 AU to 10 AU.

(c) Which distances have the greatest particle density? ______________________

6. **Light travels at 300,000 km/s. Use this piece of information to explain why last minute adjustments to New Horizon's flight past Pluto were impossible:**

______________________

______________________


**ISBN: 978-1-927309-37-7**

# 37 Studying Exoplanets

**Key Idea**: Exoplanets are planets outside our solar system. Studying them can give us clues about how our solar system formed and behaves.

For many years people have wondered if other stars have planets orbiting them. It was generally assumed that other stars must have planets but it was impossible to prove it. However since the 1990s new techniques and more precise measurements of stars have identified more than 2000 exoplanets. Studies have now found that, on average, every star has a planet orbiting it and there are possibly up to 11 billion Earth-like planets in the Milky Way.

### Finding a planet

There are many techniques for finding planets orbiting other stars. The bigger and closer a planet is to its star, the easier it is to find. Most of the early exoplanets found were massive Jupiter-like planets orbiting very close to their stars. More recently, better equipment and longer studies have been able to identify more Earth-like planets orbiting in the habitable zone of their star.

### Transit method

If a planetary system happens to be aligned with our line of sight, a planet passing in front of its star will cause the light reaching us from the star to dim. The length and amount of dimming can tell us the size and orbit of the planet.

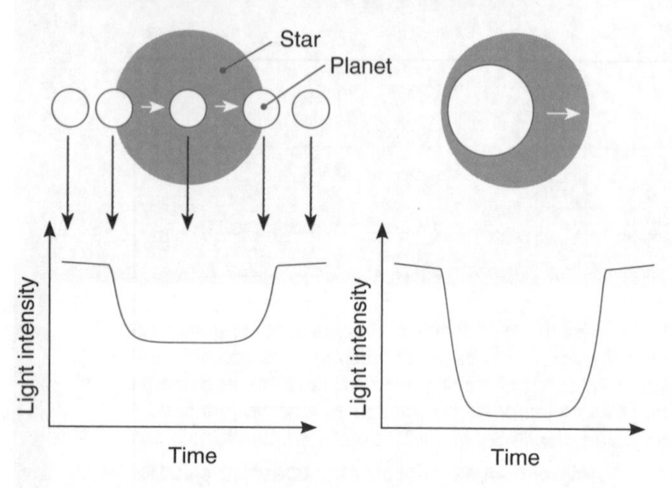


### Doppler shift

Planets and stars orbit a common center of mass, so the star orbits in a very small ellipse. As it moves away from us, its light is stretched into the red end of the spectrum. As the star moves back towards us, its light is compressed into the blue end of the spectrum. Measuring these changes allows us to determine the size and orbit of the star's planets.

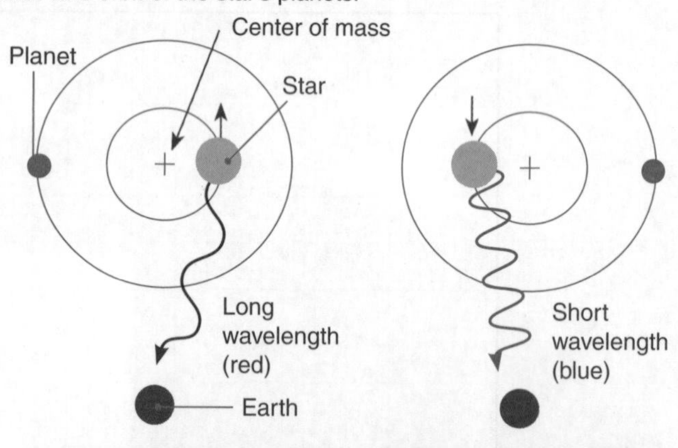


### Exoplanets and our solar system

The most common theory for the formation of our solar system is the nebula theory. This theorizes that solar systems generally form with rocky planets near the star and larger gas planets further away. However this is not what we see when we look at other planetary systems. Instead, most of the planetary systems we see have large gas planets orbiting close to their stars. Explanations for this vary but it seems that the formation of Jupiter and Saturn close together stabilized their orbits and kept them from spiraling in towards the Sun, unlike most other systems seen so far. Also the study of other systems has found that they can be many different configurations of planets including orbiting binary stars (Kepler 16 and 34) and planets orbiting on wildly different planes to each other (e.g. Kepler 56).

Kepler 186f is the most Earth-sized planet found in a star's habitable zone. It is 500 light years from Earth.

1. (a) Explain how the transit method is used to find planetary systems: ______________________________

______________________________

______________________________

(b) Explain how measuring Doppler shift can be used to find planets: ______________________________

______________________________

______________________________

(c) In what situation might these methods be unable to find planets: ______________________________

______________________________




ISBN: 978-1-927309-37-7

# 38 Gravity and Newton's Laws

**Key Idea**: Gravity is an attractive force that all mass possesses. The force of gravity decreases with increasing distance between the objects.

## What is gravity?

- Gravity is a property of mass. All mass has gravity and the more the mass the greater the gravity. Gravity is an attractive force, it pulls objects together. It is also effectively infinite in its reach, a galaxy a million light years away has enough gravity to affect the motion of our own Milky Way galaxy.
- Isaac Newton published his law of gravitation in 1687 and included the equation:

$$F = G\frac{M_1M_2}{r^2}$$

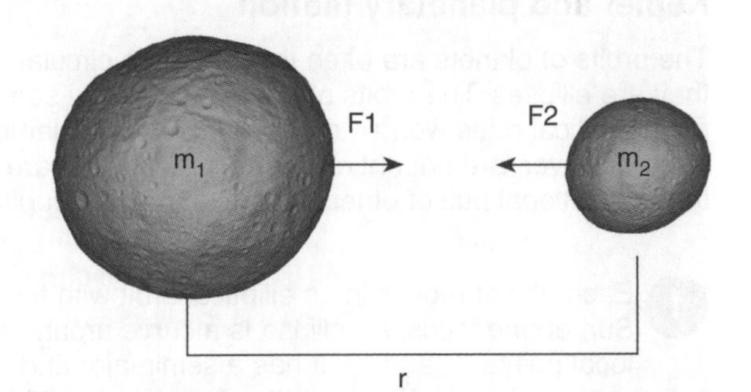


Force 1 and force 2 are equal. G is known as the **gravitational constant ($6.673 \times 10^{-11}$ Nm²/kg²)** and was not calculated until 71 years after Newton's death. M is the mass of the objects involved and r is the distance apart. The force of gravity between the objects is inversely proportional to the square of their distance (i.e. the further apart the objects are the less the force between them there is).

- The equation allows us to calculate the force of gravity acting between two objects. Using the equation we find that the force of gravity is proportional to the product of the masses and that it is inversely proportional to the square of their distance apart.

## What is an orbit?

- The Sun is the overwhelming source of gravity in our solar system. Its gravity reaches far into space and all the objects in the solar system orbit around it.
- But what is an orbit? The planets are all falling towards the Sun. However, they are also moving sideways relative to the Sun. The effect is that they trace elliptical orbits through space.
- It is important to remember the gravitational pull the planets exert on the Sun is equal to the Sun's gravitational pull on the planets, but because the Sun is so much bigger, it barely moves, instead it wobbles around it axis on a tiny orbit of its own.
- The closer the match between the mass of the orbiting objects the more pronounced this wobble is. The center of the Earth - Moon system for example is 4670 km from the center of the Earth (i.e. within the radius of the Earth).

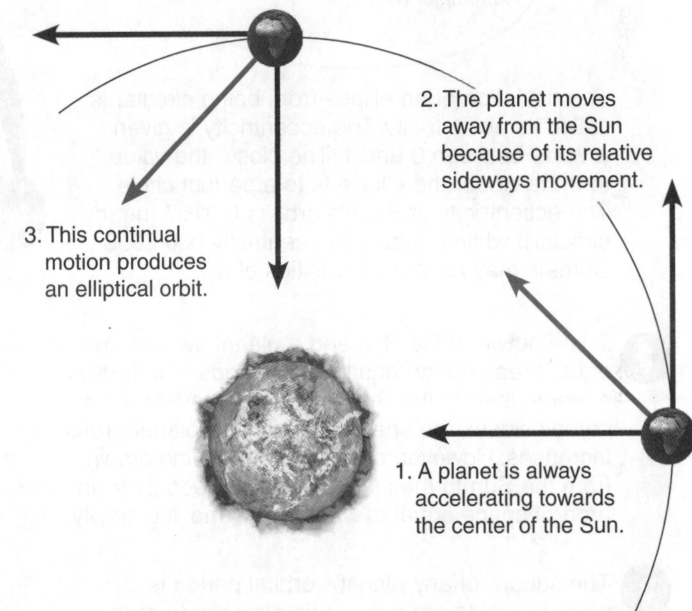


1. What is gravity? ______________________________

2. Using Newton's equation for the universal law of gravitation, what would be the effect on the force between the Earth and Moon of changing the following:

   (a) Doubling the distance between the Earth and Moon: F would (increase / decrease / remain the same) (circle one).

   (b) Doubling the mass of the Moon: F would (increase / decrease / remain the same) (circle one).

   (c) Halving the mass of the Earth: F would (increase / decrease / remain the same) (circle one).

3. The Earth exerts the same gravitational pull on the Sun as the Sun does on the Earth. Explain why the Earth orbits the Sun in a wide orbit, while the Sun only wobbles on its axis:


**ISBN: 978-1-927309-37-7**

# 39 Kepler's Laws and Orbits

**Key Idea**: Johannes Kepler developed three laws that describe planetary motion.

## Kepler and planetary motion

The orbits of planets are often thought of as circular, but in fact they are ellipses. The orbits of the planets follow some simple mathematical rules worked out by Johannes Kepler in 1609. The laws, however, are not entirely perfect as planets are affected by the gravitational pull of other planets, especially Jupiter.

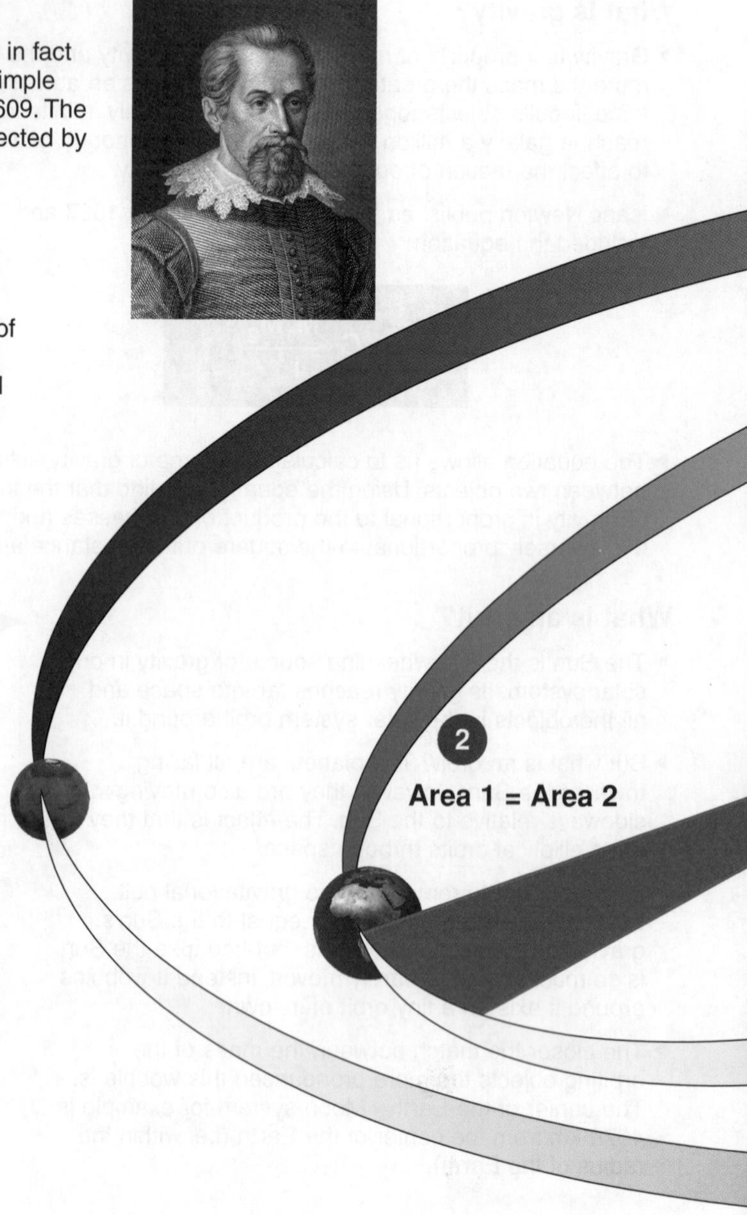


1. Each planet moves in an elliptical orbit with the Sun at one focus. An ellipse is a curve around two focal points ($f_1$ and $f_2$). It has a semimajor and a semiminor axis. Because the Sun contains 99.8% of the mass in the solar system, for any planet in our solar system both of the focal points for its elliptical orbit are very close to the center of the Sun.

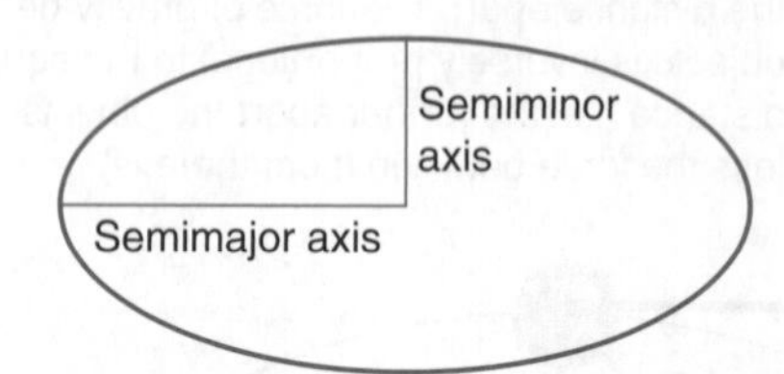


The deviation of an ellipse from being circular is called its eccentricity. The eccentricity is given a value between 0 and 1. The closer the value to 0 the closer the ellipse is to a perfect circle. The eccentricity of Earth's orbit is 0.0167 (nearly circular), while Mercury's eccentricity is 0.2056. Comets may have eccentricities of 0.7

2. A line between the Sun and a planet sweeps over equal areas during equal time periods. The further a planet is from the Sun the slower it orbits. As it comes closer, the speed at which the planet orbits increases. However, over a set time a line drawn from the Sun to the planet will still sweep over an area of space equal to when it was moving slowly.

3. The square of any planet's orbital period is proportional to the cube of its distance from the Sun along the semimajor axis of its orbit. Gravity decreases proportionally over distance, so planets further away from the Sun can take longer to travel around it. The formula $\mathbf{a^3 = p^2}$ can be used to calculate orbital periods. This equation can also be represented as:

$$\frac{p_1^2}{a_1^3} = \frac{p_2^2}{a_2^3}$$

Where **a** is the distance along the semimajor axis from the Sun to the planet and **p** is the orbital period.

The ratio for a system is constant (e.g. $p^2/a^3$ for the Earth-Sun orbit is the same as $p^2/a^3$ for Mars-Sun orbit). In this equation the units can be simplified to AUs and years.

### Examples of p and a

| Planet | a (AU) | p (Earth years) | $a^3$ | $p^2$ |
|---|---|---|---|---|
| Venus | 0.72 | 0.62 | 0.37 | 0.38 |
| Earth | 1 | 1 | 1 | 1 |
| Jupiter | 5.203 | 11.86 | 140.9 | 140.7 |
| Saturn | 9.537 | 29.46 | 867.4 | 867.9 |

**ISBN: 978-1-927309-37-7**

## Aphelion and perihelion

The elliptical shape of a planet's orbit means that at some point in its orbit the planet will make its closest approach to the Sun (perihelion) and at another point it will be at its most distant (aphelion).

Because Earth's orbit is almost circular there is very little difference between aphelion and perihelion (about 5 million km). The Northern Hemisphere's winter occurs in perihelion and the Northern Hemisphere's summer occurs during aphelion.

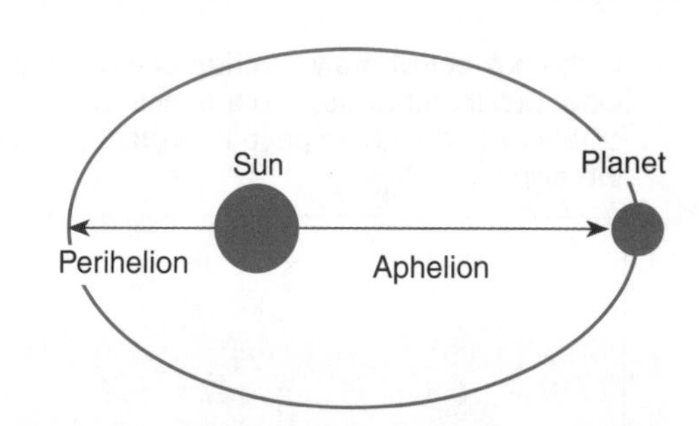


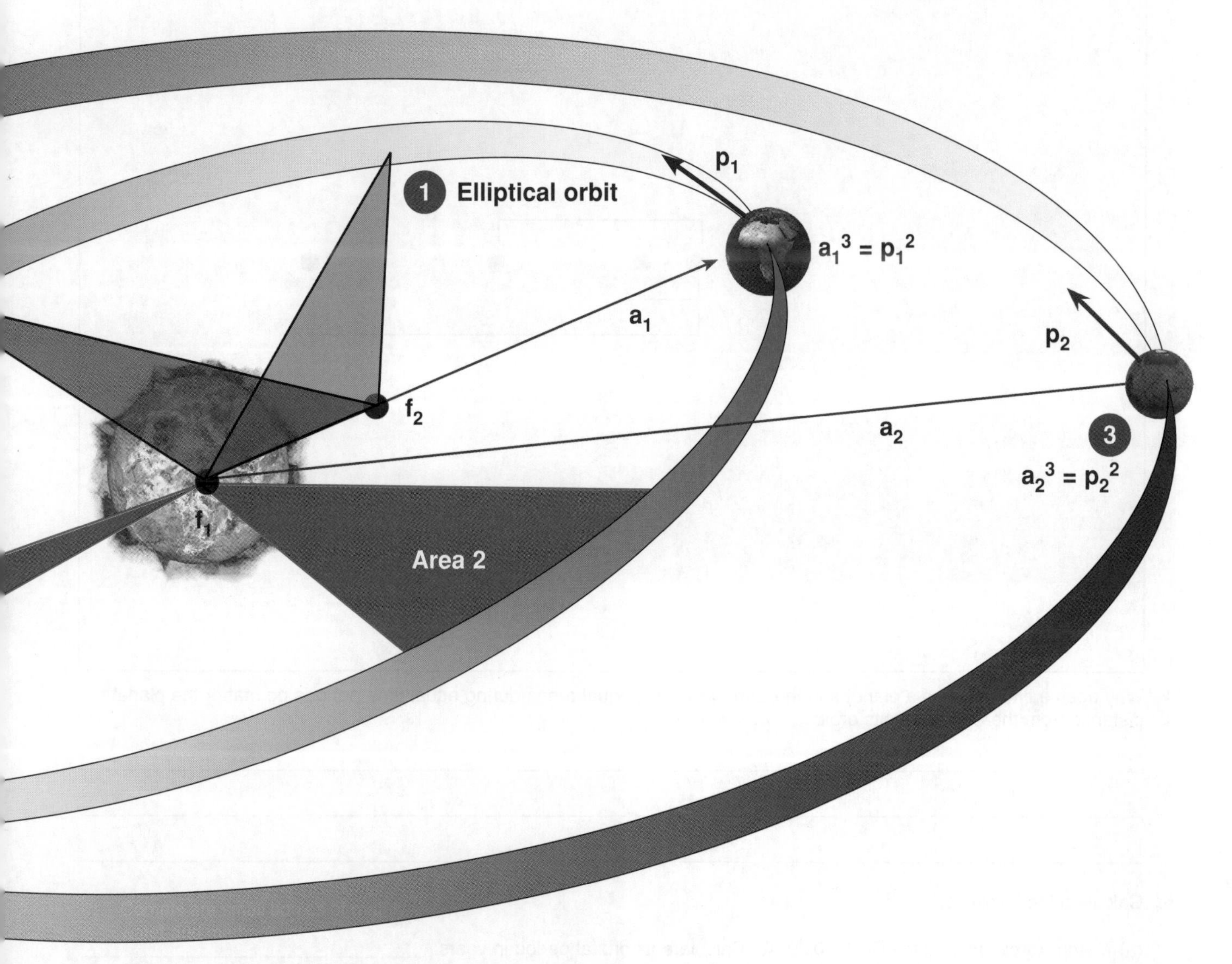


1. Label the semimajor and semi minor axis on the following ellipses:

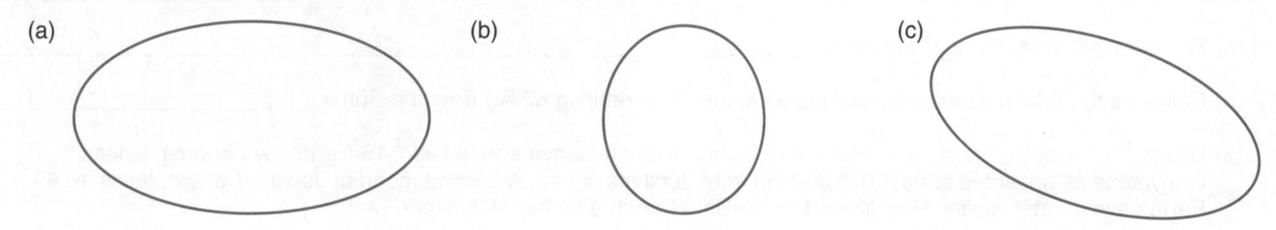


2. Complete the statement by circling the correct answers:

The Northern Hemisphere (summer/winter) occurs when the Earth is at aphelion, while its (summer/winter) occurs at perihelion. Because of this, the Northern Hemisphere's (summer/winter) is longer than its (summer/winter). Compared to the Southern Hemisphere, the Northern Hemisphere's summer is (cooler/hotter) and (shorter/longer).


**ISBN: 978-1-927309-37-7**

3. In the box below draw a series of ellipses using the pairs dots as focal points. You will need a piece of string 15 cm long, two thumbtacks, and a pencil. Tie the string into a loop. Press the thumb tacks into the dots as focal points (put corkboard behind the page to protect your book!). Loop the string around the thumbtacks and use it as a guide to draw the ellipses.

4. Why does a line between a planet and the Sun sweep out equal areas during equal time periods no matter the planet's distance from the Sun during its orbit:

5. Calculate the following:

(a) Mercury's distance to the Sun is 0.39 AU. Calculate its orbital period in years:

(b) Neptune takes 164.79 years to orbit the Sun. What is its distance from the Sun in AUs?

(c) How far is your answer to (b) in km?

(d) Calculate the orbital period of a new planet found to be orbiting 62 AU from the Sun.

6. (a) Use $p_1^2/a_1^3 = p_2^2/a_2^3$ to solve the following: Jupiter's moon Ganymede, takes 7.15 Earth days to orbit Jupiter. Ganymede is measured to be 1,070,000 km from Jupiter's center. A second moon of Jupiter, Callisto, takes 16.69 Earth days to orbit Jupiter. How far away is Callisto from the center of Jupiter?

(b) It was also noted that Jupiter's moon Io takes just 1.77 days to orbit the planet. How far away from Jupiter is Io?


**ISBN: 978-1-927309-37-7**

# 40 Applying Newton's Law to Kepler's Laws

**Key Idea**: Newton applied his law of gravitation to Kepler's work, deriving an equation that models how all orbiting objects in the universe behave.

Isaac Newton realized that the orbital period of a planet was related to the mass of the star and the mass of the planet. He was able to derive all of Kepler's laws from his law of universal gravitation, producing simple formulas which could be used to calculate the mass of anything in the universe based on the orbital period and distance of something orbiting it.

- Newton rewrote Kepler's third law as:

$$M_1 + M_2 = \frac{4\pi^2 a^3}{G p^2}$$ **Equation 1**

- In which **M** is mass in kilograms, **a** is the distance (semimajor axis) in meters, and **p** is the orbital period in seconds.
- In many orbiting systems $M_1$ and $M_2$ are very different (one is usually very big and the other is very small, e.g. a star and a planet) so $M_1 + M_2$ are often rewritten simply as M (the mass of the larger object).
- $4\pi^2$ and G are constant numbers and can be ignored (when producing estimates). Thus we can simplify the equation as:

$$M = \frac{a^3}{p^2}$$ **Equation 2**

- In which **M** is mass in solar masses, **a** is the distance (along the semimajor axis) in AUs, and **p** is the orbital period in years.
- For example the Sun takes 250 million years to orbit the center of Milky Way galaxy and is about 30,000 light years from the center. Using Newton's version of Kepler's third law we can then calculate the mass of the Milky Way galaxy!

**Example 2**: A star 4 times the size of the Sun (1.98 x$10^{30}$ kg) has a planet 5 AU from it. What is the period of the planet's orbit in years?

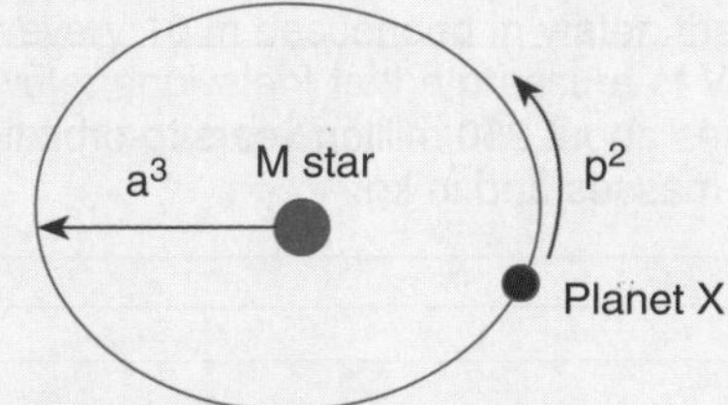


Using **equation 1**: (converting units)

M star = $4\pi^2 a^3 / G p^2$ $\quad p^2 = 4\pi^2 a^3 / MG$

$p^2 = 4 \times 3.142^2 \times (7.5 \times 10^{11})^3 / 7.92 \times 10^{30} \times 6.673 \times 10^{-11}$

$p^2 = 1.6659 \times 10^{37} / 5.2850 \times 10^{20} = 3.1578 \times 10^{16}$

p = 177,703,345.9 s = 5.6 years

And **equation 2**:

M star = $a_X^3/p_X^2$. $\quad 4 = 5^3/p^2$. $\quad p^2 = 5^3/4 = 125/4 = 31.25$.

p = 5.6 years.

**Example 1**: Planet X orbits star 1 and planet Z orbits star 2. Both planets take 2 years to orbit their respective stars. However, planet X is 2 AU from its star, while planet Z is 1 AU from its star. What is mass of each star and how much bigger is one star than the other?

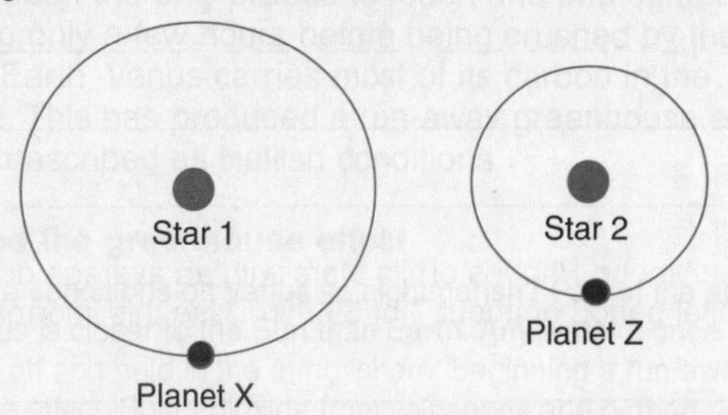


If we are to use **equation 1** we need to convert the units above into kilograms, meters, and seconds. There are 1.5 x $10^{11}$ meters in 1 AU and 31,536,000 seconds in a year.

**M star 1** = $4\pi^2 a^3 / G p^2$,

$= \frac{(4 \times 3.142^2) \times (2 \times 1.5 \times 10^{11})^3}{6.673 \times 10^{-11} \times (2 \times 365 \times 24 \times 60 \times 60)^2}$

$= \frac{39.478 \times 2.7 \times 10^{34}}{6.673 \times 10^{-11} \times 3.978 \times 10^{15}}$

$= \frac{1.066 \times 10^{36}}{265{,}457}$

$= 4.016 \times 10^{30}$ kg

**M star 2** = $4\pi^2 a^3 / G p^2$,

$= \frac{(4 \times 3.142^{2)} \times (1.5 \times 10^{11})^3}{6.673 \times 10^{-11} \times (2 \times 365 \times 24 \times 60 \times 60)^2}$

$= \frac{39.478 \times 3.375 \times 10^{33}}{6.673 \times 10^{-11} \times 3.978 \times 10^{15}}$

$= \frac{1.332 \times 10^{35}}{265{,}457}$

$= 5.019 \times 10^{29}$ kg

$4.016 \times 10^{30}$ kg / $5.019 \times 10^{29}$ kg = 8.00.

Star 1 is 8 times more massive than star 2.

We can use **equation 2** to provide a similar answer:

M star 1 = $a^3/p^2 = 2^3/2^2 = 2$ solar masses

M star 2 = $a^2/p^2 = 1^3/2^2 = 0.25$ solar masses

2 / 0.25 = 8

And checking to be sure the second equation is an accurate representation of the first:

One solar mass ~ $1.98 \times 10^{30}$ kg

So: $4.016 \times 10^{30}$ kg / $1.98 \times 10^{30}$ kg = 2.03

$5.019 \times 10^{29}$ kg / $1.98 \times 10^{30}$ kg = 0.25.

2.03 / 0.25 ~8


**ISBN: 978-1-927309-37-7**

1. (a) Observations of a star 30 light years away showed it was being orbited by a planet once every 1.5 Earth years. It was also calculated the planet was 1.75 AU from the star. What is the mass of the star in solar masses?

(b) Observations of a second star showed that its planet orbited it in 0.5 years and was 0.9 AU distant. Calculate the difference in solar masses between the two stars and state which star is larger.

2. Mars' moon Phobos orbits Mars with an average distance of about 9380 km from the center of the planet and has an orbital period of about 7 hr 39 min. Use this information to estimate the mass of Mars in kg.

3. Calculate the distance (in AU) of a planet from its star's center for a planet with an orbit of 145 years and a star 2.5 times the mass of the Sun.

4. (a) The largest star so far observed is estimated to be 315 times as massive as the Sun. Assuming it was possible for a planet to orbit this star at the same distance as the Earth is from the Sun, calculate the planet's orbital period in days:

(b) At what distance (in AU) from the star's center would the planet need to be in order to orbit it in one Earth year?

5. The Sun is about 30,000 light years from the center of the Milky Way. It takes about 250 million years to orbit the galaxy. Use this information to estimate the mass of the Milky Way galaxy in solar masses and in kg.

6. Recall that Ganymede takes 7.15 Earth days to orbit Jupiter and that Ganymede is measured to be 1,070,000 km from Jupiter's center. Use this information to calculate the mass of Jupiter:


**ISBN: 978-1-927309-37-7**

# 41 Satellites

**Key Idea**: Satellites orbit in low, medium, or high Earth orbit. They follow Newton's and Kepler's laws the same as planets orbiting the Sun.

There are hundreds of satellites orbiting the Earth. The majority are used in communications (e.g. carrying television signals) but others are used in monitoring weather or surveying the Earth's surface. Satellites orbit the Earth at different altitudes and at different inclinations, some orbit around the Earth's equator, others have a polar orbit.

## Orbiting the Earth

There are three basic orbital heights in which satellites are placed:

Geostationary orbit (GEO) is at an altitude of about 36,000 km. At this distance the satellite takes the same time to orbit the Earth as the Earth takes to complete one rotation, keeping the satellite above the same point on the Earth.

Low Earth orbit (LEO) is around 200 km to 3000 km above the surface. The International Space Station is in LEO.

Medium Earth orbit (MEO) is at an altitude of about 3000 km to 30,000 km. It is used for communications and GPS satellites.

## Orbital speed

- The closer a satellite is to the Earth, the faster is has to move in order to stay in orbit. Satellites that orbit too close to the Earth encounter drag from the Earth's atmosphere and slow down, falling to a lower orbit. Occasionally these need to be manoeuvred to maintain their altitude. We can use Kepler's laws to estimate the speed and orbital period of a satellite.

## Calculating orbital speed and period

The international space station (ISS) orbits at about 400 km above the Earth. How long does it take to complete one orbit and at what speed is it moving?

- First we must add the Earth's radius to the altitude of the ISS to determine the distance of the ISS from the center of the Earth: $6.371 \times 10^6$ m + $4 \times 10^5$ m = $6.771 \times 10^6$ m. The mass of the Earth is $5.972 \times 10^{24}$ kg. Applying Kepler's third law we find that $p^2$ = 30,752,275 and p = 5,545 s = 92 minutes, or about 1.5 hours.
- The ISS orbit is almost circular, so calculating the circumference of a circle the size of its orbit using $c = 2\pi r$ = 42,543,448 m. Dividing distance by time = 42,543,448 m / 5545 seconds = 7672 m/s

We can also use the equation $\mathbf{v = \sqrt{(GMe/r)}}$ to calculate orbital speed:

Where v is the velocity (orbital speed) in m/s, G is the gravitational constant, Me is the mass of the Earth (kg), and r is the radius of the orbit (m).

1. Explain why a geostationary satellite is useful for carrying a continuous television signal from one side of the planet to the other.

______

______

______

2. (a) The Hubble Space Telescope orbits 559 kilometers above the surface of the Earth. Calculate the time (in hours) it takes to orbit the Earth:

______

______

(b) Calculate the HST's orbital speed: ______

______


**ISBN: 978-1-927309-37-7**

# 42 Moving Planets

**Key Idea**: The relative motion of the planets in their orbits causes them, from our point of view, to trace loops and zig zags across the sky.

- When a planet is observed in the night sky, night after night and at the same time each night the planet is seen to move across the sky (the term "planet" from Greek meaning "wandering star"). The planet does not move in a uniform way, however. Planets move east to west across the sky then reverse direction for days or weeks, then move east to west across the sky again.
- The geocentric model explained this by having each planet performing epicycles (cycles within cycles) as it apparently orbited the Earth. The heliocentric model explained this **apparent retrograde motion** as an illusion caused by the motions of the planets relative to each other.

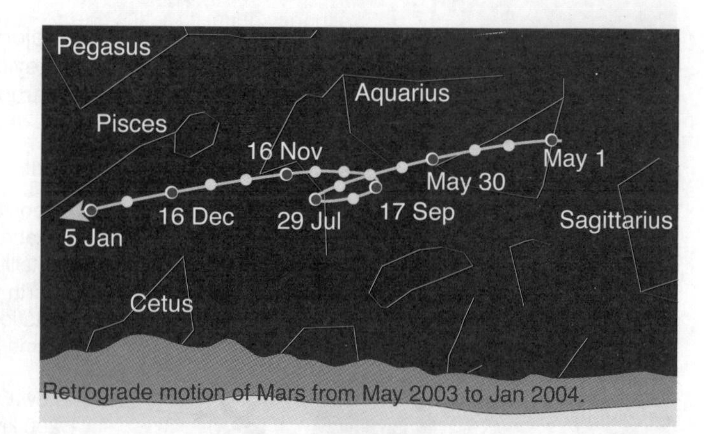

Retrograde motion of Mars from May 2003 to Jan 2004.

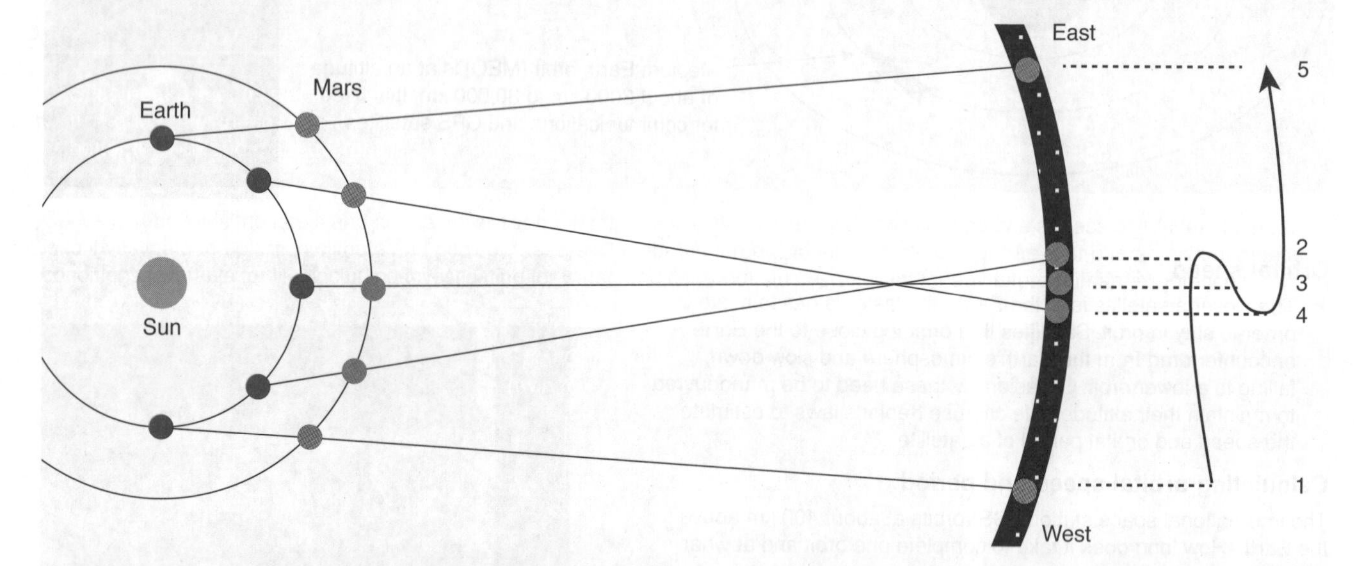


- Earth orbits the Sun faster than planets further away from the Sun, e.g. Mars. It is constantly catching up with, then overtaking Mars. From our point of view Mars appears as a red point of light against the background stars that moves across the night sky from west to east as the Earth approaches it. Then as the Earth begins to catch up with Mars, the planet begins to move backwards across the sky to the west. The Earth then passes Mars and it appears to start moving east once more.
- A similar apparent motion occurs with the inner planets Venus and Mercury from our viewpoint except that it is the inner planets that are catching up with and passing the Earth.
- Depending on the orientation of the planets as they pass each other the retrograde motion may take the form of a loop or a zig zag in the sky.

1. Complete the model below of the retrograde motion of Venus (use the diagram above as a guide):

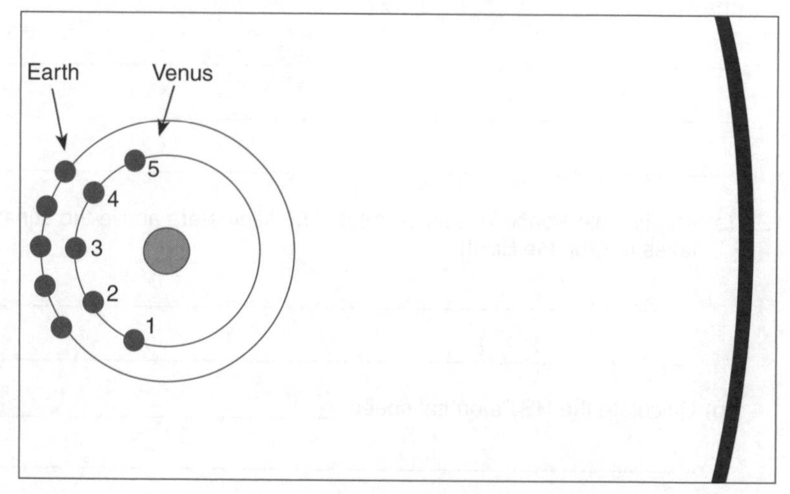


2. Use this model to explain why Venus usually can't be seen at position 3.

3. Explain why the retrograde motion of planets in the sky is an illusion:

**ISBN: 978-1-927309-37-7**

# 43 The Earth and The Moon

**Key Idea**: The Earth and the Moon influence each other in many ways including the Moon causing tides on Earth and the Earth affecting the rotation of the Moon.

## Formation of the Moon

The Moon is the Earth's only natural satellite. It formed about 4.5 billion years ago, just a few million years after the solar system condensed into a swirling disk. A Mars sized proto-planet (commonly called Theia) smashed into the Earth. The debris that was flung off eventually condensed as the Moon. Evidence for this hypothesis includes:

- The Moon has a lower density than the Earth (consistent with the density of the upper layers of the Earth).
- The Earth and Moon have similar isotope ratios of their elements suggesting they formed at the same time.
- The Moon's angle of orbit (5°) is different to the Earth's tilt (23.4°) (the Earth was tilted over due to the collision).

## The Moon's orbit

The Moon is 384,400 km from the Earth. It completes an orbit of the Earth once every 27.3 days, the same time it takes to complete one rotation. This means the same side of the Moon always faces the Earth. It is a phenomenon called tidal locking. The Earth's gravity pulls on the Moon's crust as the Moon rotates. This causes friction that slows the Moon's rotation and keeps one side of the moon facing the Earth.

The Moon is the largest natural satellite in comparison to its planet in the solar system. Its mass is large enough to move the center of the Earth-Moon orbit (the barycenter) 4670 km from the center of the Earth.

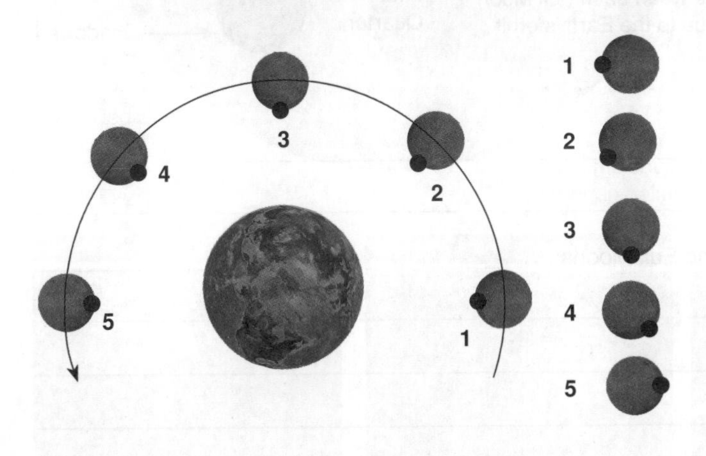


The Moon may appear to not rotate from our point of view but this diagram shows how a point on the moon does rotate.

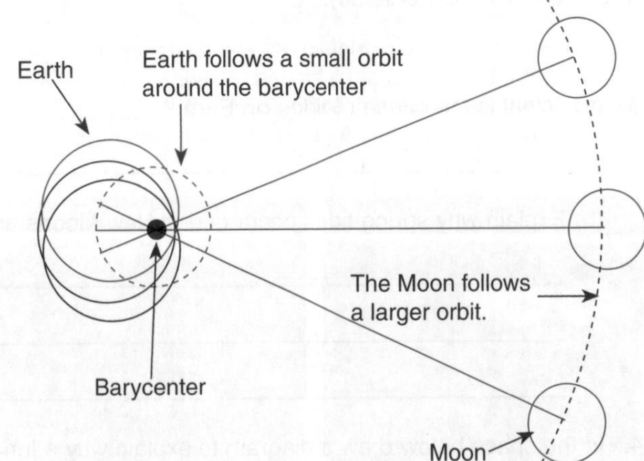


The Earth and Moon orbit a common center called the barycenter. This is inside the radius of the Earth.

1. Use the mass of Earth ($5.972 \times 10^{24}$ kg) and the distance of the Moon from Earth to show that the Moon takes 27.3 days to orbit the Earth and not 29 days are it appears to by the phases of the moon:

2. Explain why even though the Moon rotates once every 27.3 days only one side of the moon ever faces the Earth.


ISBN: 978-1-927309-37-7

## Tides

One of the most noticeable effects of the Moon on Earth are tides. The Moon's gravity is much less than the Earth's but it is still strong enough to pull the water on Earth towards the Moon causing tides. In addition, the Sun's gravity also pulls on the water and modifies the size of the tides.

### Spring tides and neap tides.

Spring tides occur when the Moon, Earth, and Sun and arranged in a line. The gravity of the Moon and Sun combine to create extra-high high tides and extra-low low tides. When the Moon, Earth, and Sun form a 90° angle, their gravities act against each other and result in only a small difference between high and low tide (a neap tide).

## Phases of the Moon

As the Moon orbits the Earth it reflects light from the Sun to the Earth. The amount of the Moon's surface that is visible (the phase of the Moon) depends on the position of the Moon relative to the Earth and Sun.

A Full Moon occurs when the Moon is on the opposite side of the Earth from the Sun, whereas a New Moon occurs when the Moon is between the Earth and Sun. Because the Moon's orbit around the Earth is not in the same plane as the Earth's orbit around the Sun, the Earth does not often cast a shadow on the Moon during a Full Moon (a lunar eclipse), nor does the Moon often pass in front of the Sun during a New Moon (a solar eclipse).

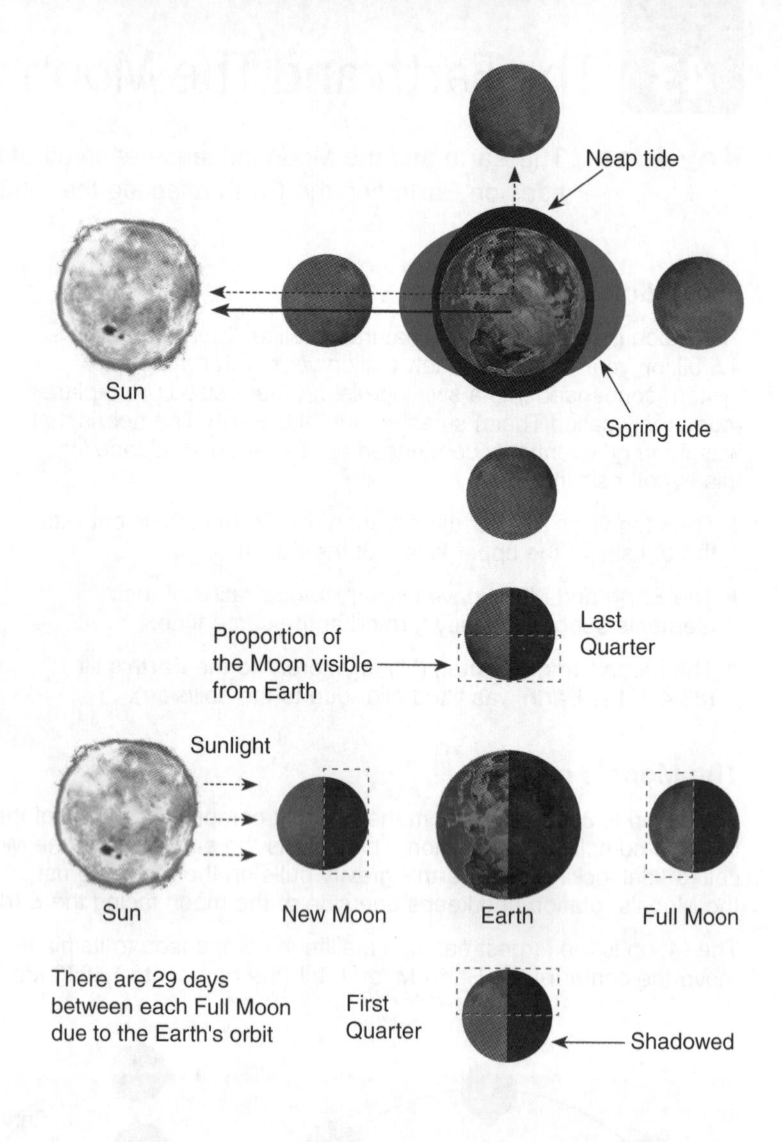


3. (a) What is the cause of tides on Earth? ___

(b) Explain why spring tides occur during New Moons and Full Moons: ___

4. In the space below draw a diagram to explain why a lunar eclipse will only occur during a Full Moon and a solar eclipse will only occur during a New Moon:


**ISBN: 978-1-927309-37-7**

# 44 Cyclical Changes

**Key Idea**: There are long term changes to the Earth's orbit, tilt, and rotation. These affect the level of sunlight on the Earth and therefore the climate.

Earth is not stable in space. It is affected by the gravity of the Sun, the planets, and the Moon. These influences affect the tilt of the Earth's axis and the shape of its orbit. The changes in orbit and tilt can combine to cause extreme changes in the Earth's climate, e.g. producing ice ages. The Earth experiences three main orbital and rotational cycles:

### Axial tilt (obliquity)

Relative to its orbital axis, the Earth's rotational axis at an angle of about 23.4°. This angle is responsible for seasonal changes. The angle is not constant and changes between 22.1° and 24.5° over a period of about 41,000 years. The greater the degree of tilting the greater the difference between the summer and winter seasons.

| Planet | Axial tilt | Planet | Axial tilt |
|---|---|---|---|
| Mercury | 0.1° | Jupiter | 3° |
| Venus | 177° | Saturn | 27° |
| Earth | 23.4° | Uranus | 98° |
| Mars | 25° | Neptune | 30° |

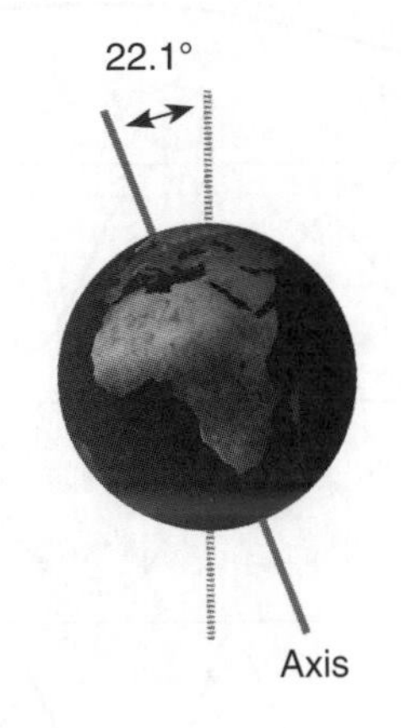


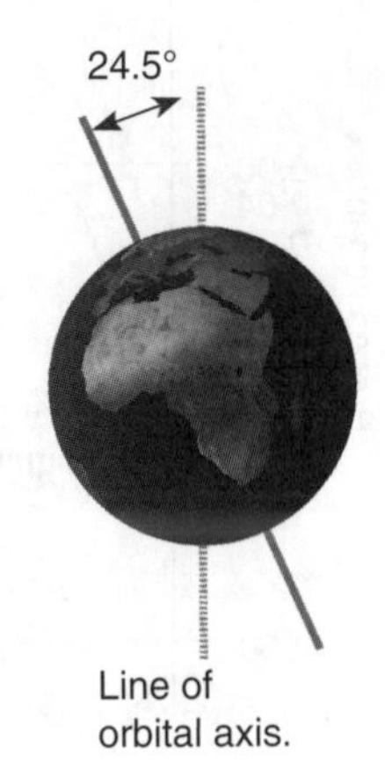


### Precession

Like a spinning top, Earth wobbles about its axis. This causes the Earth's axis to describe a cone in space. An entire cycle takes about 26,000 years. Precession alters the direction the Earth's axis is pointing. For example, during June, the Northern Hemisphere is pointed towards the Sun. However 13,000 years ago the occurrence of the seasons will be opposite to what they are today.

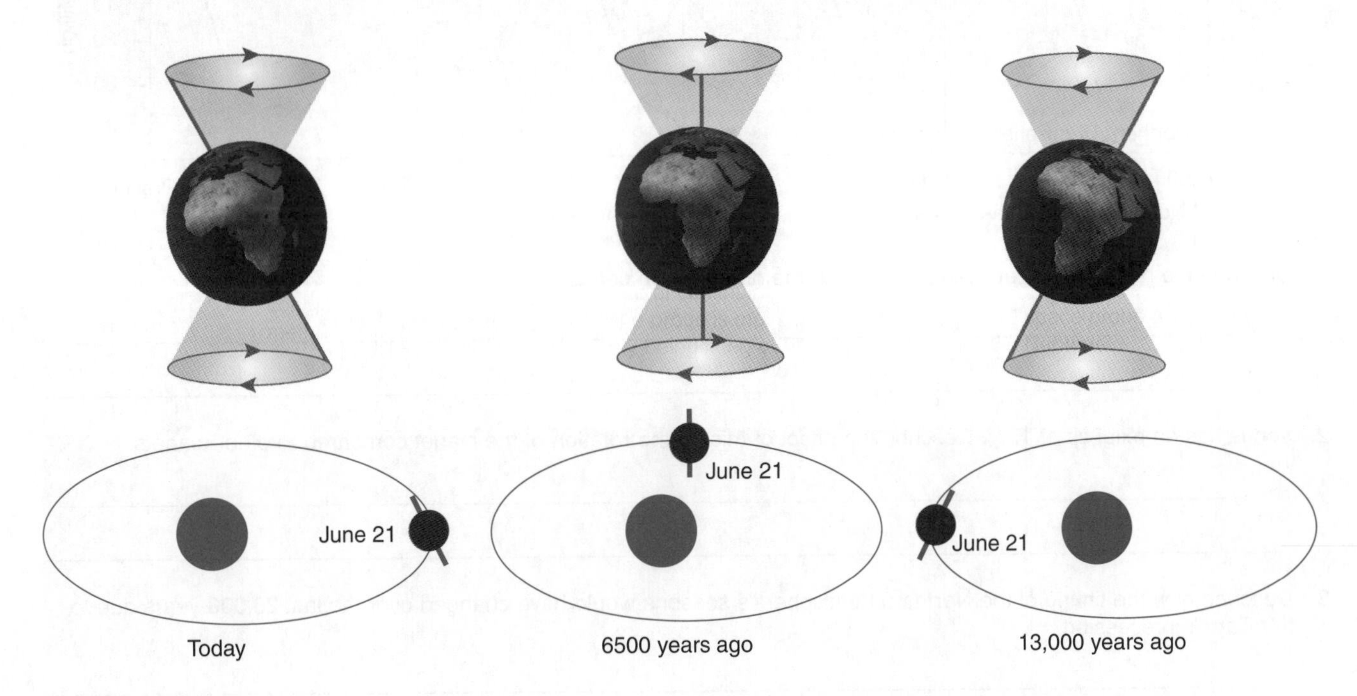


### Orbital eccentricity

The Earth's elliptical orbit changes its eccentricity from very nearly circular (eccentricity of 0.000055) to slightly elliptical (eccentricity 0.0679). Circular orbits tend to make the differences between the seasons rather mild, while more elliptical obits exacerbate the difference. The changes in eccentricity have a cycle of about 100,000 years.

Of the planets in our solar system Mercury has the most eccentric orbit, while Venus has the least eccentric orbit.

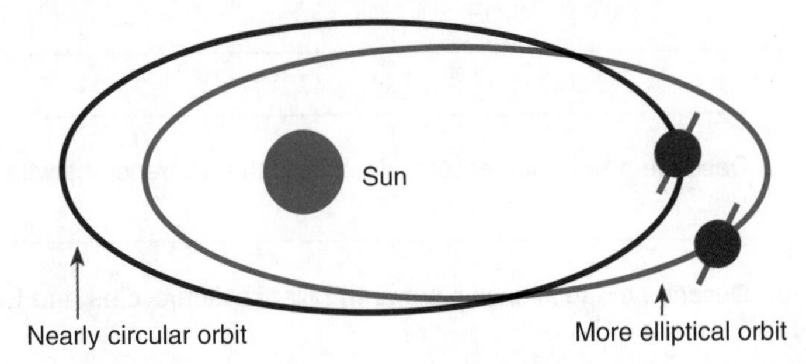



**ISBN: 978-1-927309-37-7**

## Effect of cycles

- The combination of obliquity, precession, and eccentricity change (called Milankovitch cycles) have major effects on the Earth's climate, especially if the extremes of each coincide.
- Obliquity, precession, and eccentricity change can all be calculated for hundreds of thousands of years in the past and future. Their effects can then be combined to produce a graph showing the maximum solar radiation received by Earth. Ice cores can then be compared to see if there are any correlations.

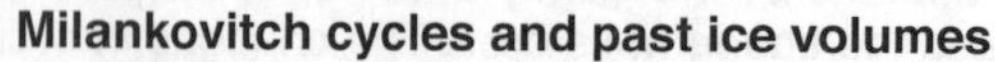

**Milankovitch cycles and past ice volumes**

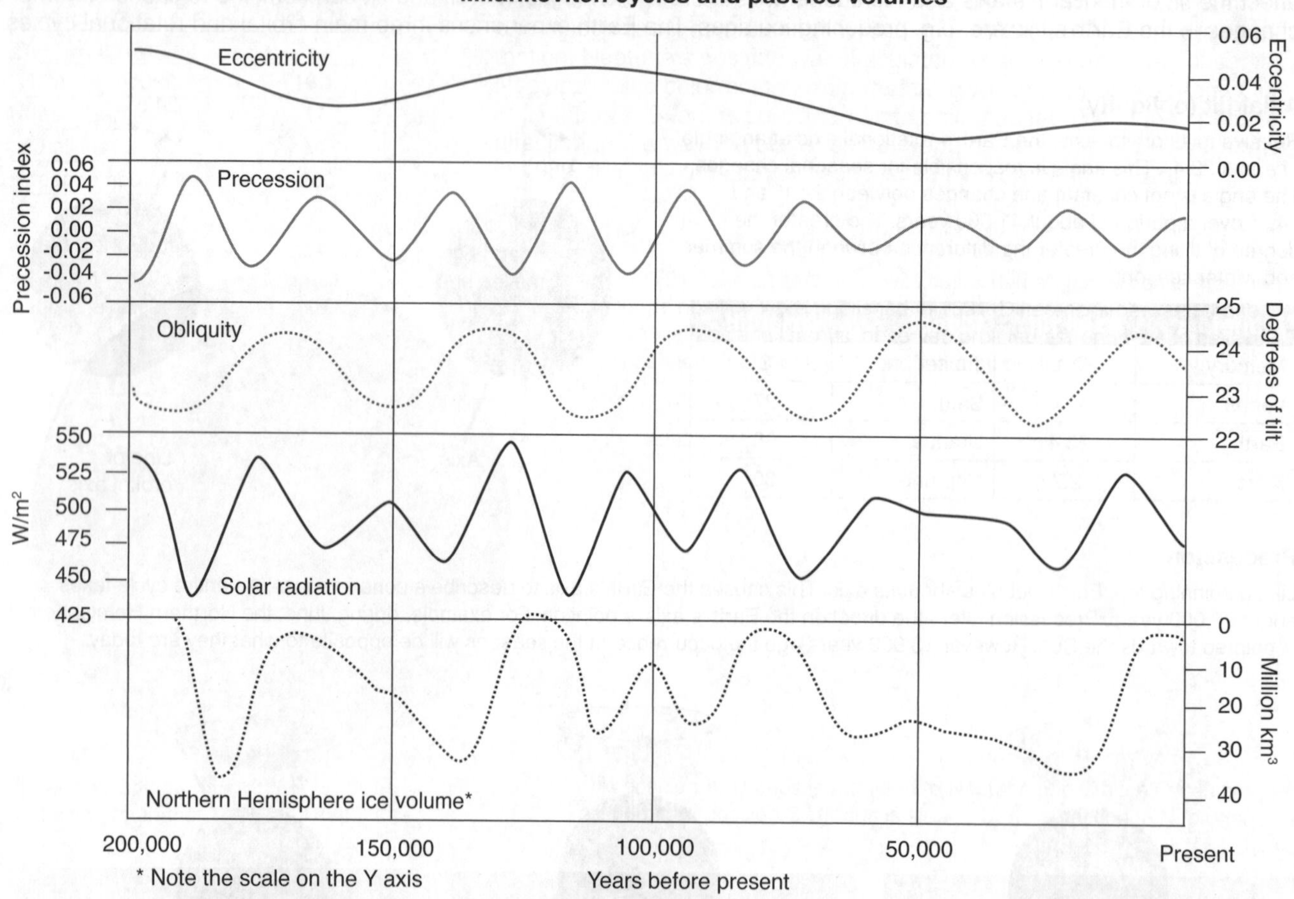


1. Describe the relationship between axial tilt and the difference in seasons: ____________

2. Venus has an axial tilt of 177°. Describe the effect of this on the rotation of the planet compared to other planets:

3. Describe how the timing of the Northern hemisphere's seasons would have changed over the last 26,000 years due to the Earth's precession:

4. Describe how orbital eccentricity affects the difference between seasons: ____________

5. Describe the relationship between Milankovitch cycles and Earth's climate: ____________


**ISBN: 978-1-927309-37-7**

# 45 Chapter Review

Summarize what you know about this topic so far under the headings provided. You can draw diagrams or mind maps, or write short notes to organize your thoughts. Use the introduction and the images and hints included to help you:

The solar system

HINT: Key steps in the formation of the solar system. The planets and technologies used to explore them.

Kepler and Newton

HINT: Kepler's laws of planetary motion and how Newton's laws of motion apply to them. Calculating orbital motion and using Kepler's third law.

Planets and cycles

HINT: Influences of gravity on short and long term cycles on Earth and other planets.


ISBN: 978-1-927309-37-7

# 46 KEY TERMS AND IDEAS: Did You Get It?

1. Test your vocabulary by matching each term to its definition, as identified by its preceding letter code.

|  | |  |
|---|---|---|
| eccentricity | A | The gravitationally-bound system comprising the Sun at the center and the objects that orbit it, either directly or indirectly. |
| exoplanet | B | Any object orbiting any larger object. In astronomy these are commonly divided into natural and artificial (man made). |
| gravity | C | German astronomer who formulated three laws of planetary motion around 1610. |
| Kepler | D | The gravitationally induced elliptical pathway of an object about a point in space. |
| newton (N) | E | A planet that orbits a star other than our Sun. |
| Newton | F | A measure of the extent that an ellipse differs from a circle. |
| orbit | G | The unit of force, named after Isaac Newton. One unit is the force required to accelerate one kilogram of mass by one meter per second squared. |
| planet | H | An object orbiting a star that has enough gravity to form a spherical shape and clear its orbit of other orbiting bodies or debris. |
| satellite | I | An attractive force between masses in which the attractive force is proportional to the product of the masses and inversely proportional to the square of the distance between them (Newtonian definition). |
| solar system | J | English mathematician who (along with many other important scientific contributors) formulated the laws of motion and universal gravitation. |

2. Venus takes 0.615 Earth years to orbit the Sun. Calculate its distance from the Sun in AUs: ______

3. A space probe reaching Neptune was placed into orbit 45,000 km from the center of the planet. Calculate the orbital period in hours of the space probe (note Neptune has a mass of $1.024 \times 10^{26}$ kg):

4. The exoplanet Kepler-186f orbits the red dwarf star Kepler-186. It takes 129.9 days to orbit the star and is about 0.40 AUs from it. Calculate the mass of the star Kepler-186 in solar masses.

5. The diagrams below show two orbiting systems and their relative masses and their barycenters (black dot). Describe how the systems would orbit and explain why (diagrams may be useful).

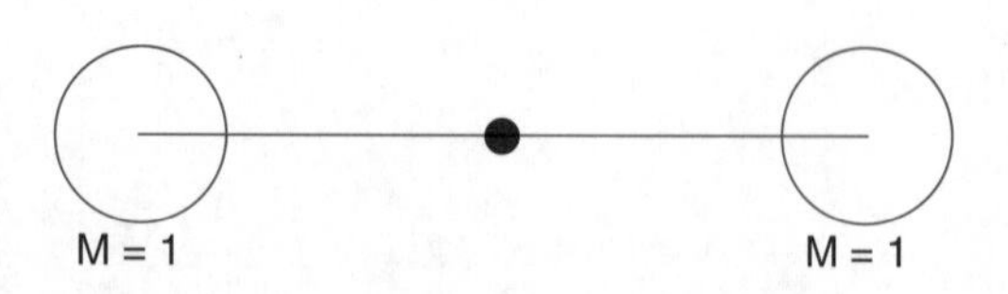


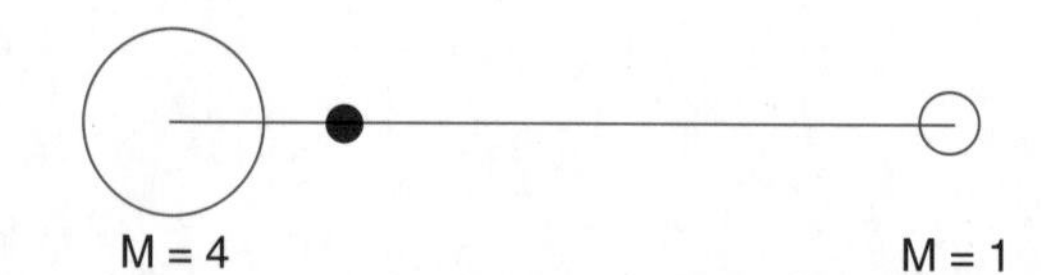



ISBN: 978-1-927309-37-7

# 47 Summative Assessment

1. Below is a model of a distant solar system. The star in the center is one solar mass (it is the same size as our Sun). The distances from the center of the star to the orbits shown is to scale. It has been determined that the inner most planet takes 123 Earth days to orbit the star. Use this data to calculate the distance from the star to the other four planets and the comet along their semimajor axis, and the time they take to orbit the star. Write the answer to your calculations next to the appropriate planet.

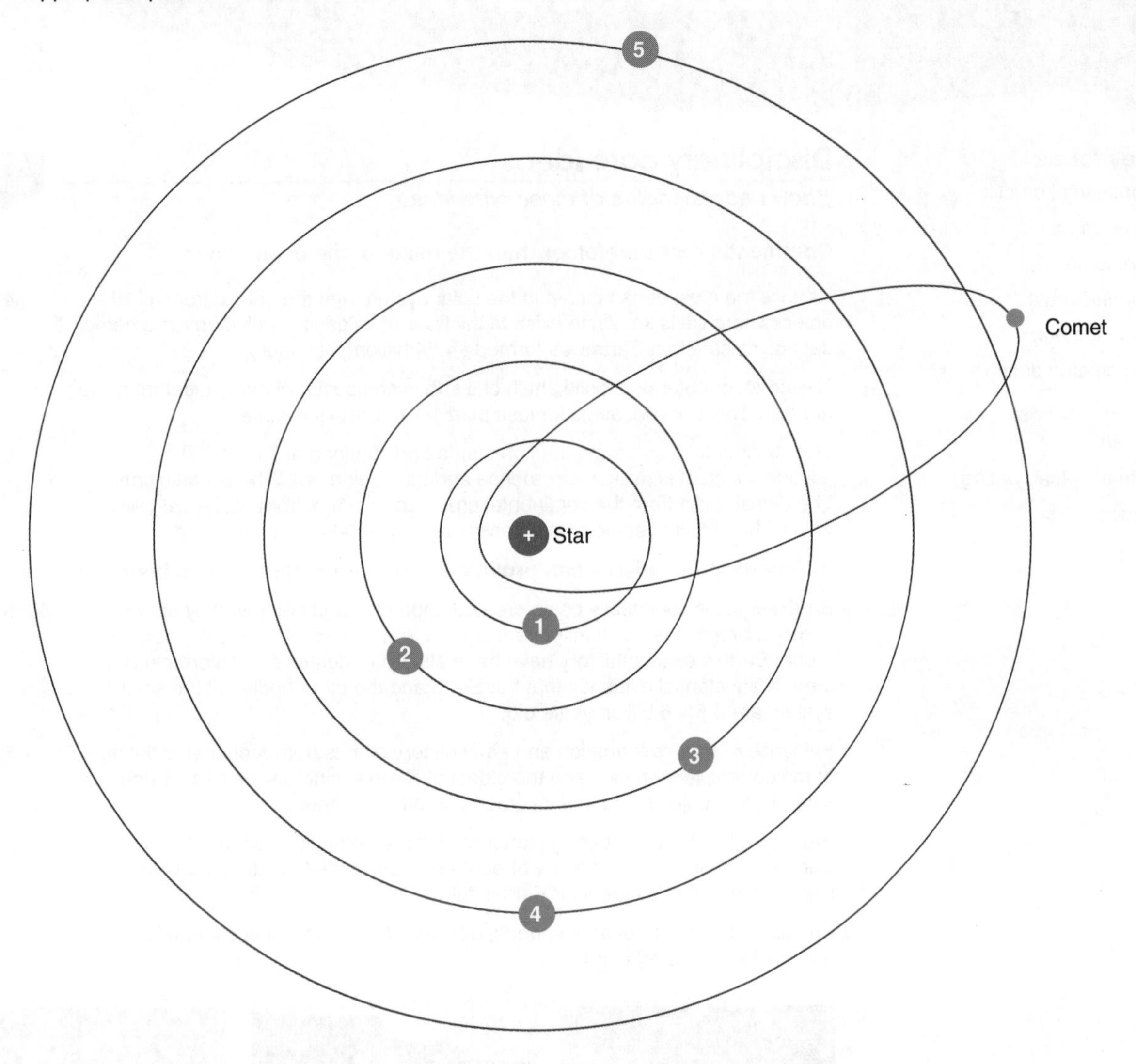


2. The diagram below shows the path of a comet with an eccentric orbit about a star. Use the diagram to explain the motion of the comet as it moves around its orbit:

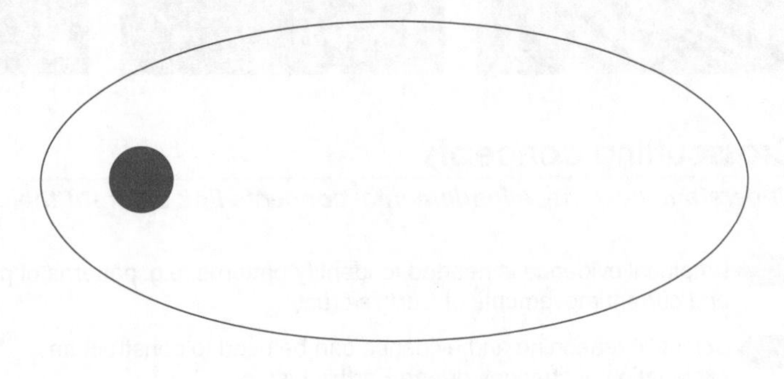


ISBN: 978-1-927309-37-7

ESS1.C

# The History of Planet Earth

**Key terms**

continental crust
lunar rocks
meteorite
oceanic crust
plate tectonics
radiometric dating
terrestrial rocks
Theia
uranium–lead dating
zircon

## Disciplinary core ideas

*Show understanding of these core ideas*

Activity number

### Continental rocks are older than the rocks of the ocean floor

| | | Activity number |
|---|---|---|
| ☐ | 1 Earth is the most dense planet in the solar system and the only astronomical object where life is known to exist. Many lines of evidence, including radiometric dating, indicate that Earth was formed ~4.54 billion years ago. | 48 54 |
| ☐ | 2 The crust, or outermost solid shell, of Earth is composed of many different types of rocks. The crust accounts for less than 1% of Earth's volume. | 49 |
| ☐ | 3 The continental crust forms the continents and continental shelves. The continental crust is thicker, less dense and much older than the oceanic crust. The oldest rocks from the continental crust range in age from ~3.7-4.28 billion years. The oldest oceanic crustal rocks are only ~200 million years old. | 48 49 |

### Extraterrestrial objects can provide information about Earth's history

| | | Activity number |
|---|---|---|
| ☐ | 4 Earth's crust is continually being created, modified, and destroyed by active geologic processes, such as plate tectonics and erosion. As a result, rocks that record Earth's earliest history have been altered or destroyed. Nevertheless, there is substantial evidence that the Earth and the other bodies of the solar system are 4.5-4.6 billion years old. | 48 51 52 |
| ☐ | 5 Evidence of Earth's formation and early history comes from radiometric dating of meteorites, lunar rocks, and the oldest of Earth's minerals, as well as from solar system objects and impact craters on other planets. | 50-53 |
| ☐ | 6 Studies of Earth's neighboring planet Mars have potential to shed light on Earth's early history. The history of geological and climate changes on Mars may also serve as a model for Earth's future. | 51 55 |
| ☐ | 7 An immense amount of atmospheric, geological, and biological change has occurred since Earth's formation. | 53 |

Meteorite Recon cc 3.0

## Crosscutting concepts

*Understand how these fundamental concepts link different topics*

Activity number

| | | Activity number |
|---|---|---|
| ☐ | 1 **P** Empirical evidence is needed to identify patterns, e.g. patterns of past and current movements of Earth's crust. | 49-50 53 |
| ☐ | 2 **SC** Scientific reasoning and evidence can be used to construct an explanation of changes during Earth's history. | 49-53 |

## Science and engineering practices

*Demonstrate competence in these science and engineering practices*

Activity number

| | | Activity number |
|---|---|---|
| ☐ | 1 Evaluate the evidence for explanations of the ages of crustal rocks. | 49-53 |
| ☐ | 2 Use scientific reasoning to link evidence to accounts of Earth's formation and early history and assess the extent to which the account is supported. | 49-53 58 |

# 48 The Earth

**Key Idea**: The Earth is 4.5 billion years old. During its long history, the Earth's geological activity has continually reshaped its surface.

▶ The Earth formed around 4.5 billion years ago along with the rest of the solar system. The interior is still molten, reaching up to 4700°C in the core. This heat drives crustal movements that have produced the surface we see today.

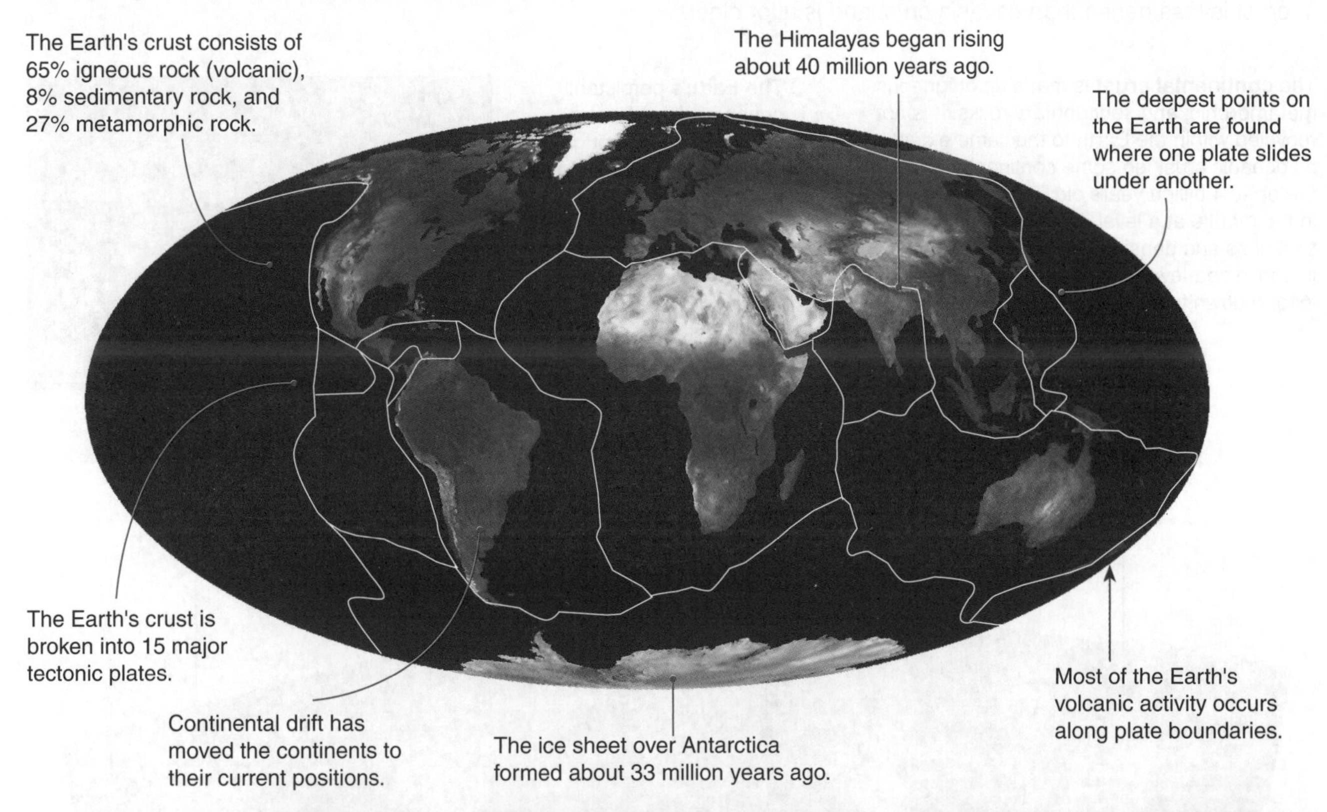

The Earth is geologically active. Convection currents in the mantle below the crust drive continuing continental drift and help form volcanoes and mountains.

Earth's liquid outer core surrounds an iron inner core producing a strong magnetic field. Research shows that the field's polarity has reversed many times.

Earth's atmosphere is 78% nitrogen, 20.9 % oxygen. The oxygen was produced by photosynthetic cyanobacteria about 2.3 billion years ago.

1. What process moves the continents across the surface of the Earth? ______________________

2. What is a consequence of the convection currents in the mantle? ______________________

3. What cause the Earth's magnetic field? ______________________

4. Where did the Earth's atmospheric oxygen come from? ______________________


**ISBN: 978-1-927309-37-7**

# 49 The Earth's Crust

**Key Idea**: The Earth's crust can be divided into continental crust and oceanic crust, each having specific properties.

- The Earth's crust is thin compared to the bulk of the Earth, averaging just 25-70 km thick below the continents and about 10 km thick below the oceans. The crust can be divided into continental crust and oceanic crust. Continental crust is less dense than oceanic crust and is a lot older.

The **continental crust** is made up of igneous, metamorphic, and sedimentary rocks. It is not recycled within the Earth to the same extent as oceanic crust, so some continental rocks are up to 4 billion years old. The crust "floats" in the mantle at a level determined by its thickness and density. The more mass there is above sea-level, the deeper the crust must extend down to support it.

The Earth's persistent oceans of liquid water cycle moisture through the atmosphere to the land and back again.

Water precipitated from the atmosphere forms rivers and lakes, which flow back to the ocean eroding the landscape in the process.

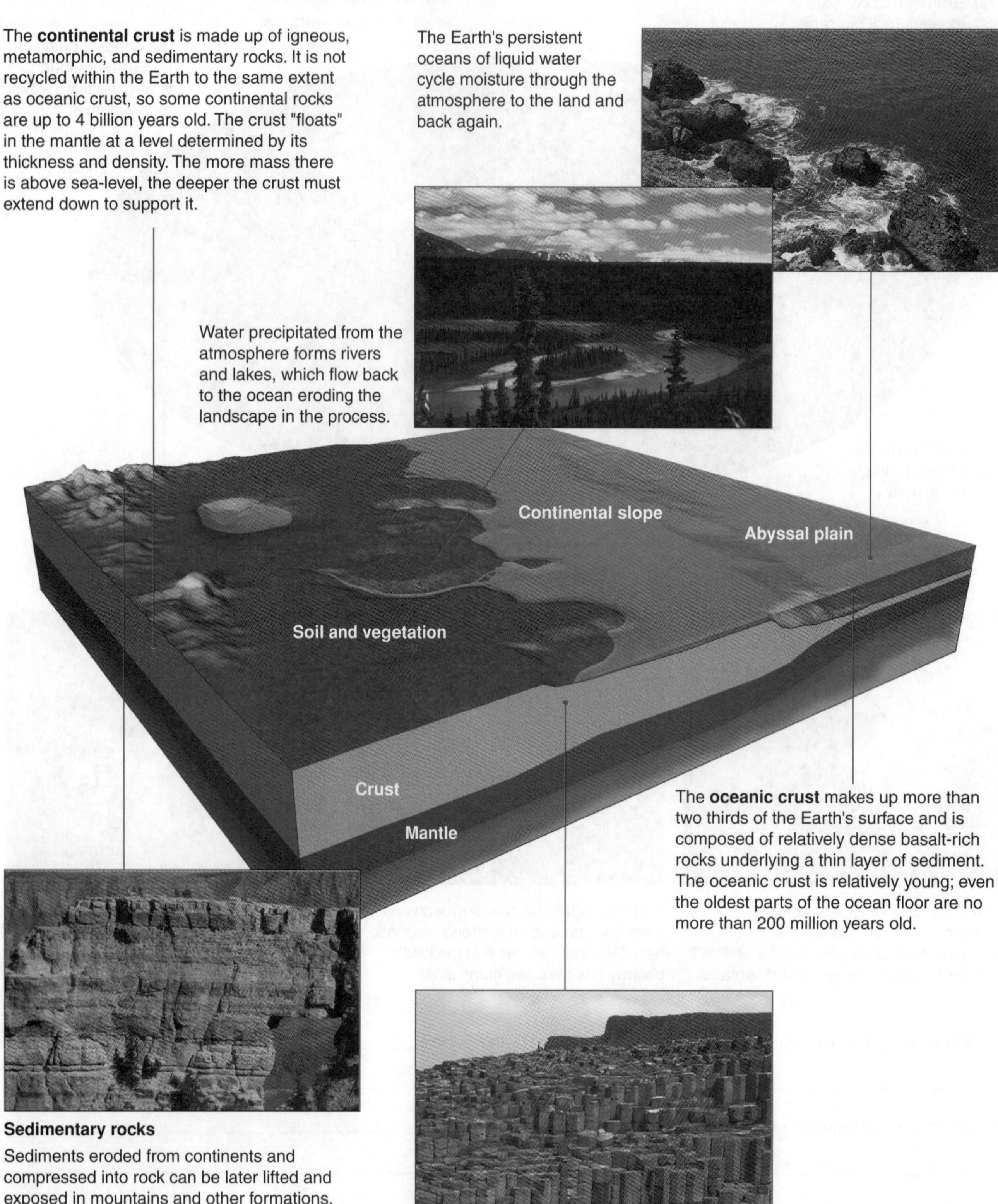


The **oceanic crust** makes up more than two thirds of the Earth's surface and is composed of relatively dense basalt-rich rocks underlying a thin layer of sediment. The oceanic crust is relatively young; even the oldest parts of the ocean floor are no more than 200 million years old.

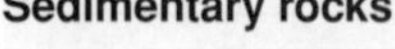

**Sedimentary rocks**

Sediments eroded from continents and compressed into rock can be later lifted and exposed in mountains and other formations.

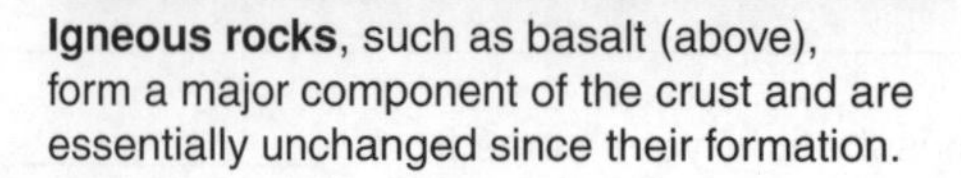

**Igneous rocks**, such as basalt (above), form a major component of the crust and are essentially unchanged since their formation.

**ISBN: 978-1-927309-37-7**

## Oldest and newest rocks

Earth is dynamic, the surface is always changing. Rocks on the surface are constantly being eroded, buried, melted, and regenerated. However some parts of the continental crust are relatively stable and have survived repeated cycles of rifting and collision. These areas contain the oldest rocks on Earth. On the ocean floor new rock is constantly being formed at the mid ocean ridges.

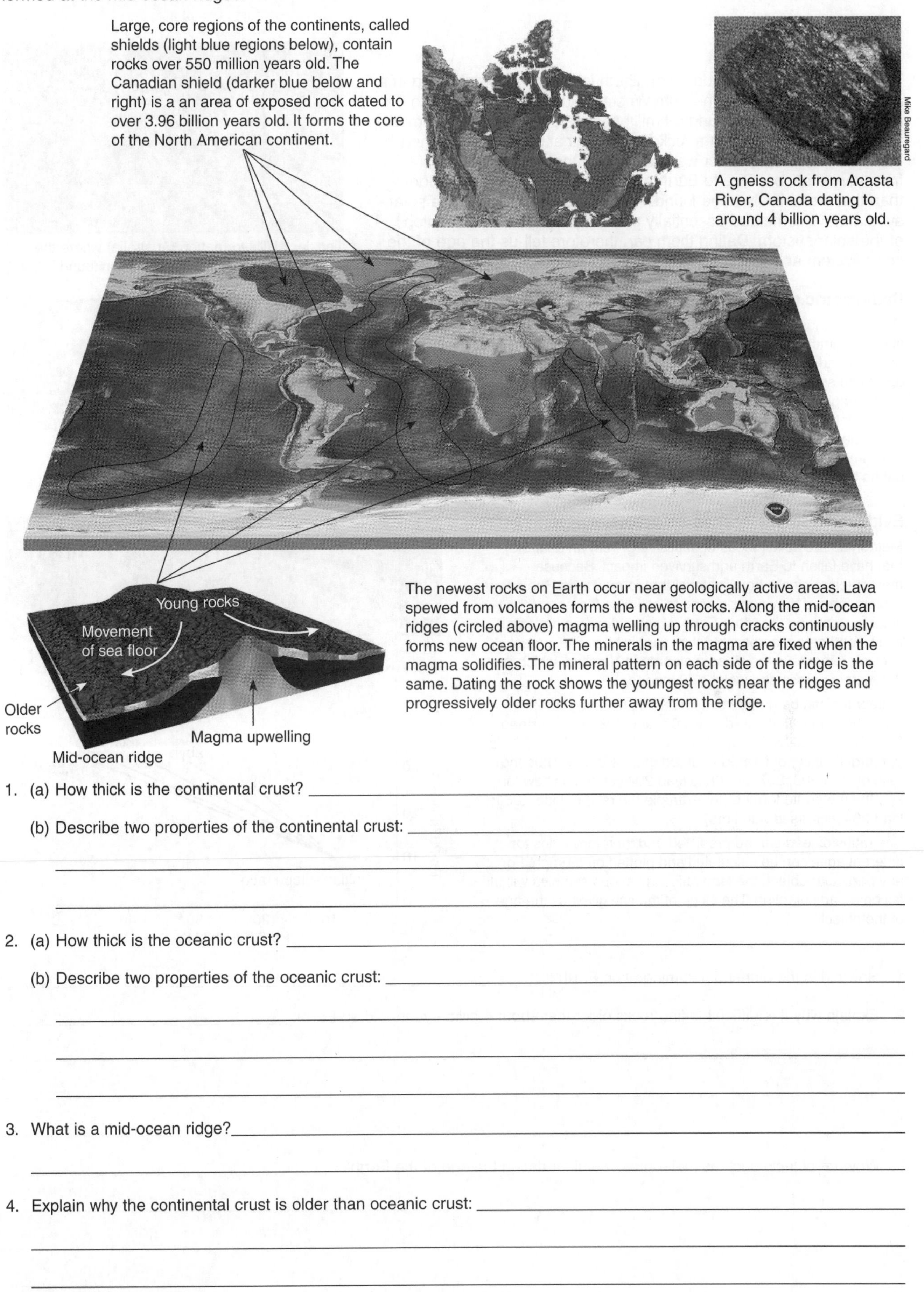


Large, core regions of the continents, called shields (light blue regions below), contain rocks over 550 million years old. The Canadian shield (darker blue below and right) is a an area of exposed rock dated to over 3.96 billion years old. It forms the core of the North American continent.

A gneiss rock from Acasta River, Canada dating to around 4 billion years old.

The newest rocks on Earth occur near geologically active areas. Lava spewed from volcanoes forms the newest rocks. Along the mid-ocean ridges (circled above) magma welling up through cracks continuously forms new ocean floor. The minerals in the magma are fixed when the magma solidifies. The mineral pattern on each side of the ridge is the same. Dating the rock shows the youngest rocks near the ridges and progressively older rocks further away from the ridge.

1. (a) How thick is the continental crust? ____________________

   (b) Describe two properties of the continental crust: ____________________

2. (a) How thick is the oceanic crust? ____________________

   (b) Describe two properties of the oceanic crust: ____________________

3. What is a mid-ocean ridge? ____________________

4. Explain why the continental crust is older than oceanic crust: ____________________


**ISBN: 978-1-927309-37-7**

# 50 Dating the Earth

**Key Idea**: The Earth can be dated by comparing minerals on Earth with minerals found on extraterrestrial bodies.

## Dating the Earth

The very oldest material dated on Earth is a zircon crystal found in a metamorphosed sandstone from Western Australia. It is dated to 4.4 billion years old, just a hundred million years after the Earth formed. The mineral was found in a rock formation that suggests it formed in water, suggesting water was present on the surface of Earth from very early on. Due to Earth's dynamic history, minerals older than this are unlikely to be found. However other bodies in the solar system have remained essentially unchanged since the formation of the solar system. Dating them can therefore tell us the age of the solar system and therefore how long ago Earth first formed.

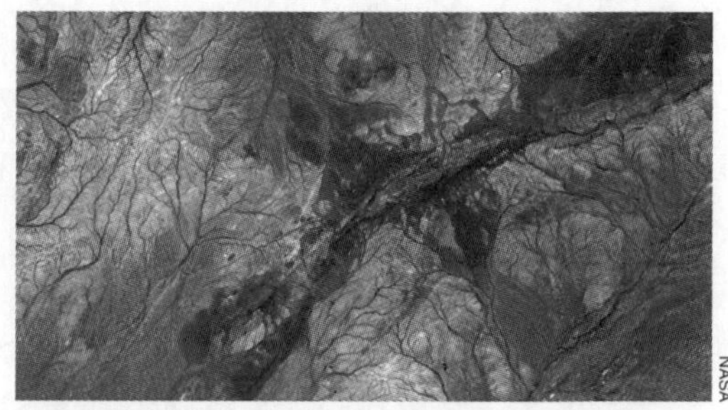


The Jack Hills formation (Australia) where the oldest minerals on Earth have been found.

### Radiometric dating

One of the most accurate ways of dating a mineral sample is radiometric dating. Many of the heavier elements (e.g. uranium and thorium) decay over time (their atoms break apart into smaller atoms). The rate of decay depends only on the original element's isotope. Uranium-238 ultimately decays into lead-206. The ratio of uranium-238 to lead-206 in a sample can therefore be used to determine the time since the sample formed. There are other ways of using radiometric dating but all depend on decay of heavy radioactive elements.

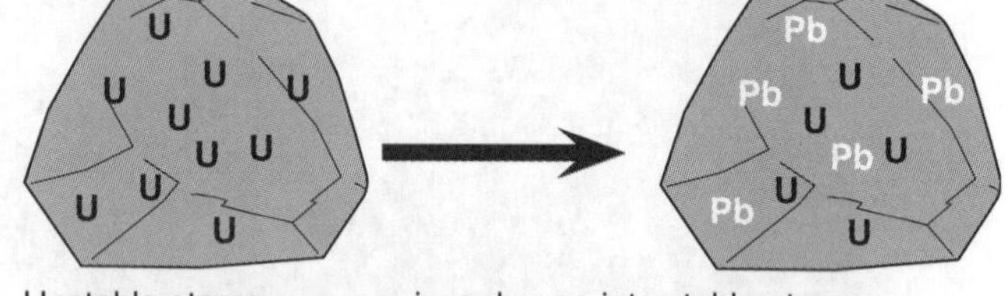


Unstable atoms, e.g. uranium change into stable atoms, e.g. lead, following a long but predictable decay series involving many radioactive elements.

### Evidence from meteorites

Meteorites are solid pieces of debris (e.g. from an asteroid), that have fallen to Earth and survived impact. Because meteorites originate from material that formed when the solar system formed they can be used to estimate the age of the solar system and therefore Earth. Most meteorites are small, a few centimeters across, but they vary in size. The largest known meteorite is the Hoba meteorite in Namibia, which is 2.7 meters across and weighs 60 tonnes.

Meteorites can be dated using the ratios of lead-206, 207, and 208 to lead-204. Lead 206, 207, and 208 are all formed from the radioactive decay of uranium or thorium. Lead 204 is primordial - it is not formed from radioactive decay. Thus the ratio of lead-206, 207, or 208 to lead 204 can tell us how long ago the meteorite formed (for example the ratio of lead-206 to lead-204 increases with time).

The ratios of lead can be predicted and therefore ratios for different ages can be calculated and plotted on a graph. For any particular object, the lead ratios in various samples will all fall on a particular line. The slope of the line gives us the age of the object.

The Hoba meteorite

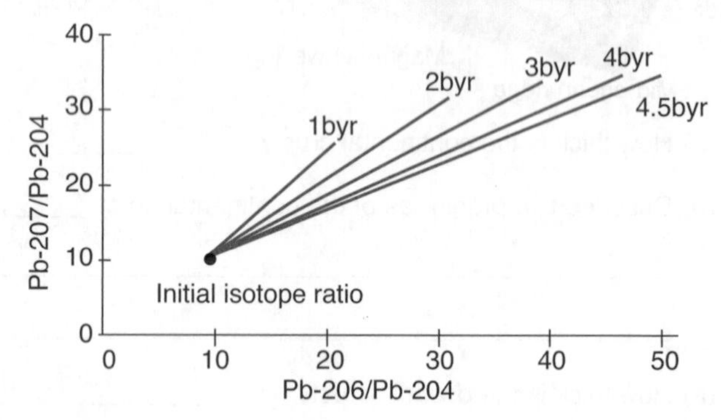


1. How old is the oldest dated mineral on Earth? ______________________

2. Explain why it is difficult to find rocks older that about 4 billion years old on Earth: ______________________

3. Why are objects such as meteorites useful for dating the age of the Earth? ______________________

KNOW

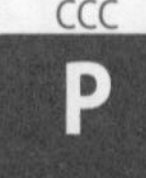

ISBN: 978-1-927309-37-7

Collecting moon rocks (Apollo 17)

Basalt moon rock (Apollo 15)

## Evidence from moon rocks

Moon rocks here on Earth come from lunar meteorites and samples collected by the manned American Apollo and unmanned Soviet Luna missions. The Apollo missions brought back a total of 380.96 kg of moon rock from the various landing sites on the moon.

The Moon has little active geology so most of the surface rocks are almost as old as the Moon. The Moon was formed when the protoplanet Theia smashed into the Earth early in the development of the solar system. The full age of the Earth cannot therefore be dated using moon rocks, but moon rocks can help confirm the current calculated age of the Earth based on their age.

The Moon has also suffered major impacts, some so large they exposed the lunar mantle. Thus lunar rocks taken from the lowlands (the darker areas seen from Earth) are younger than those of the highlands (the brighter areas). Radiometric dating places the age of highland rocks at around 4.5 billion years old and lowland rocks at about 3.16 billion years old. Also the lunar highlands are more heavily cratered than the lowlands, indicating they are older.

## Crater counting

The amount of impact cratering created by meteorites can be used to date a surface. At its very simplest, a young surface has very few craters, an old surface has more. Different sized craters form at different rates (there are more smaller craters compared to larger craters). Therefore we can expect that ancient surfaces not only have a larger number of craters but a larger number of any particular sized crater compared to young surfaces. Generally craters around 1 km in diameter are used as a reference point when using this method. Using the rocks brought back from the Moon, the number of craters at the sample site and the age of the surface can be compared and a plot produced.

On a planet such as Earth only large or very recent craters are visible on the surface. Smaller and more ancient craters have been eroded away by geological processes.

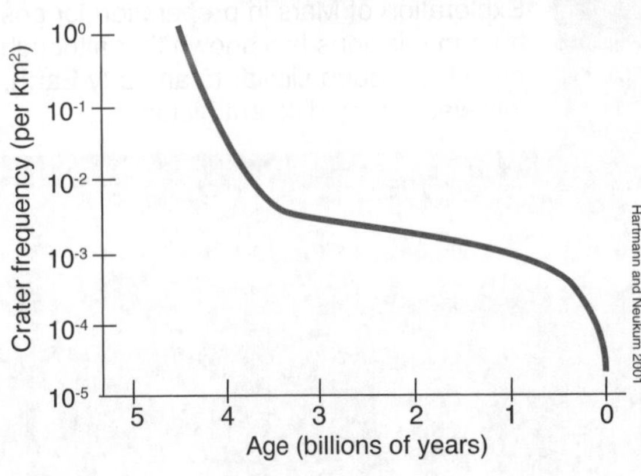


4. Give a brief description of how radiometric dating works: ______________________

5. Why is lead-204 useful when radiometrically dating a meteorite? ______________________

6. Explain how dating rocks from the Moon can help in dating the age of the Earth: ______________________

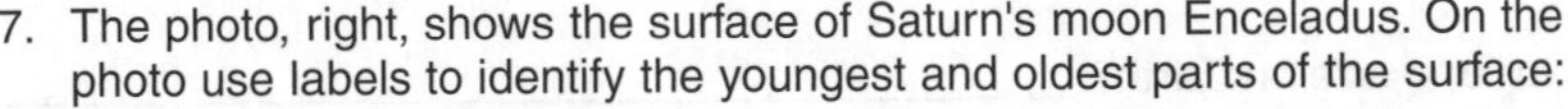

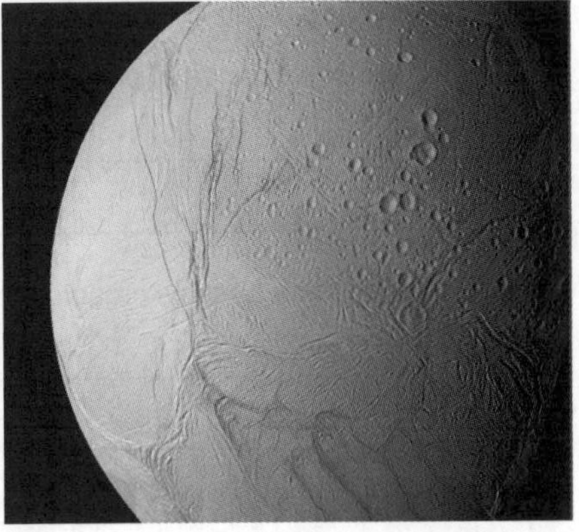


7. The photo, right, shows the surface of Saturn's moon Enceladus. On the photo use labels to identify the youngest and oldest parts of the surface:

8. Study the crater frequency graph above carefully. From what you know about the formation of the solar system, explain why the number of craters increases rapidly from 4 billion years ago to 5 billion years ago:

______________________


**ISBN: 978-1-927309-37-7**

# 51 What Can Mars Tell Us?

**Key Idea**: Robotic exploration of Mars has helped us understand why it is different to Earth.

Mars has some similarities to Earth in inclination, structure, composition, and even the presence of water on their surfaces. However Mars is also very different. Its surface is dry and cold and its gravity is only a third of Earth's. Yet Mars is the most explored of all the planets except Earth because it is the most hospitable of the other planets. Mercury is far too close to the Sun and Venus is far too hot and its atmosphere too dense. Mars offers the best possibility of all the planets for human colonization.

Exploration of Mars in preparation for possible future human missions has shown that although Mars may once have been similar to an early Earth, it has followed a very different history.

Earth is loaded with organic molecules (molecules that contain both carbon and hydrogen). Most of these are related to biological activity. Mars has few organic materials. Using Curiosity's Sample Analysis at Mars (SAM) laboratory, simple organic molecules were detected in significant (but still small amounts). However these molecules could have formed by simple reactions in the rocks.

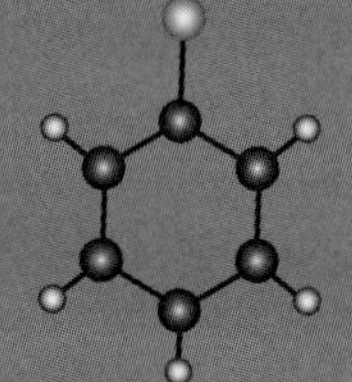

Chlorobenzene is one of the simple organic molecules found on Mars.

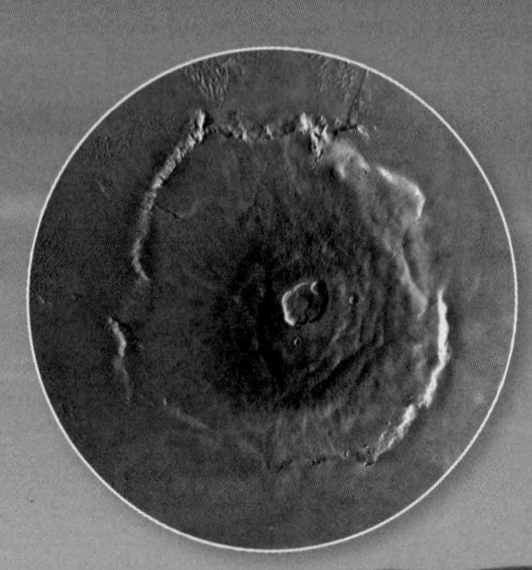

Olympus Mons is the largest volcano in the solar system. It is a basalt shield volcano that formed over tens of thousands of years. Olympus Mons and Mars' other volcanoes formed from hot spots under the crust. Unlike Earth, Mars has no plate tectonics, so the crust remained above the hot spot, allowing enormous volcanoes to form.

All images NASA

1. What is the significance of plate tectonics to volcano size on Earth? ______________________________

______________________________

______________________________

2. Explain why Mars has few organic molecules in its surface soils: ______________________________

______________________________

______________________________

______________________________

**ISBN: 978-1-927309-37-7**

One of the instruments aboard Curiosity is the ChemCam. This fires high intensity laser pulses at a target rock and the resulting spark at the rock surface is analyzed by spectrometers. The different wavelengths of light indicate different types of elements and molecules in the rock.

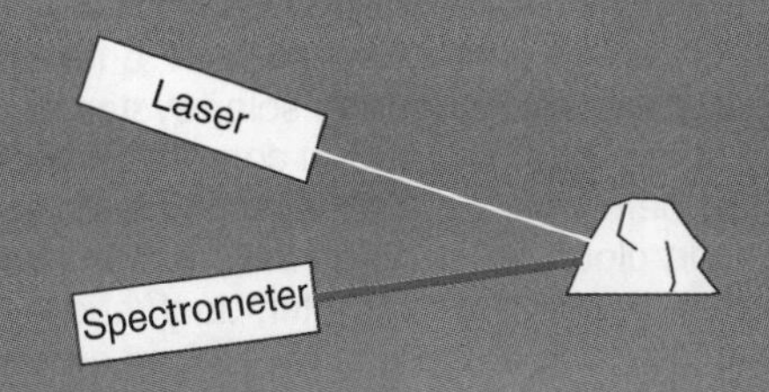


ChemCam has confirmed that much of the dust and soil on Mars is hydrated (it has water bonded to it). Other analysis has shown that most of the hydrogen in Mar's water is deuterium (hydrogen with a neutron). The water containing lighter hydrogen (protium) has evaporated into space.

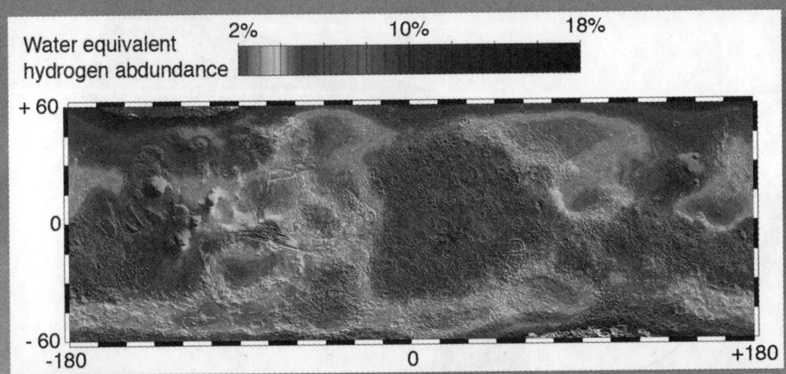


This map shows the distribution of Mar's subsurface water. Water is present in the highest concentrations near the poles and on the highlands.

Images of dry river beds, canyons, and grooves on sedimentary rock (above) indicate that Mars had plentiful running water at some time in the distant past. The grooves on the rock are caused by water currents.

The confirmation that Mars had plentiful water in the past (and probably a fairly hospitable climate) and possibly subsurface water today leads us to the ultimate question: If Mars was hospitable, did life evolve there and, if so, is it still there today?

We have yet to answer this question. It has been tested many times but the answers so far have been negative or ambiguous.

One clue for life on Mars is the detection of methane in the atmosphere. On Earth, methane is produced predominantly through biological processes. Methane seems to be being continuously made on Mars, but is just as quickly being removed. Methane's origin could be biological or geological.

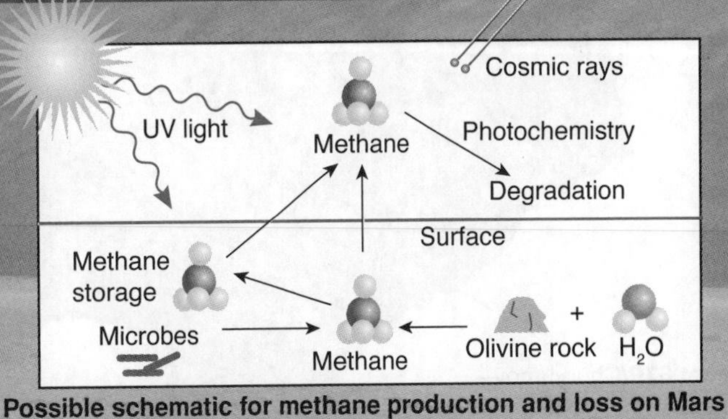


**Possible schematic for methane production and loss on Mars**

3. What evidence is there that Mars once had large amounts of flowing water? ______

4. What happened to Mars' water? ______

5. Why does the detection of methane not necessarily indicate life on Mars? ______


**ISBN: 978-1-927309-37-7**

# 52 Information from Probe Missions

**Key Idea**: Information from a large number of space probes has helped us understand the earliest history of Earth.

## Comets

Comets are chunks of frozen rock, dust, water, and gas that move through the solar system in highly eccentric orbits. Short term comets originate in the Kuiper belt whereas long term comets originate in the Oort cloud. Passing by the Sun causes them to heat up, vaporizing the water, dust, and gas, and creating a long tail that always faces away from the Sun. Comets represent material from the early solar system. Eight space probes have carried out missions to nearby comets.

Comet McNaught (also known as the Great Comet of 2007) as seen from New Zealand just before sunset.

## Rosetta

The **Rosetta** space probe was launched by the ESA in 2004. It made a gravity assist flyby of Mars, and flybys of the comets 2867 Šteins and 21 Lutetia. Rosetta then entered hibernation for 31 months before reawakening in January 2014. In August 2014 the Rosetta space probe reached comet 67P and entered orbit around it.

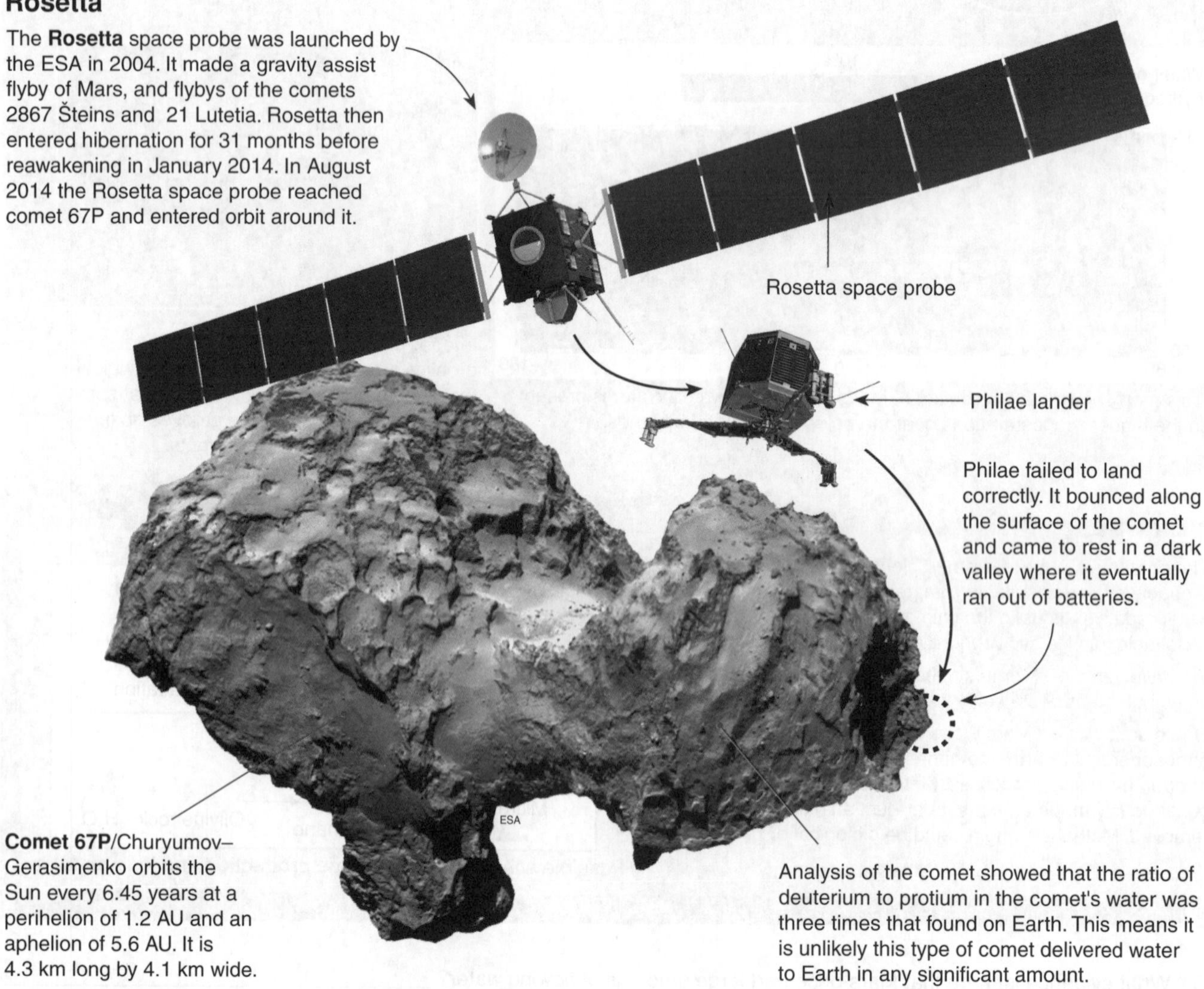


Philae failed to land correctly. It bounced along the surface of the comet and came to rest in a dark valley where it eventually ran out of batteries.

**Comet 67P**/Churyumov–Gerasimenko orbits the Sun every 6.45 years at a perihelion of 1.2 AU and an aphelion of 5.6 AU. It is 4.3 km long by 4.1 km wide.

Analysis of the comet showed that the ratio of deuterium to protium in the comet's water was three times that found on Earth. This means it is unlikely this type of comet delivered water to Earth in any significant amount.

### Detecting water

Hydrogen atoms can be found as both a single proton with an electron (protium) or as a proton and neutron with an electron (deuterium). Deuterium is heavier than protium and is very rare (0.0156% of all hydrogen). In Earth's oceans there is about one atom of deuterium in 6420 of hydrogen. Changes in this ratio in water molecules in comets or asteroids can be used to help answer questions about the origin of Earth's water.

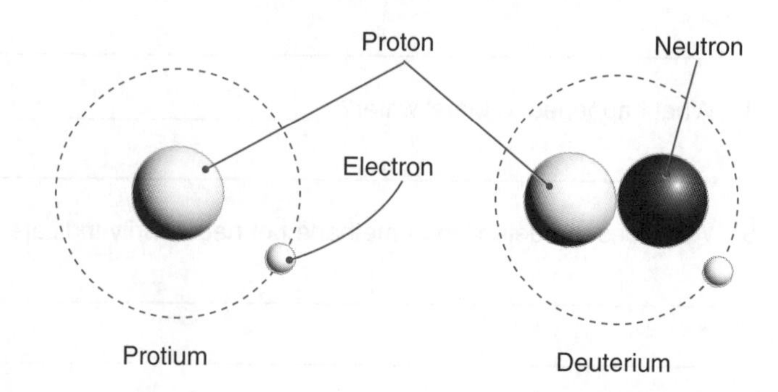

**ISBN: 978-1-927309-37-7**

## Stardust

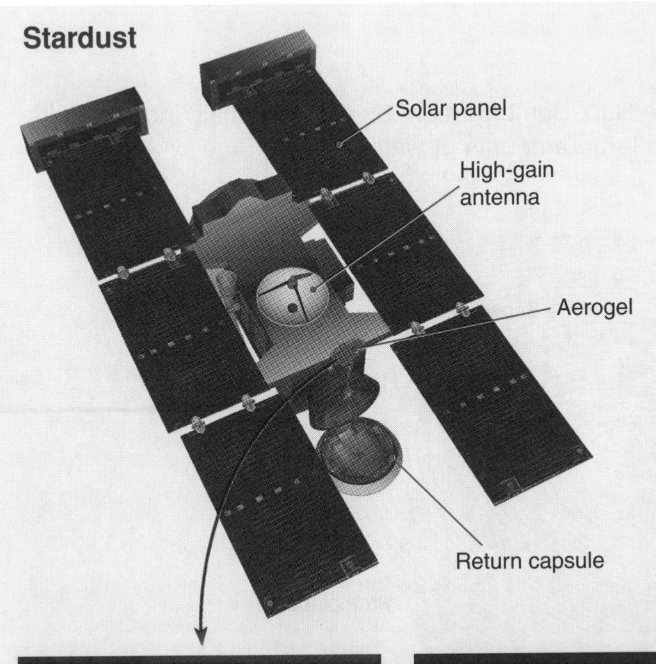


- **Stardust** was a space probe launched by NASA in 1999. Its primary mission was to rendezvous with comet **Wild-2**, collect dust samples from the coma (the dust cloud surrounding the comet nucleus), and return them to Earth. It was the first time samples from a comet were returned to Earth.
- As the probe neared the comet, a capsule opened and deployed a tennis racket sized collector holding blocks of aerogel. The dust particles streaming from the comet buried themselves into the gel.
- The capsule was then released as the probe passed by Earth in 2006.
- Analysis of the dust particles gathered by the probe suggested that material from the inner solar system was transferred to the outer solar system after its formation.
- The amino acid glycine was also detected in the material ejected from the coma, adding weight to the idea that comets brought at least some of the building blocks of life to Earth.

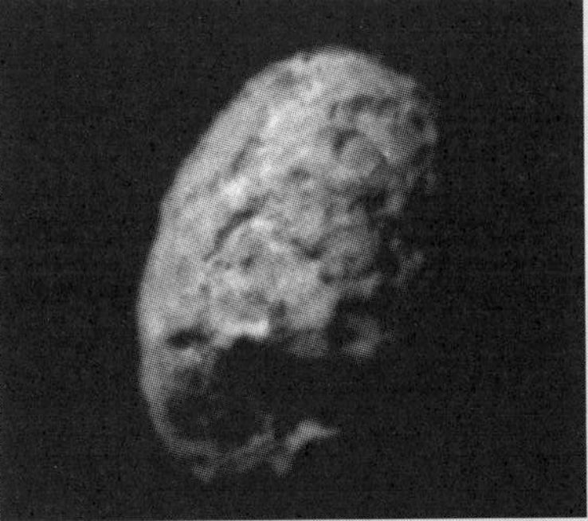

Comet Wild-2 (left) has a diameter of about 5 km. Until 1974 it orbited in the outer solar system, taking about 43 years per orbit. In September 1974 it passed within 1 million km of Jupiter. Jupiter's massive gravity pulled the comet into the inner solar system and changed its orbit to about 6 years.

Andrzej Mireck

The space probe **Giotto** flew past **Halley's Comet** in 1986, becoming the first space probe to study a comet. It found the comet to be 4.5 billion years old and 80% water.

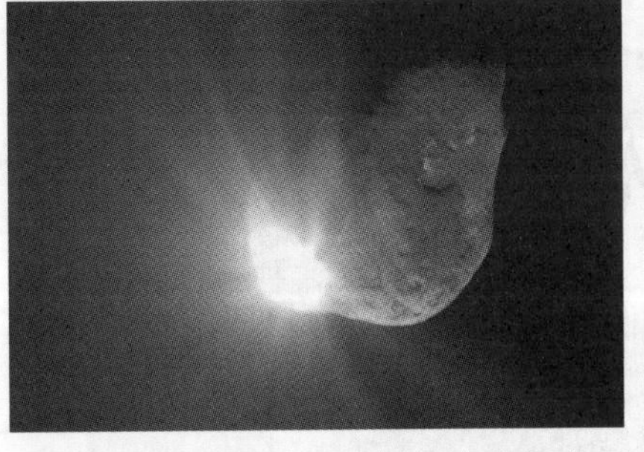

The **Deep Impact** space probe visited comet Tempel-1 in July 2005. It released a copper impactor that created a 100 m wide crater (above) so that cameras could photograph the interior of the comet.

All images NASA

Deep Impact also visited Comet **Hartley 2**. In 2011 it was found that the comet contained water very much like Earth's, supporting the argument that comets brought at least some water to Earth.

1. (a) What evidence is there that comets may have delivered water to Earth early in its history? ______

   (b) What evidence is there that only certain types of comet delivered the water? ______

2. What is the evidence that the building blocks of life were carried to Earth on comets? ______

3. Why was an impactor deployed by Deep Impact at Tempel-1 and what might it help discover? ______


**ISBN: 978-1-927309-37-7**

# 53 The Earth's History

**Key Idea**: Earth formed at the same time as the Sun. Bombardment by planetesimals increased its size, formed the Moon, and delivered large amounts of water.

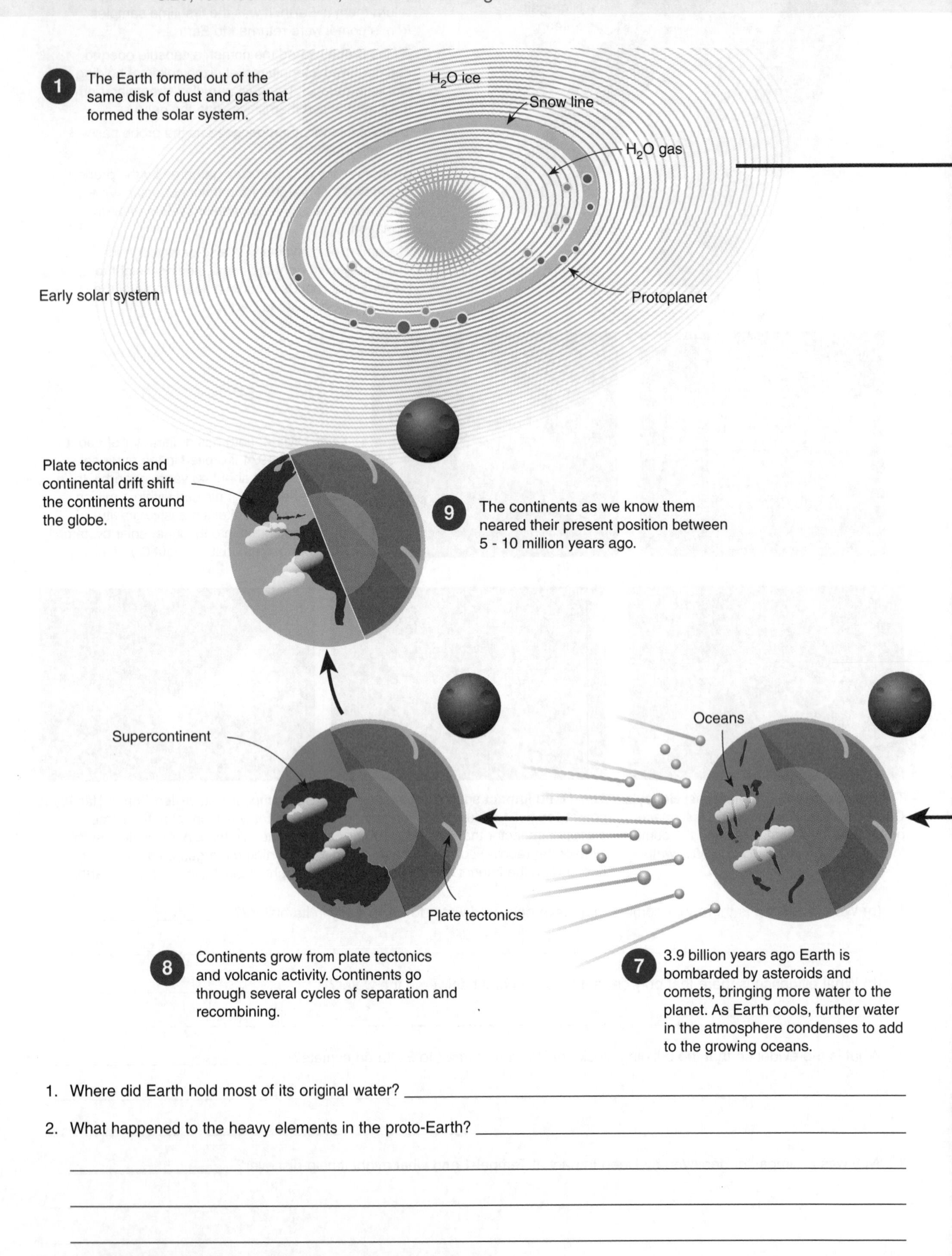


1. Where did Earth hold most of its original water? ___

2. What happened to the heavy elements in the proto-Earth? ___

___

___

___

**ISBN: 978-1-927309-37-7**

2. 4.55 billion years ago the Earth is a ball of molten rock. The heavier elements such as iron and nickel sink towards the center of the Earth. Water is present in the hot mantle.

3. The proto-Earth is bombarded with asteroids and planetesimals (small rocky bodies that could coalesce to form a planet). Hydrated minerals in asteroids and comets bring more water to the Earth.

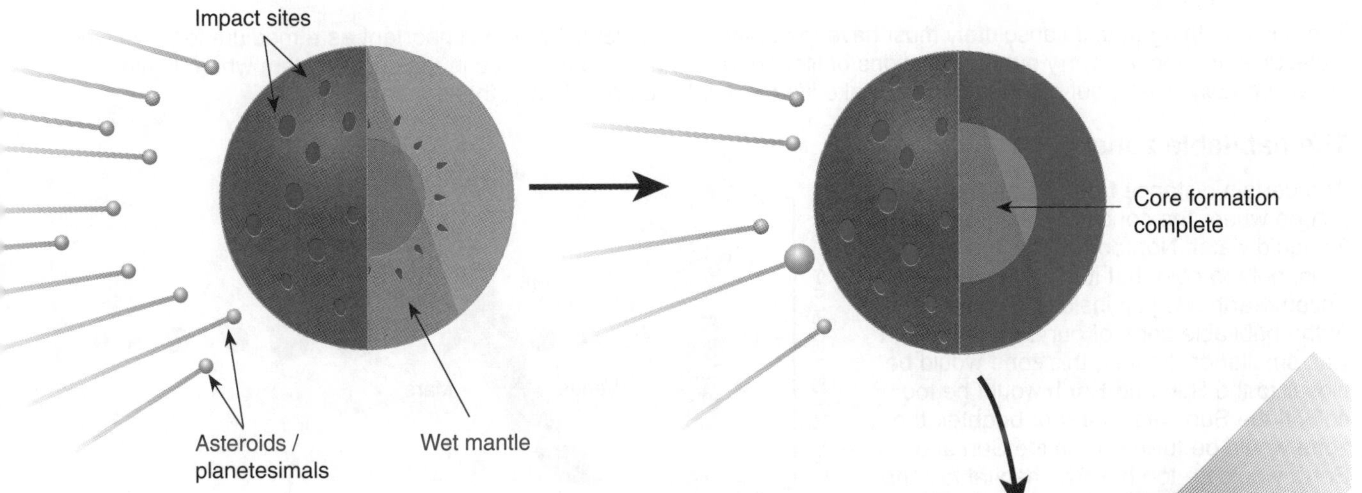

Impact sites

Asteroids / planetesimals

Wet mantle

Core formation complete

4. 4.4 billion years ago a Mars sized protoplanet called Theia smashes into the Earth in a low speed collision. Much of Theia was added to the Earth with about 20% being thrown into orbit around the Earth.

It is believed Theia formed in the L5 Lagrangian position in the Earth's orbit. Changes to Theia's mass destabilized its orbit and caused the collision.

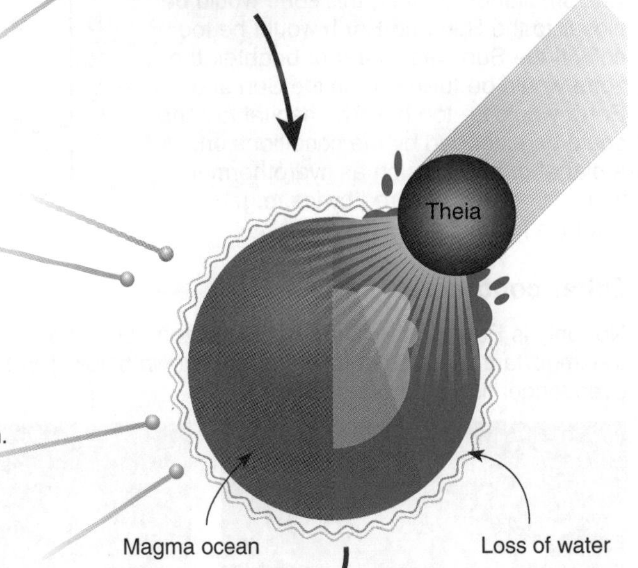

Theia

Magma ocean

Loss of water

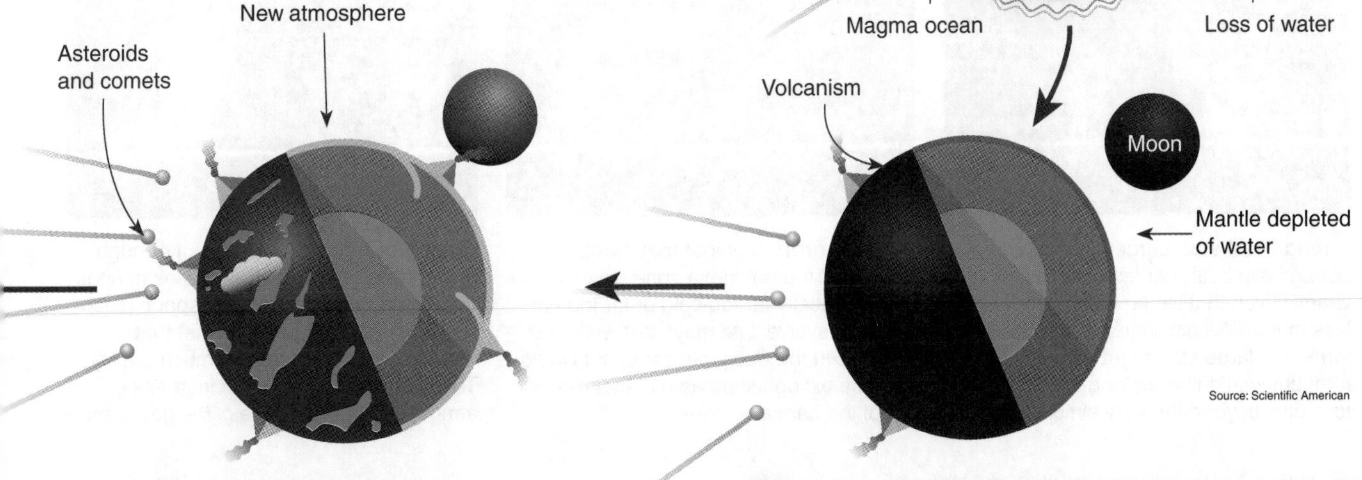

New atmosphere

Asteroids and comets

Volcanism

Moon

Mantle depleted of water

Source: Scientific American

6. As the Earth cools, liquid water appears on the surface. Plate tectonics begin as the crust hardens around the molten mantle.

5. The impact of Theia knocks Earth into its present 22° - 24° angle of rotation. The debris splashed into orbit around the Earth would have taken only around 100 years to coalesce into the Moon. Much of Earth's atmosphere and water is lost due to the impact of Theia. Volcanic activity and impactors (e.g. comets) gradually replace these.

3. How was Earth's atmosphere replaced after its collision with Theia? ______________________________________

4. What mechanisms created and shifted the continents? ______________________________________


**ISBN: 978-1-927309-37-7**

# 54 The Goldilocks Planet

**Key Idea**: As far as we know, Earth is the only place in the Universe where life exists. The Earth's distance from the Sun, its size, and the Sun itself are probable factors.

There is one thing that life absolutely must have to survive: liquid water. Water is important as a medium for dissolved molecules and ions to carry out the reactions of life. There are numerous places in the solar system where liquid water probably exists, but other conditions make life (certainly complex life) in those places unlikely.

## The habitable zone

At a certain distance from a star there is a zone where the conditions are just right for liquid water. Not too hot that the water boils, not too cold that it is permanently frozen. Earth sits just inside the inner edge of the habitable zone of our Sun. If the Sun was smaller or dimmer, this zone would be closer to the Sun and Earth would be too cold. If the Sun was bigger or brighter, the zone would be further from the Sun and Earth would be too hot. The habitable zone could be extended by the conditions on a planet or moon, such as hydrothermal vents in oceans covered by ice, e.g. Saturn's moon Enceladus.

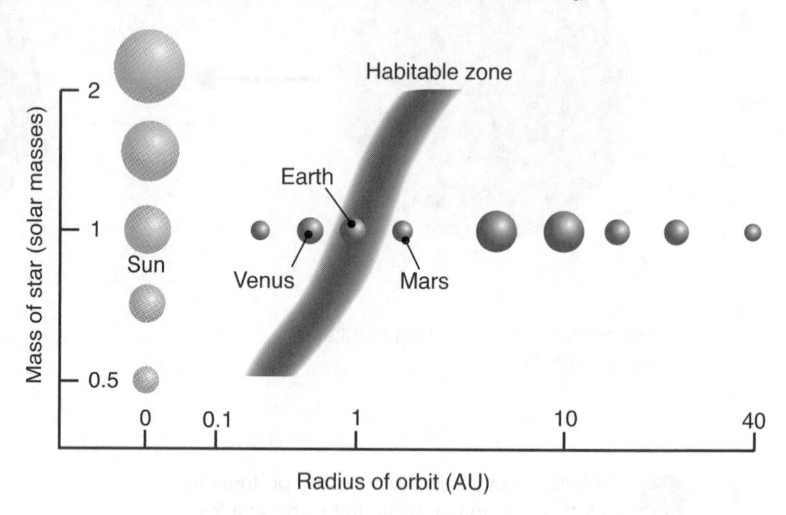


## Other conditions

Not only is Earth in the Sun's habitable zone but it has several other properties that make it suitable for life. Some of the important requirements for life are shown below. For the development of complex life, it is likely that there is an even longer list of requirements.

**A long lived star**: Large stars can burn out and explode after less than a billion years. Given that life on Earth appeared less than half a billion years after Earth formed, a large star might allow life to form but wouldn't exist long enough for life to evolve beyond the very simplest forms.

**A stable orbit**: A planet that moves in and out of the habitable zone would either be periodically far too cold or far too hot for life to evolve. Life may start while the planet is in the habitable zone, but would likely be extinguished as a planet moved out of the habitable zone.

All images NASA

**Sufficient mass**: If there is not enough mass, water and important gases cannot be retained and escape into space (such as occurred on Mars). A planet that is too massive generates too much gravity. This can have the effect of increasing atmospheric pressure (e.g. the gas planets).

1. Why is water needed for life? ___

2. What are some factors that help maintain liquid water on Earth's surface? Explain: ___

3. Why is a long lived star necessary for life to arise and evolve? ___


ISBN: 978-1-927309-37-7

# 55 Models of the Earth's Age

**Key Idea:** Scientists have been modeling the Earth's age for centuries. As techniques have become more sophisticated the models have become more refined and accurate.

## Early models and estimates

- Before the discovery of radioactive decay, calculating the age of the Earth was difficult. Most models focused on the cooling of the Earth from a molten ball of rock, but did not take into account the effect of heat produced by radioactive decay (because there was no knowledge of it).
- Georges Louis Leclerc (later Comte de Buffon) developed one of the earliest models for dating the Earth. In 1778 he used the cooling rate of iron spheres to model the cooling of the Earth from a ball of molten rock. Extrapolating his data he proposed the Earth was 70,000 years old, far older than the age of a few thousand years that many people believed at the time.
- In 1862, William Thomson (Lord Kelvin) calculated possible ages of between 20 million and 400 million years old.
- In 1895, John Perry argued that Lord Kelvin had not included convection currents in a molten mantle. He included convection currents in his calculations and arrived at an age of 2 to 3 billion years.
- In 1900, John Joly calculated the rate that the seas should have accumulated salt to reach their current level of salinity. He used this to calculate the age of the Earth at 100 million years old.

## Radioactivity offers another solution

- The discovery of radioactivity and the heat it produces in 1896 added another factor to the models of Earth's age. Heat in the Earth was being continually replaced and so the assumption that Earth simply cooled down after its formation was incorrect.
- Early attempts to use radioactive decay as a way of dating the Earth were filled with errors due to the many unknowns of the new science, but they did begin to show that the Earth had to be at least hundreds of million of years old.
- Using the theory that uranium decays to lead following a precise decay chain, Arthur Holmes published calculations in 1927 that the Earth was between 1.6 to 3.0 billion years old.
- Refinements of the decay series for various radioactive isotopes and ways of measuring isotope ratios have helped date the age of the Earth to 4.54 billion years, plus or minus a few tens of millions of years.

**Radioactive decay of potassium into calcium and argon**

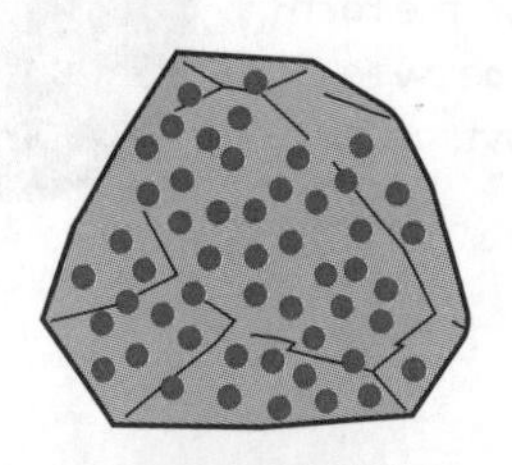


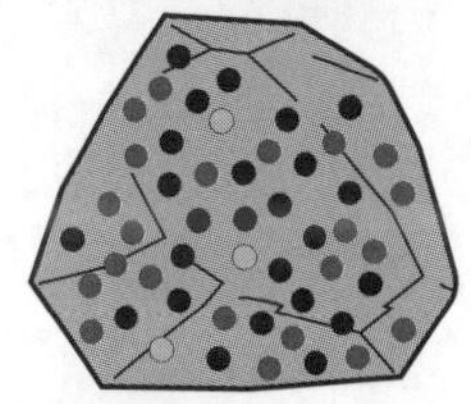


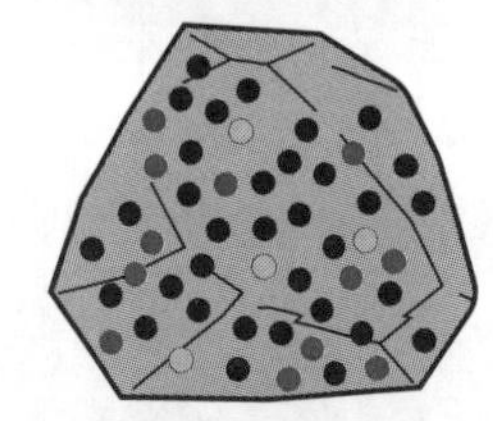


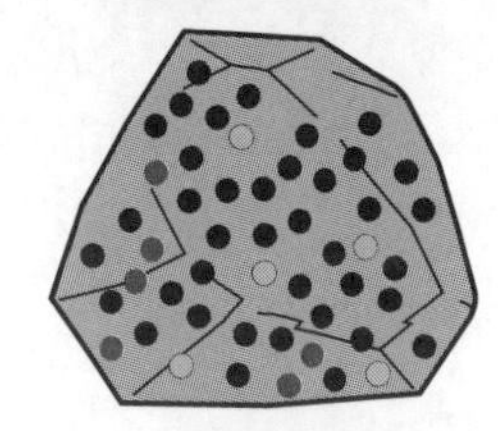


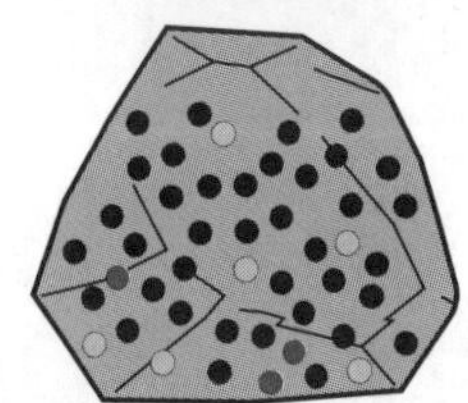


Potassium-argon dating is a commonly used radiometric dating technique. Potassium decays into calcium 89% of the time and argon about 11% of the time. Argon is inert and so is not contained in most minerals. Its mass in minerals is therefore related to the decay of potassium.

1. What assumption did many of the early models for dating the Earth make about the cooling of the Earth? ____________

2. How did the discovery of radioactive decay invalidate many of the early models? ____________

3. Explain why using the decay series of several different radioactive isotopes to date the Earth will provide a more robust, accurate answer to the Earth's age:


ISBN: 978-1-927309-37-7

# 56 Chapter Review

Summarize what you know about this topic so far under the headings provided. You can draw diagrams or mind maps, or write short notes to organize your thoughts. Use the introduction and the images and hints included to help you:

**The Earth's crust**

HINT: Outline the make up of the Earth's crust, including differences between continental and oceanic crust.

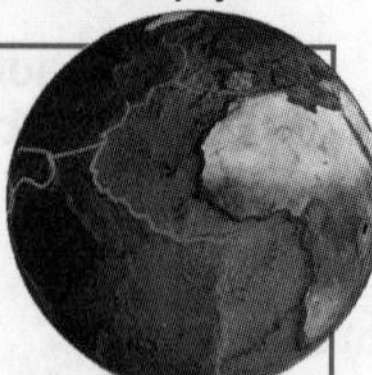

**Dating the Earth**

HINT: Describe methods for dating the Earth, including radiometric dating.

**Formation of the Earth**

HINT: Describe how the Earth formed and the influence of extraterrestrial bodies on its formation.




ISBN: 978-1-927309-37-7

# 57 KEY TERMS AND IDEAS: Did You Get It?

1. Test your vocabulary by matching each term to its definition, as identified by its preceding letter code.

|  | |  |  |
|---|---|---|---|
| continental crust | | A | The theoretical Mars-sized protoplanet that struck Earth during the formation of the solar system, producing the Moon from the debris. |
| lunar rocks | | B | The part of the Earth's crust that "floats" on the mantle and forms the continents. |
| meteorite | | C | The scientific theory which describes how the Earth's crust is divided into plates that move about due to convection currents in the mantle. |
| oceanic crust | | D | Rocks that have originated on Earth. |
| plate tectonics | | E | Part of the Earth's crust that is basalt rich, relatively dense and geologically young. It tends to underlie the ocean basins. |
| radiometric dating | | F | The common name for the mineral containing zirconium silicate and trace amounts of uranium and thorium. Radiometric dating has shown some of these minerals to be the oldest minerals on Earth. |
| terrestrial rocks | | G | Dating method that uses isotopes or measurement of radioactivity to produce an absolute time since the formation of a substance, +/- a specific error. |
| Theia | | H | Fragment of rock or iron from outer space that has survived both passage through the Earth's atmosphere and impact with the Earth's surface. |
| uranium–lead dating | | I | Radiometric dating technique that uses the ratios of uranium isotopes and lead isotopes to calculate the age of a mineral. |
| zircon | | J | Rocks that have been collected from the Moon by manned or unmanned missions. |

2. When using radiometric dating to date rocks on Earth, why is it better to use rocks from continental shields than elsewhere on Earth?

3. The image right shows Uranus' moon Miranda. Note that some parts are cratered and other parts have long streaks across them which may have formed from tidal stretching. Study the area in the white box. Which event occurred first, the tidal stretching or the crater impacts? Explain your answer:

4. When studying other planets and extraterrestrial objects (e.g. comets) identifying the ratio of hydrogen isotopes in any water found is an important task. Describe the information the hydrogen isotope ratio can tell us about an object's water in relation to Earth:


**ISBN: 978-1-927309-37-7**

# 58 Summative Assessment

1. The diagram below shows the South Atlantic ocean. Samples taken from the seabed were dated using radiometric dating. The positions of the sample and their age are shown in the table below:

| Sample number | Latitude | Longitude | Age of rock (millions of years) |
|---|---|---|---|
| 1 | 14° S | 23° W | 62 |
| 2 | 35° S | 43 W | 100 |
| 3 | 25° S | 7° E | 80 |
| 4 | 41° S | 6° W | 40 |
| 5 | 11° S | 11° W | 20 |
| 6 | 26° S | 15° W | 10 |
| 7 | 24° S | 23° W | 55 |
| 8 | 25°S | 13°W | 10 |
| 9 | 50°S | 19°W | 40 |

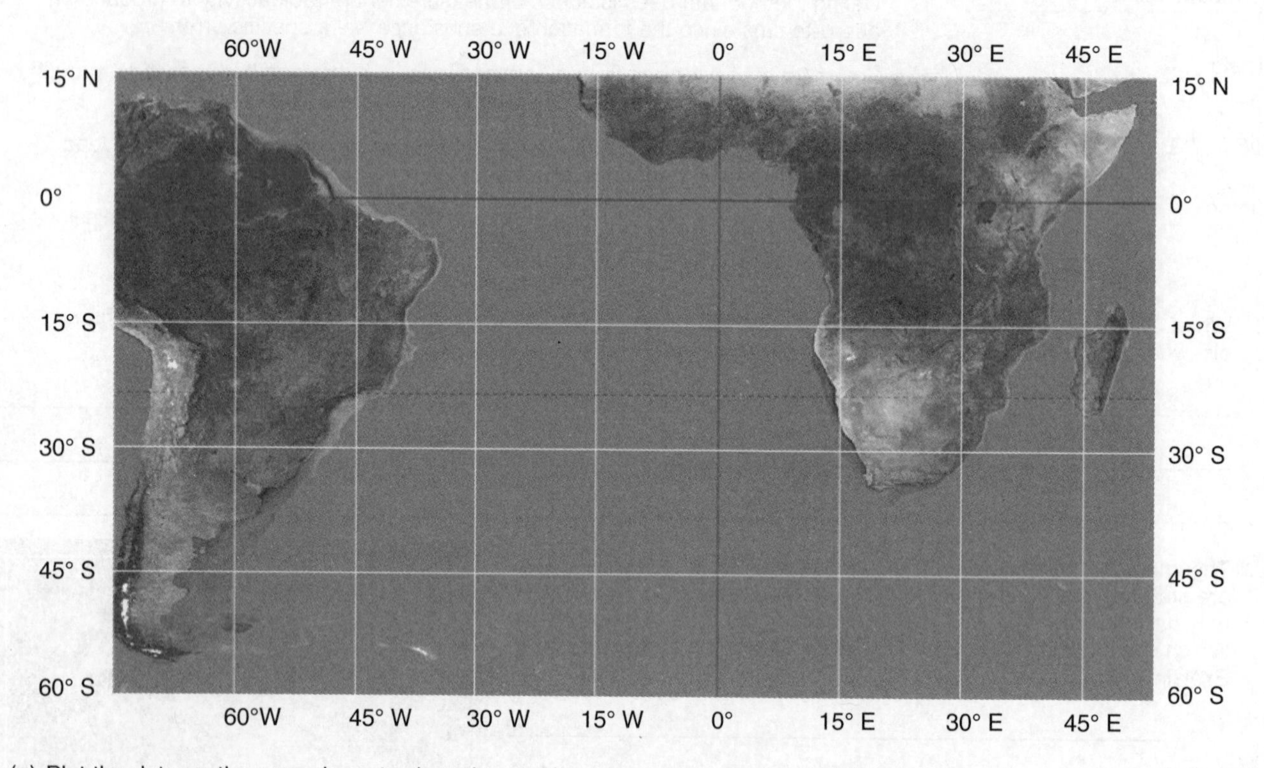


(a) Plot the data on the map above to show the position where the sample was taken:

(b) What happens to the age of the rock as you move from west to east across the South Atlantic?

(c) Draw a line on the map to indicate where rocks of the age 0 million years would be found:

(d) Explain what causes the production of new rocks on the seabed:

2. Explain why rocks in the continental crust are older than rocks in the oceanic crust:

TEST CCC SC PRACTICES

**ISBN: 978-1-927309-37-7**

3. The data below shows the age of various rocks and materials from the Earth and other bodies of the solar system. The material was dated using radiometric dating:

| Material | Age (millions of years) | Error (+/-) (millions of years) | Dating method |
|---|---|---|---|
| Chondrite meteorite | 4568.2 | 0.3 | Pb-Pb |
| Chondrite meteorite | 4566 | 0.7 | Pb-Pb |
| Chondrite meteorite | 4565.1 | 0.9 | Pb-Pb |
| Mars meteorite | 4070 | 40 | Ar-Ar |
| Mars meteorite | 4040 | 100 | U-Pb |
| Moon rock highland | 4426 | 65 | U-Pb |
| Moon rock highland | 4339 | 5 | U-Pb |
| Moon rock lowland | 3800 | 20 | Ar-Ar |
| Moon rock lowland | 3770 | 70 | Ar-Ar |
| Earth - Australia | 4404 | 68 | Pb-Pb |
| Earth - Australia | 4341 | 6 | Pb-Pb |
| Earth - Canada | 3939 | 31 | Pb-Pb |
| Earth - Greenland | 3809 | 7 | U-Pb |

(a) Which of the types of rocks would give the best indication of the time the Earth began to form? ____________

____________

(b) Explain why: ____________

____________

____________

____________

(c) Explain why the rocks from the lunar lowlands are younger than the rocks from the lunar highlands: ____________

____________

____________

____________

(d) Use the data above to estimate the time it took for planetary bodies to accrete material and cool. Explain the evidence for your choice:

____________

____________

____________

____________

4. Discuss the development of models for dating the Earth and how these models changed as techniques developed and new information was discovered.

____________

____________

____________

____________

____________

____________

____________

____________


**ISBN: 978-1-927309-37-7**

# Earth's Systems

**Concepts and connections**

Use arrows to make your own connections between related concepts in this section of this workbook

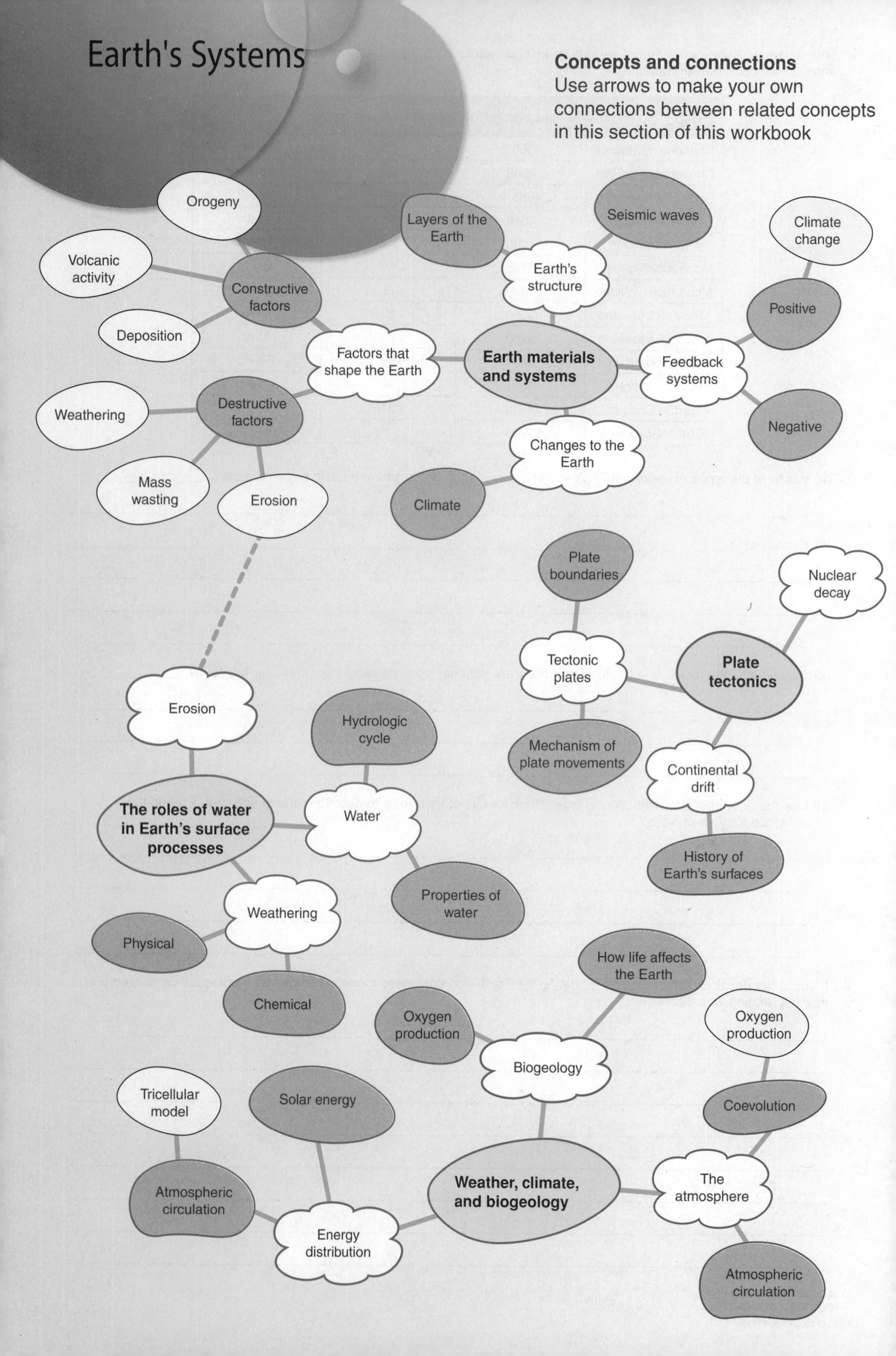

# ESS2.A Earth Materials and Systems

**Key terms**

albedo
asthenosphere
constructive mechanisms
convection
crust
destructive forces
glaciation
ice age
inner core
interglacial
lithosphere
mantle
negative feedback
outer core
P-wave
positive feedback
S-wave

## Disciplinary core ideas

Activity number

***Show understanding of these core ideas***

### Earth's systems cause feedback effects

| | | Activity number |
|---|---|---|
| ☐ | 1 Earth's systems (the atmosphere, geosphere, biosphere, and hydrosphere) are dynamic and interacting. The constantly changing interactions between these different systems cause feedback effects. Feedback can be negative and stabilize systems against changes, or positive, and amplify changes. | 63 64 66 67 |
| ☐ | 2 The Earth's surface is shaped by both constructive forces, such as volcanism and tectonic uplift, and destructive mechanisms, such as mass wasting, weathering, and erosion. These processes operate at different scales of space and time. | 62 68 |

### Empirical evidence lead to a model of the Earth's structure

| | | Activity number |
|---|---|---|
| ☐ | 3 A model of the Earth's structure includes a solid inner core, a liquid outer core, and a solid mantle and crust (including the lithosphere and asthenosphere). | 59 |
| ☐ | 4 The model of the Earth's internal structure is based on several lines of evidence. | 60 |
| ☐ | 5 Motions of the mantle and its plates occur mainly through thermal convection. | 61 |

### The geological record shows changes to global and regional climates

| | | Activity number |
|---|---|---|
| ☐ | 6 The geological record shows that changes to global and regional climate can be caused by interactions between the changing energy output of the Sun or the orbit of the Earth, tectonic events, ocean circulation, volcanism, vegetation, glaciers, and human activity. These changes can occur on a variety of time scales. | 69 70 |

NOAA

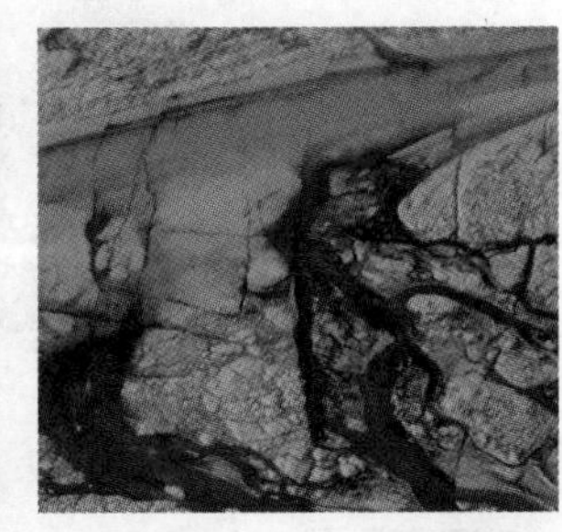

## Crosscutting concepts

Activity number

***Understand how these fundamental concepts link different topics***

| | | Activity number |
|---|---|---|
| ☐ | 1 SC Change and rates of change in the Earth's surface features over short and long periods of time can be quantified and modeled. | 62 66-70 |
| ☐ | 2 SC Feedback between the Earth's systems can stabilize or destabilize those systems, e.g. in the regulation of climate. | 63 64 66 |
| ☐ | 3 EM Energy drives the cycling of matter by thermal convection in the mantle. | 59 61 |
| ☐ | 4 CE Empirical evidence can be used to determine causes of climate change on different time scales. | 65 68 69 70 |

## Science and engineering practices

Activity number

***Demonstrate competence in these science and engineering practices***

| | | Activity number |
|---|---|---|
| ☐ | 1 Use a model to illustrate how the Earth's surface features are created. | 61 62 66 67 |
| ☐ | 2 Use an evidence-based model to describe the structure of the Earth. | 59 60 |
| ☐ | 3 Use a model to describe the multiple causes of climate change. | 63 64 68-70 |
| ☐ | 4 Analyze data to make a valid claim that change to one of Earth's systems can create feedbacks that cause changes to other Earth systems. | 65 68-70 |
| ☐ | 5 Conduct an investigation to produce data as evidence of the ice albedo effect. | 65 |

# 59 Structure of the Earth

**Key Idea**: The Earth has several layers, each with different characteristics.

The Earth is layered due to the density of different materials in it. The Earth's crust has a density of about 3 g/cm$^3$ while the core has a density of about 12 g/cm$^3$. Movement of convection currents in the mantle shifts the plates of the Earth's crust, while movement of the outer core produces the Earth's magnetic field.

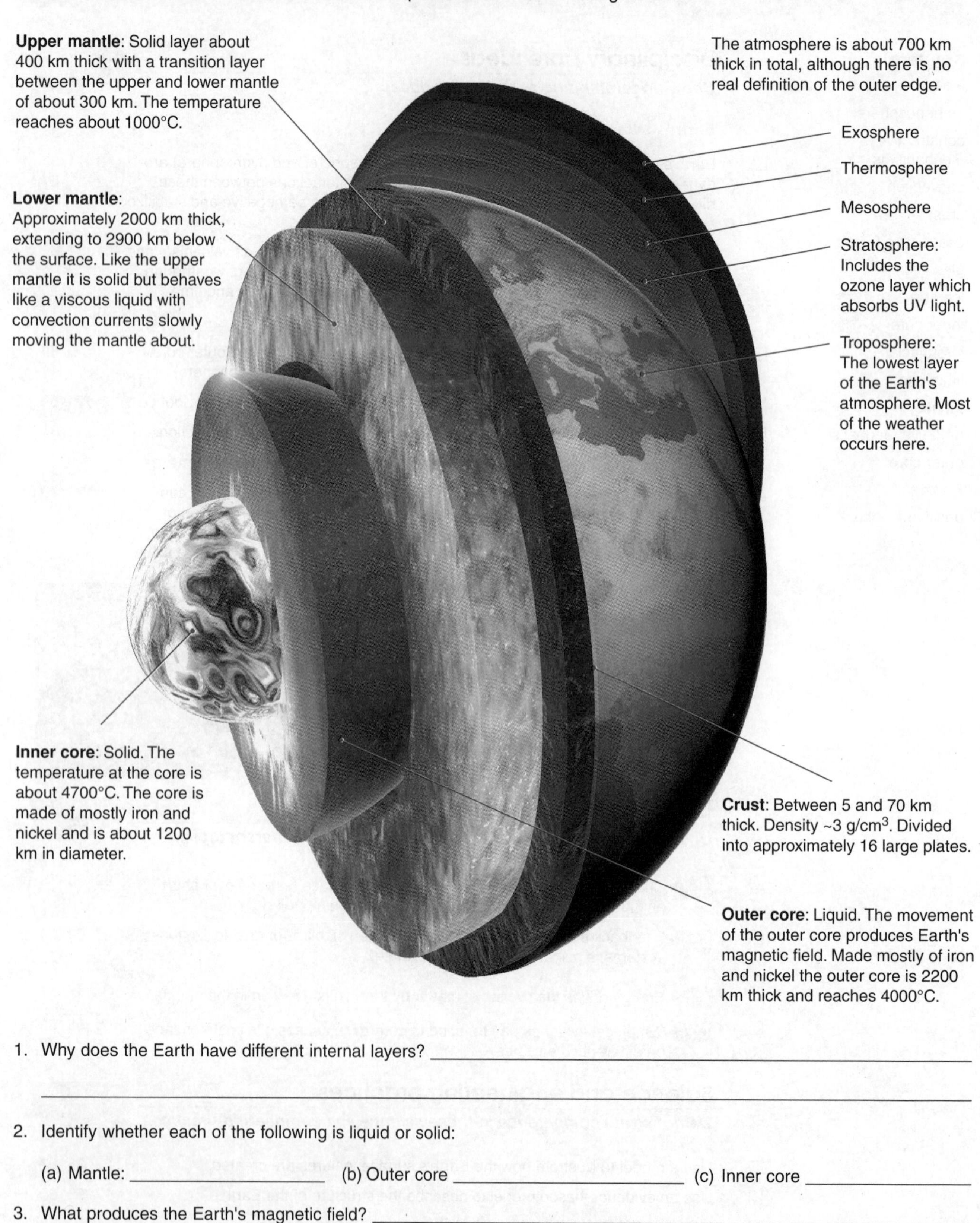


1. Why does the Earth have different internal layers? ______________________________

______________________________________________________________

2. Identify whether each of the following is liquid or solid:

(a) Mantle: ______________ (b) Outer core ______________ (c) Inner core ______________

3. What produces the Earth's magnetic field? ______________________________

______________________________________________________________

KNOW

WEB 59 CCC EM PRACTICES 

**ISBN: 978-1-927309-37-7**

# 60 Evidence for Earth's Structure

**Key Idea**: Seismic evidence and measurement of magnetic changes in rocks suggests the Earth has many layers.

## Seismic waves

Movement of the ground along fault lines in the Earth's crust causes earthquakes. During an earthquake, two types of ground wave are produced; compressional, or **P-waves** and shear waves, or **S-waves**.

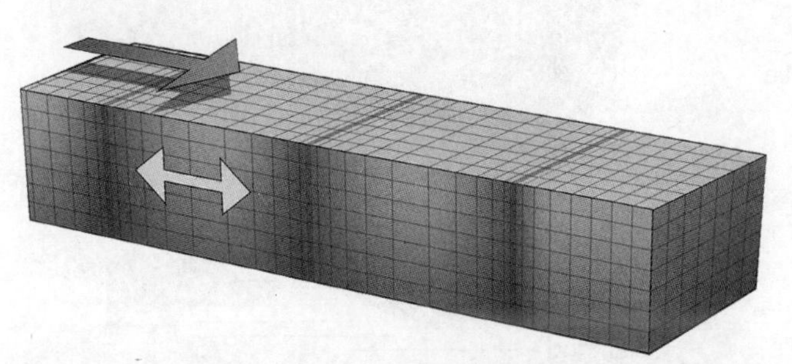

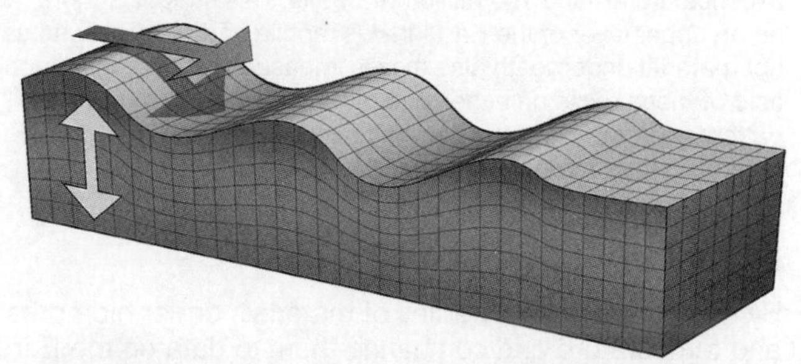

**P-waves** are compression waves (similar to sound waves in air). P-waves are the fastest moving wave from an earthquake and so are the first to arrive at a seismograph. P-waves can travel through all media and so can pass through all layers of the Earth.

**S-waves** are transverse waves. They move the ground perpendicular to their direction of travel. They are unable to travel through liquids.

Changes in density cause changes in the velocity of P and S-waves (velocity increases with greater density and pressure). The internal characteristics of the Earth can be deduced by measuring the time waves take to reach seismographs around the world.

Seismic waves refract (bend) as they pass through the layers of the Earth. Because of this, there is a shadow between 103° and 142° from the earthquake's focus (origin) where no P-waves are detected.

S-waves do not travel through liquids. No S-waves are detected opposite where they would have to pass through the Earth's outer core. We can therefore conclude that the outer core must be liquid.

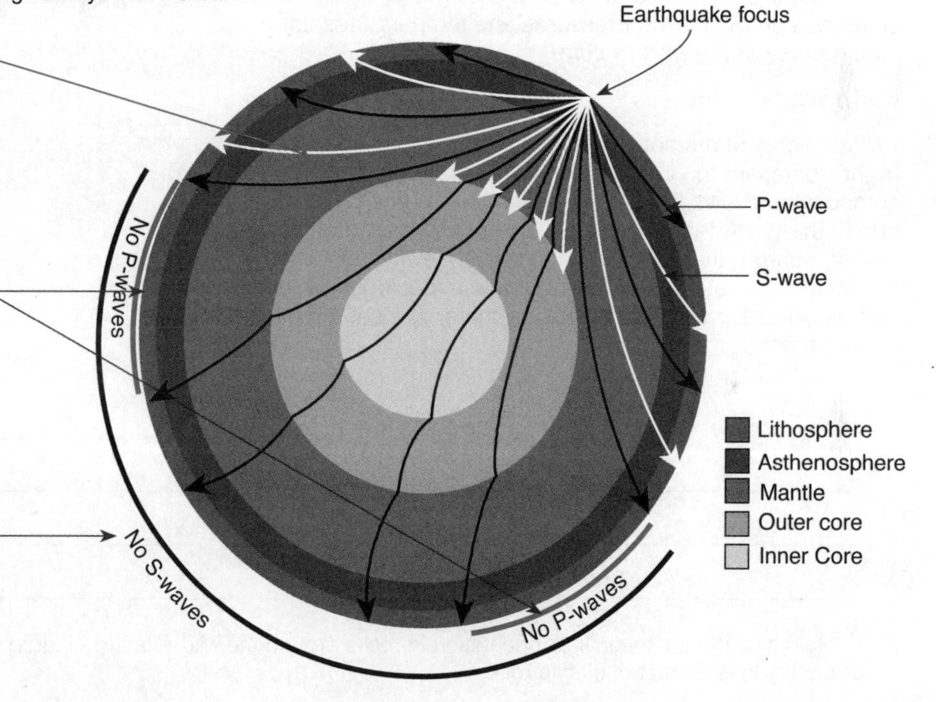


1. Describe two differences between P-waves and S-waves: ____________________

2. What causes the P-wave shadow between 103° and 142° from an earthquake's focus? ____________________

3. How do we know that the mantle is a solid that behaves like a viscous fluid rather than liquid? ____________________


**ISBN: 978-1-927309-37-7**

## Magnetic evidence

The fact that the Earth has a layered structure can be deduced by measuring the direction of the magnetic field in rocks about the globe. Indeed the simple fact that the Earth has a strong magnetic field is evidence to suggest there must be a liquid outer core surrounding a iron inner core, producing a dynamo effect.

### Mid ocean ridges:

This magnetic field aligns magnetic minerals in rocks when they form (e.g. cooling lava). The best example of this is along the Mid-Atlantic Ridge, although it occurs along all the mid ocean ridges. The rocks on either side of the ridge can be dated using radiometric dating. Dating shows that the youngest rocks are close to the mid ocean ridge, while those close to the continent of Africa and South America are around 100 million years old. This suggests there must be an upper layer of the Earth that is mobile and moves about as hot material underneath also moves. Measurements of the magnetic field of these rocks on either side of the ridge show identical patterns. Interestingly, the patterns show that the magnetic field reverses every few million years.

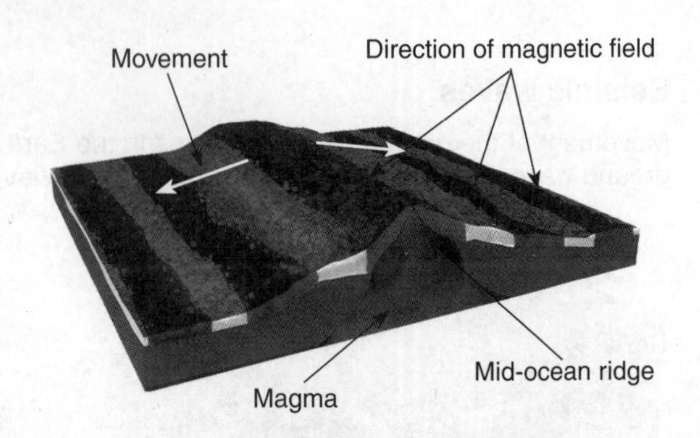


## Laboratory experiments

By measuring the properties of materials under high pressure and temperature and comparing them to data on the Earth's core it is possible to identify the materials that make up the inner and outer cores and the mantle. Experiments like this are done in small devices called diamond anvil cells (right). The sample is squeezed between two diamond anvils while being heated by a laser. Pressures and temperatures can match those of the Earth's core.

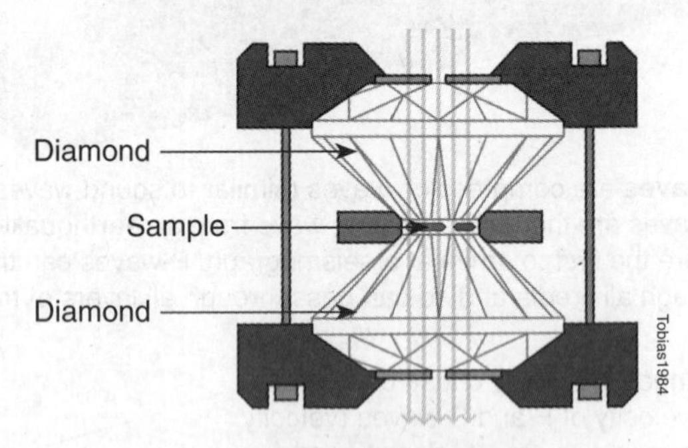


## Meteorites

Certain types of meteorites called carbonaceous chondrites (right) represent the composition of the early Earth. Their composition shows that there should be more nickel and iron in the Earth than we see on the surface or in magma welling up from the mantle. The speed of seismic waves in the mantle do not match what would be expected if more iron and nickel was present. The missing material must therefore be in the core.

4. What evidence is there that the Earth's outer core is liquid? ______________________________

5. Explain how the magnetic field orientation in rocks around the Mid-Atlantic Ridge provides evidence that the outer layer of the Earth is a crust of mobile rock:

______________________________

6. How do laboratory experiments help us understand the structure of the Earth? ______________________________

7. How does evidence from meteorites help our understanding of the Earth's composition and structure? ______________________________


**ISBN: 978-1-927309-37-7**

# 61 Lithosphere and Asthenosphere

**Key Idea**: The upper mantle is made up of the solid lithosphere on top of a more fluid asthenosphere. The properties of these layers can be determined by the measurement of seismic waves.

## The lithosphere

- The **lithosphere** (*lithos* = "stone") is made up of the crust and the uppermost part of the upper mantle. It is both rigid and solid, and broken up into sections called tectonic plates.
- The lithosphere can be divided into continental lithosphere, which contains relatively light minerals, and oceanic lithosphere, which contains much denser minerals. The lithosphere ranges from 400 km thick over the continents to 70 km thick in the oceans.

## The asthenosphere

- The **asthenosphere** (*asthenes* = "weak") lies below the lithosphere. This layer of rock is viscous and plastic (semi-fluid) in its behavior. It changes through plastic deformation, slowly moving about and so allowing for movement of the tectonic plates above.
- The asthenosphere is relatively thin, around 100 km thick. The boundary between the lithosphere and asthenosphere is thermal. The lithosphere conducts heat out to the surface whereas the asthenosphere retains its heat.

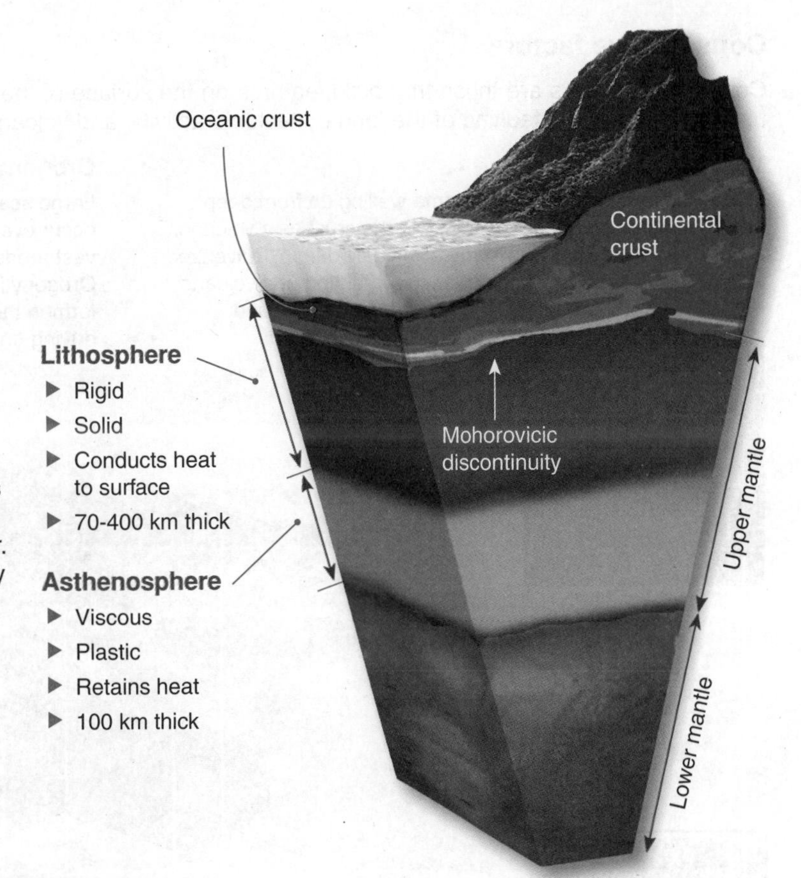


### How do we know?

- The Mohorovicic discontinuity marks the boundary between the crust and the mantle. It was discovered by Andrija Mohorovicic in 1909 while he was studying seismic waves from an earthquake near Zargreb, Croatia. Mohorovicic noticed that the seismographs 200 km or more away from the earthquake epicenter were reading the seismic waves as arriving earlier that he expected. He realized that this was because the waves were traveling through a denser layer of material and so speeding up, arriving earlier than if they traveled through material that was of uniform density. Because this change in speed happened at a specific distance from the epicenter, the change in material must have been very sudden.
- The lithosphere - asthenosphere boundary can be detected by measuring seismic waves. In a recent experiment in New Zealand, geologists exploded charges of TNT to produce ground waves and measured their echoes. This revealed the lithosphere-asthenosphere boundary is a thin jelly-like layer of rock.

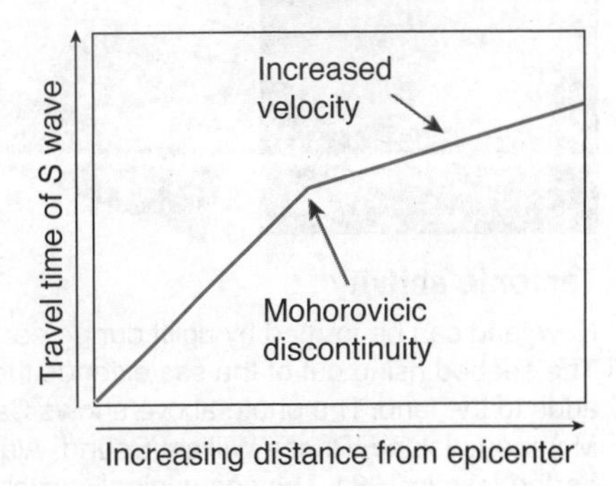


1. (a) Describe the structure of the lithosphere: ____________________

____________________

____________________

(b) Describe the structure of the asthenosphere: ____________________

____________________

____________________

2. Describe the general method that scientists use to study the interior of the Earth: ____________________

____________________

____________________


**ISBN: 978-1-927309-37-7**

# 62 Factors That Shape the Earth

**Key Idea**: The Earth's surface is shaped by the interaction between constructive factors (e.g. tectonic uplift) and destructive factors (e.g. weathering).

## Constructive factors

Constructive factors are those that build features on the surface of the Earth. They include deposition of sediments, uplifting, folding and faulting of the land by tectonic activity, and volcanic activity.

### Volcanic activity

Volcanic activity builds land. **Magma** welling up from deep below the crust bursts out of cracks in the crust as a volcano. When magma reaches the surface it is called **lava**. Pulverized rock and lava thrown into the air by the eruption of a volcano is called ash. Together lava and ash can build new land.

### Orogeny

Large scale tectonic uplifts are called orogenies and may occur over millions of years. Orogenies are capable of lifting vast areas the Earth's crust above the surface. The Laramide Orogeny in North America, ending about 35 million years ago, formed the Rocky Mountains due to oceanic tectonic plates driving under the western edge of the North American Plate.

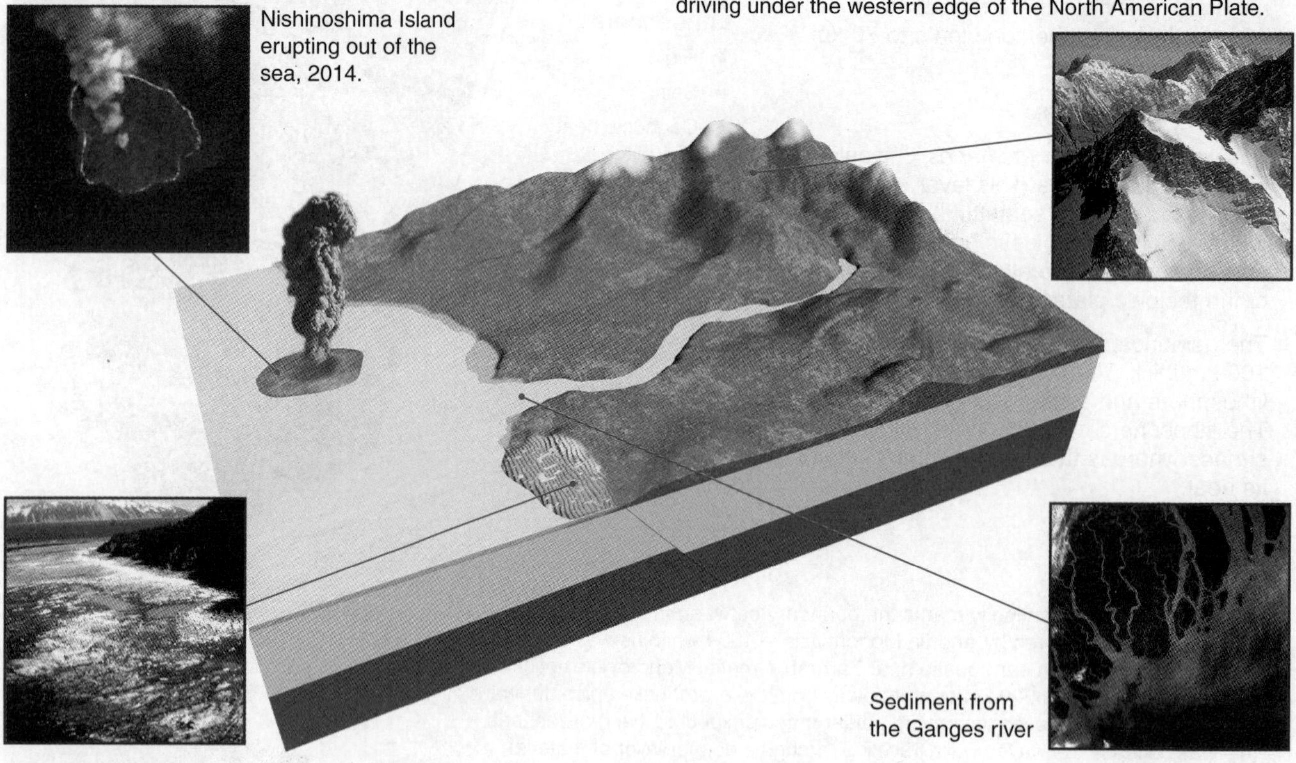


Nishinoshima Island erupting out of the sea, 2014.

Sediment from the Ganges river

### Tectonic activity

New land can be formed by uplift during earthquakes. The seabed rising out of the sea extends the beach and adds to the land. The photo above shows Cape Cleare, Montague Island, Prince William Sound, Alaska after a large earthquake in 1964. The sea originally extended to the base of the cliffs. However the earthquake lifted the seabed by an estimated 10 m, creating a new beach (white area).

### Sediment deposition

Deposition is the process of sediments being added to a land mass. Sediments, such as sand, may be carried by streams or rivers, blown by the wind, carried in ice, or slide down hills as landslides. New land can be formed when sediments are deposited in river deltas or near the shore. If the sediments are held in place by tree roots, e.g. mangroves, then the land may become permanent and be slowly extended out to sea.

1. Describe three constructive factors that shape the Earth's surface: ______________________

______________________

______________________

______________________

______________________

2. Describe three destructive factors that shape the Earth's surface (see next page): ______________________

______________________

______________________

______________________

ISBN: 978-1-927309-37-7

## Destructive factors

Destructive factors are those that remove features on the Earth's surface. They include mechanisms such as weathering, mass wasting, and erosion.

### Weathering

Weathering is caused by physical, chemical, and biological processes. Together they break down rocks into finer particles. Physical weathering includes factors such as heat and pressure changes. Chemical weathering involves chemical reactions that may dissolve minerals in the rock. Biological weathering occurs when rocks are exposed to the actions of living organisms, such as the organic acids and enzymes produced by microorganisms.

### Mass wasting

Mass wasting, or mass movement, is the sudden movement of large volumes of rock and material, as in landslides. All rocks have a finite strength above which they will fail. The pull of gravity combined with weathering and erosion will eventually cause a hillside to collapse. High cliffs and steep slopes are particularly susceptible to mass wasting. A high cliff undercut by a river or wave action can quickly collapse, causing the edge of the cliff to retreat.

### Erosion

Erosion is the loosening and removal of weathered material. A key part of erosion is the transport of materials away from their origin so it lowers the mean level of the land. Erosion may occur through the action of water, wind, or glaciers. It is often linked to deposition as materials are transported and deposited elsewhere (e.g. river deltas).

3. Use the information presented in this activity to draw and label a model of land formation and destruction:


**ISBN: 978-1-927309-37-7**

# 63 Feedback in Earth's Systems

**Key Idea**: Changes in one system may cause changes in another system forming a circuit of cause and effect. On Earth the climate is the end result of many of these feedback systems.

## Feedback on Earth

Feedback occurs when the output of a system is used as input in that system. On Earth there are many feedback systems, both negative and positive, operating at the same time. Negative feedback systems tend to stabilize a system around a mean (average condition) whereas positive feedback tends to increase a departure from the mean.

### Negative feedback in nature

Feedback systems can be complex and the result of many interacting factors. The diagram below illustrates a simplified negative feedback system involving the production of clouds. Clouds reflect incoming sunlight back into space so have the effect of lowering the Earth's surface temperature.

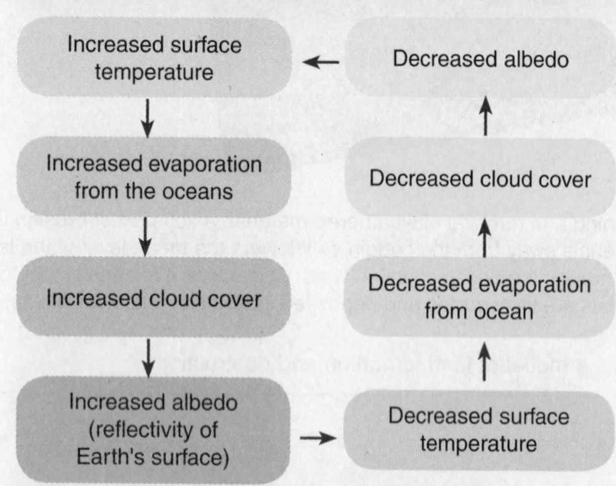


Negative feedback systems help to stabilize the Earth's climate. The evaporation of water from the oceans is affected by temperature, which may be influenced by an increase in solar output or carbon dioxide. The negative feedback of cloud production keeps the cloud cover of the Earth relatively constant.

### Positive feedback in nature

Positive feedback systems on Earth tend to drive large scale changes to environments and the climate. The current increase in $CO_2$ in the atmosphere is driving numerous positive feedback systems. The diagram below illustrates the effect of methane (a greenhouse gas) release from permafrost. As the Earth warms, the permafrost melts, releasing methane which in turn causes the Earth to warm further.

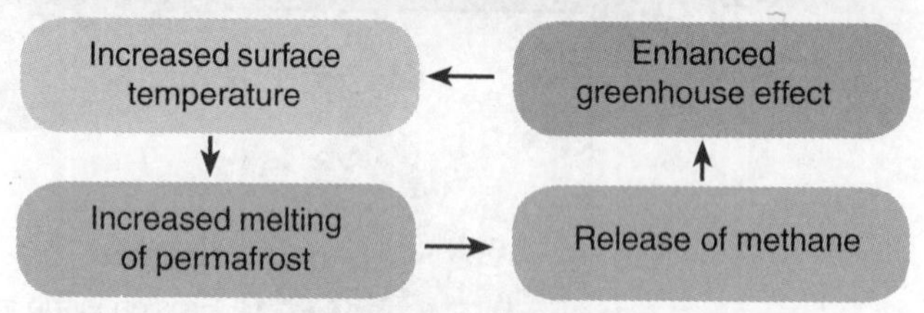


Increased surface temperatures also increase the amount of ice melting and so decreases the Earth's albedo:

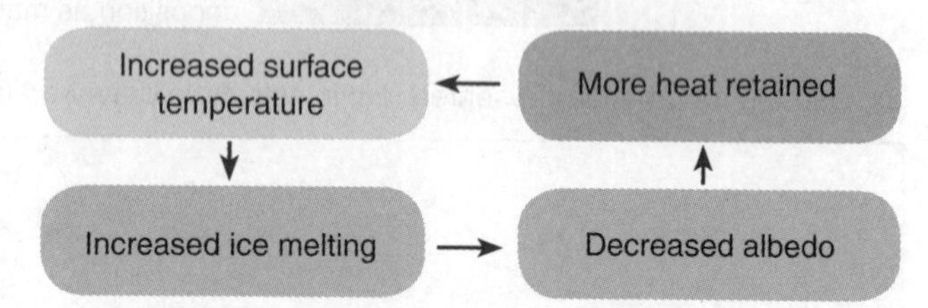


Several positive feedback systems acting at the same time can cause large changes to the climate. Although these are balanced to some extent by negative feedback systems, it is likely there will eventually be a "tipping point" at which a runaway climate change event will occur.

1. What is the difference between positive feedback and a negative feedback? ______________________

2. On Earth, negative feedback systems tend to have what effect on the climate? ______________________

3. What effect do positive feedback systems have on Earth's climate? ______________________

ISBN: 978-1-927309-37-7

# 64 Ice Sheet Melting

**Key Idea**: The melting of the ice sheets can cause a positive feedback loop that exposes more heat absorbing surfaces and increases ice sheet melting.

## Changes in polar sea ice

- The surface temperature of the Earth is partly regulated by the amount of ice on its surface, which reflects a large amount of heat into space. However, the area and thickness of the polar sea-ice is rapidly decreasing. From 1980 to 2008 the Arctic summer sea-ice minimum almost halved, decreasing by more than 3 million $km^2$. The 2012 summer saw the greatest reduction in sea-ice since the beginning of satellite recordings.
- This melting of sea-ice can trigger a cycle where less heat is reflected into space during summer, warming seawater and reducing the area and thickness of ice forming in the winter. At the current rate of reduction, it is estimated that there may be no summer sea-ice left in the Arctic by 2050.

### Arctic air temperature* changes

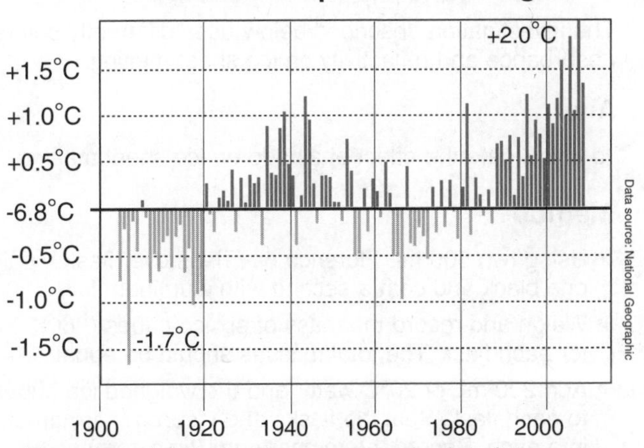


*Figure shows deviation from the average annual surface air temperature over land. Average calculated on the years 1961-2000.

Arctic sea-ice summer minimum 1980: 7.8 million $km^2$

Arctic sea-ice summer minimum 2012: Record low, 3.41 million $km^2$

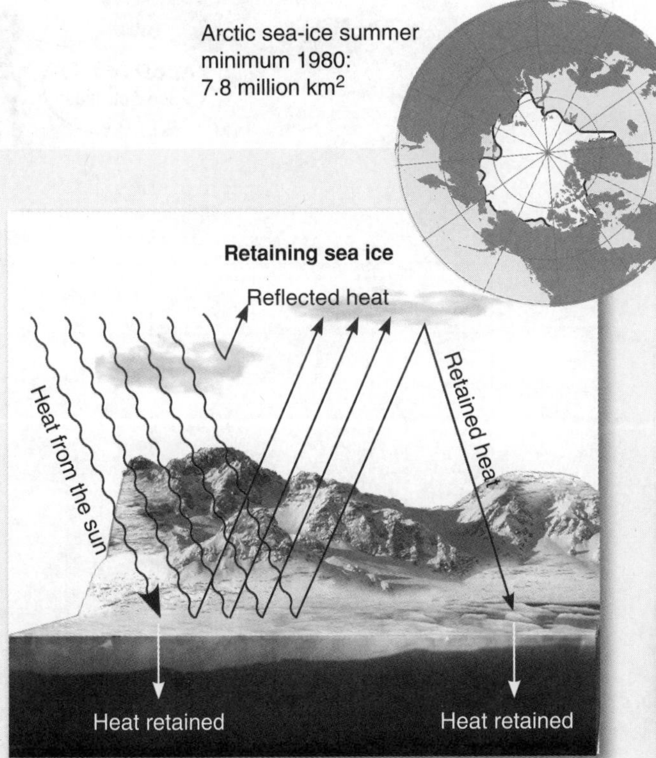


The high **albedo** (reflectivity) of sea-ice helps to maintain its presence. Thin sea-ice has a lower albedo than thick sea-ice. More heat is reflected when sea-ice is thick and covers a greater area. This helps to reduce the sea's temperature.

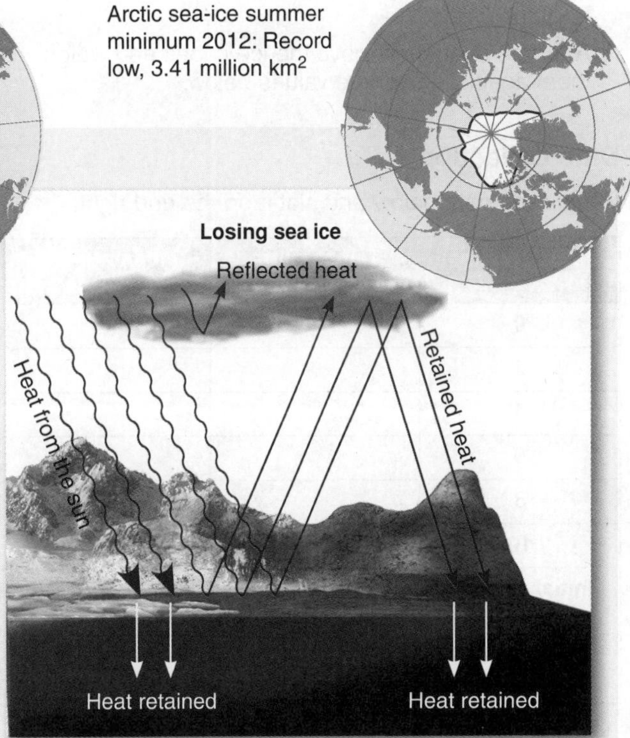


As sea-ice retreats, more non-reflective surface is exposed. Heat is absorbed instead of reflected, warming the air and water and causing sea-ice to form later in the fall than usual. Thinner and less reflective ice forms, perpetuating the cycle.

1. Calculate the difference in summer sea-ice area between 1980 and 2012: ____________________

2. How does low sea-ice albedo and volume affect the next year's sea-ice cover? ____________________

____________________

____________________

____________________

3. What type of feedback system is operating here? ____________________


**ISBN: 978-1-927309-37-7**

# 65 Modeling Ice Sheet Melting

**Key Idea**: Color and surface reflectivity affect the amount of heat absorbed by an object.

## Investigating heat absorbance

The investigation described below uses differently colored flasks to allow you to understand the importance of heat absorbance and reflectivity on ice sheet melting.

### Aim

To investigate the effect of albedo on ice sheet melting.

### Method

- Using two 500 mL Florence or Erlenmeyer flasks, paint one black and coat a second with aluminum foil.
- Weigh and record the mass of six ice cubes (~60-90 g) for each flask. The total masses should be equal.
- Add 200 mL of 20°C water and the weighed ice cubes to each flask. Seal the flasks and insert a thermometer into each. Record the temperature (time zero).
- Leave the flasks in a sunlit area and record the temperature every two minutes for ten minutes. You could also use a 60W tungsten lamp placed 15 cm from the flasks.
- After ten minutes remove the ice cubes and weigh them again. Record the values below.

Thermometer
Aluminum foil
500 mL Florence flask painted black
200 mL of water + 6 ice cubes

NEED HELP? See Activities 7 & 9

## Data collection

Record the data below and plot it on the grid right:

| Time (minutes) | Temperature black flask (°C) | Temperature foil coated flask (°C) |
|---|---|---|
| 0 | | |
| 2 | | |
| 4 | | |
| 6 | | |
| 8 | | |
| 10 | | |
| Initial mass of ice (g) | | |
| Final mass of ice (g) | | |

1. Which flask has the greater albedo?
2. Calculate the change in mass of the ice cubes for both the black and foil covered flasks:
3. Why is it important to start with the same total mass of ice in each flask?
4. Write a conclusion for the investigation:




ISBN: 978-1-927309-37-7

# 66 Dams and Erosion

**Key Idea**: Dams, especially very large dams, can have severe "knock-on" effects on rivers due to blocking the water flow and disrupting sediment transport.

## Effects of dams on natural systems

### Sediment transport

Sediment carried from the upper part of the river is released when the river reaches a lake behind a dam. In the absence of a dam, this sediment may have been laid down on the river bed further downstream or it may have been carried to the sea and added to the estuary and surrounding beaches.

### Fish populations

Dams prevent migratory fish from moving up or down river, either to spawning areas as adults or out to sea as fry (juveniles). Some dams may have inbuilt fish ladders to allow fish to bypass the dam but many do not and so disrupt life cycles of migratory fish such as salmon.

### Flood plain fertility

Natural rivers flood and deposit nutrient-rich sediment over the floodplain, as well as modifying the floodplain and river channel (e.g. changing direction or producing layered terraces). The natural flood cycle is disrupted by dams, which are often used for flood control.

### Salinity

Reduced water discharge from the river allows seawater to intrude into the delta or estuary. This causes levels of salt in the surrounding soil to rise. In some river deltas, e.g. the Nile delta, once fertile soil can no longer be used due to high salt levels.

### Erosion

Without the sediment load, waters down river of the dam tend to pick up more sediment, eroding the riverbed (instead of building it with sediment transported from upstream). Without sediment depositing at the estuary, the sea can quickly erode the foreshore. Sediment is often carried by the sea and deposited on beaches. Without the sediment the beaches are also quickly eroded.

1. Explain how damming a river to produce a reservoir for irrigation or hydroelectric power can have several unintended "knock-on effects down river from the dam:

2. Use the following list to complete the diagram of the downstream effects of damming a river: *reduced sediment, reduced flooding, coastal erosion, ecosystem deterioration, dam, saltwater intrusion, reduced nutrients*:

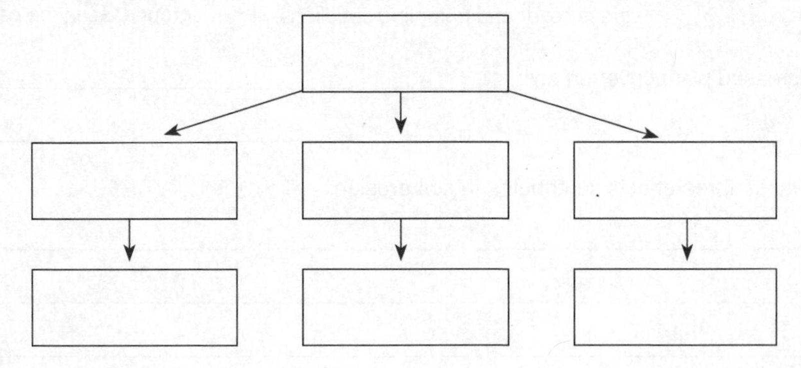


**ISBN: 978-1-927309-37-7**

# 67 Soil Erosion and Feedback

**Key Idea**: Earth's natural systems can be easily disrupted by over-exploitation or removal of a particular aspect of the system. This can then lead to further disruption.

## Soil and water loss

- Soil is important for plant growth. Soil contains various minerals and materials that help it to retain water and remain bound together. The loss of vegetation can reduce the water in soil by increasing water runoff and the effect of wind action. This in turn reduces the growth of vegetation.
- This positive feedback can quickly cause land to become unproductive, especially in harsh environments such as the semi-arid rangelands of the south and mid-western United States. Here, overgrazing by livestock removes plant cover and increases evaporation from exposed soils. Rangelands can quickly be turned into dry and sandy deserts as the soils dry out and plant growth is reduced.

USDA

Livestock eat vegetation cover → Decreased plant cover

Decreased plant cover → Increased evaporation from exposed soils; Increased exposure to wind; Reduced water infiltration rate

Increased exposure to wind → Increased evaporation from exposed soils; Increased surface erosion

Increased evaporation from exposed soils → Increased surface erosion

Reduced water infiltration rate → Increased water runoff → Increased surface erosion

Increased surface erosion → Loss of soil → Decreased plant production → Decreased litter cover → Reduced water retention in soil → Decreased plant cover

USDA

The Dust Bowl in the 1930s was a result of drought and the removal of vegetation. Soils lying fallow (without crop cover) dried out and blew away, ruining over 14 million hectares of farmland.

Overgrazing has detrimental effects on grasslands, opening space for invasion by weeds and increasing susceptibility to erosion by removing grass cover and trampling seedlings.

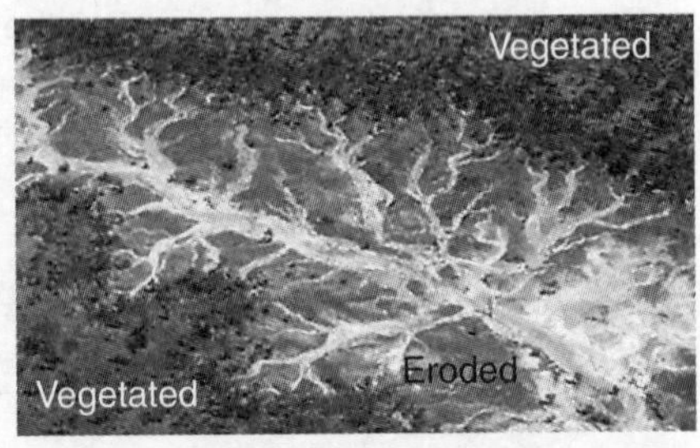



Erosion of soil in forests can result from deforestation. The tree roots maintain soil structure. Without them, the soil washes away leaving bare earth and unstable ground, as in the catchment pictured above.

1. List three effects of decreased plant cover on soil: ______________________________

______________________________

2. Describe how each of these three effects contributes to soil erosion:

(a) ______________________________

(b) ______________________________

(c) ______________________________

KNOW | CCC: SC | PRACTICES


ISBN: 978-1-927309-37-7

# 68 Short and Long Term Changes

**Key Idea**: The Earth is dynamic (constantly changing). Some of these changes occur over a short time, others take many years or even millennia.

Environmental changes come from three sources: the biosphere, geological forces (crustal movements and plate tectonics), and cosmic forces (the movement of the Moon around the Earth and the Earth and planets around the Sun).

- All three forces can cause cycles, steady states, and trends (directional changes) in the environment. Environmental trends (such as climate cooling) cause long term changes in communities.
- Some short term cycles, such as tides or day and night, may influence local environmental patterns. Others, e.g. seasons, can cause large scale environmental changes such as the advance and retreat of the polar sea-ice every winter and summer.

Every winter, the freezing of the sea around Antarctica almost doubles the continent's effective size.

**Time scale and geographic extent of environmental change**

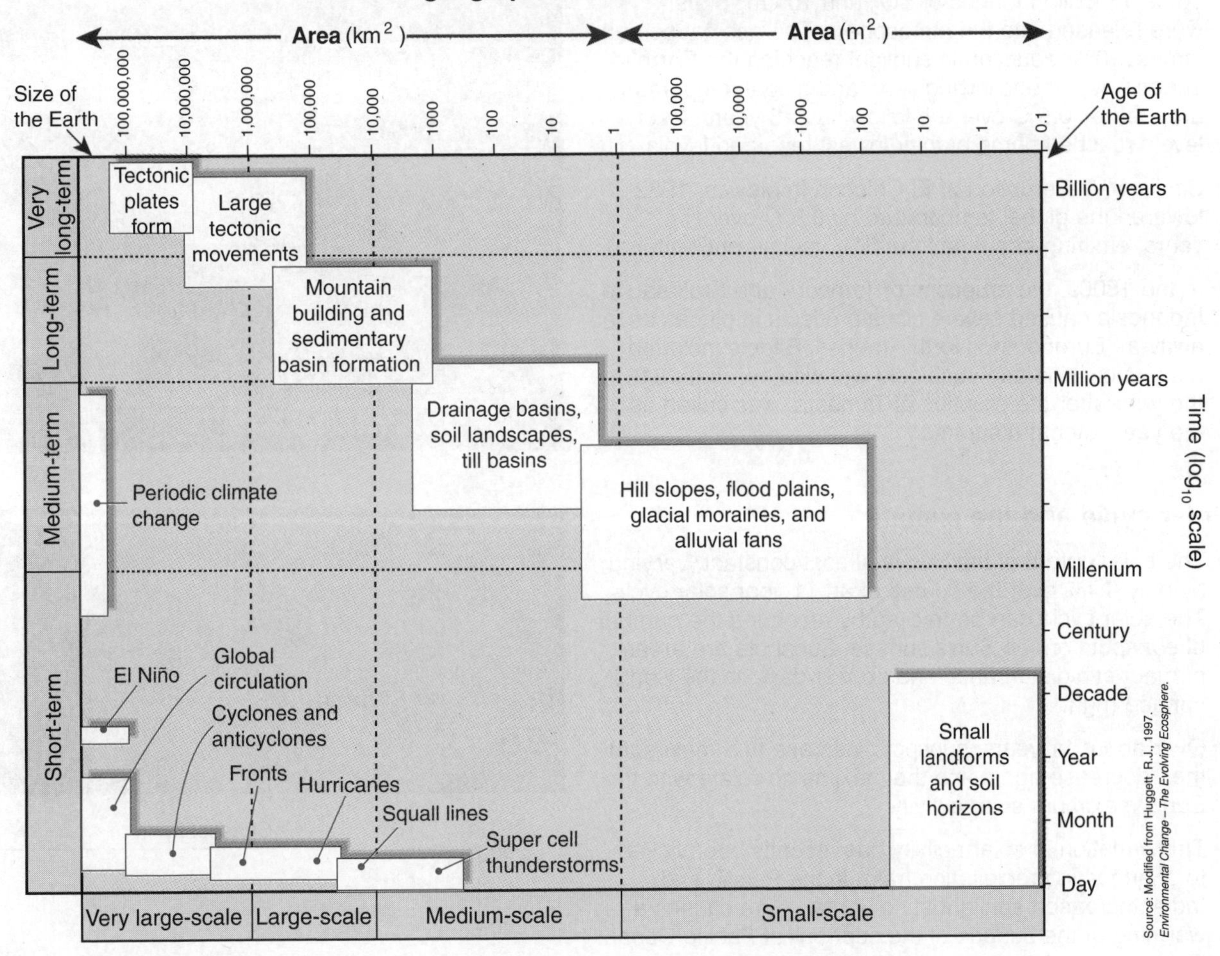


1. Identify the geologic event that is the largest and takes the longest amount of time: ______________________

______________________

2. Identify two very large scale events that take a relatively short amount of time to occur: ______________________

______________________

3. How long does it take for a soil landscape to form? ______________________

______________________


ISBN: 978-1-927309-37-7

# 69 Short-Term Changes to the Earth

**Key Idea**: Short term changes to the Earth include cyclic changes and periodic sporadic changes such as volcanic eruptions.

The Earth is constantly changing. Some of these changes occur in short term cycles, such as the El Niño Southern Oscillation and seasonal changes over the year. Others are sudden but short lived events that cause sudden but non-cyclic changes, such as volcanic eruptions.

## Volcanic eruptions and climate

- Large volcanic eruptions can affect the global climate by blasting ash and aerosols into the atmosphere.
- A well documented example of this is the eruption of Mount Pinatubo, located on the island of Luzon in the Philippines. Eruptions began on June 3 1991, after almost 500 years of virtual inactivity. On June 15, after many large explosions, Mount Pinatubo entered its final eruptive phase, blasting 10 $km^3$ of ash 34 km into the atmosphere. The Earth's climate was severely affected by Pinatubo's eruption. Over the course of the eruption, some 17 million tonnes of $SO_2$ and 10 $km^3$ of ash were released into the atmosphere. The ash caused an almost 10% reduction in sunlight reaching the Earth's surface over the following year, and global temperatures dropped by 0.5°C over the following 2-3 years. Ozone levels reached some of their lowest recorded levels.
- Similarly, the eruption of El Chichon in Mexico, 1982, lowered the global temperature by 0.5°C over two years, emitting about half the $SO_2$ volume of Pinatubo.
- In the 1800s, the eruptions of Tambora and Krakatau in Indonesia caused severe climate effects in places as far away as Europe and North America. Effects included early snow falls and such cold wet weather that 1816, the year after the eruption of Tambora, was called as the year without a summer.

Mount Pinatubo eruption 1991

USGS

## Solar cycle and the climate

- The solar output of the Sun is almost constant, varying by only 0.1% over the course of an 11 year solar cycle. The solar cycle can be tracked by recording the number of sunspots on the Sun's surface. Sunspots are areas of magnetic disturbance and appear dark on the Sun's surface (right).
- Over about 11 years, sunspots increase to a maximum then decrease again and the maxima correlate with the Sun's maximum solar activity.
- This variation in solar activity has recently been linked to changes in precipitation (rain) in the Pacific and India. Increased sunlight at solar maximum causes a warming of the surface of the subtropical Pacific Ocean. This warming increases evaporation and so increases the amount of water vapor in the atmosphere. The water vapor is carried by the trade winds to normally rainy areas, fuelling heavier rains. The effect continues a few years after the solar maximum and may affect climatic cycles such as El Niño.
- A link between sunspot activity and climate change is often made in popular media, citing a correlation between the Little Ice Age and a reduction in sunspots between the 1600s and 1700s. However, there is no clear evidence and solar activity is currently decreasing even though global temperatures are increasing.

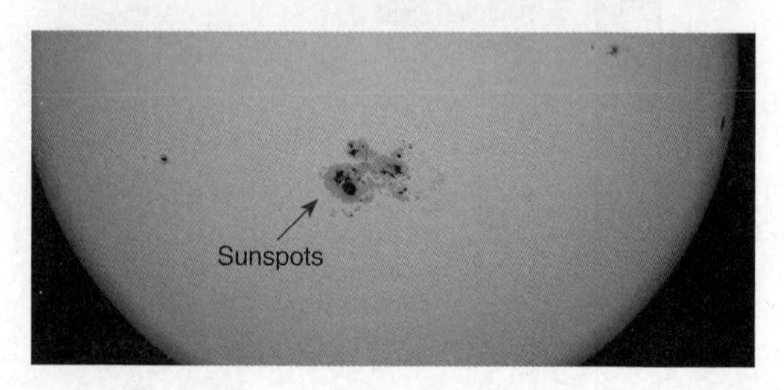


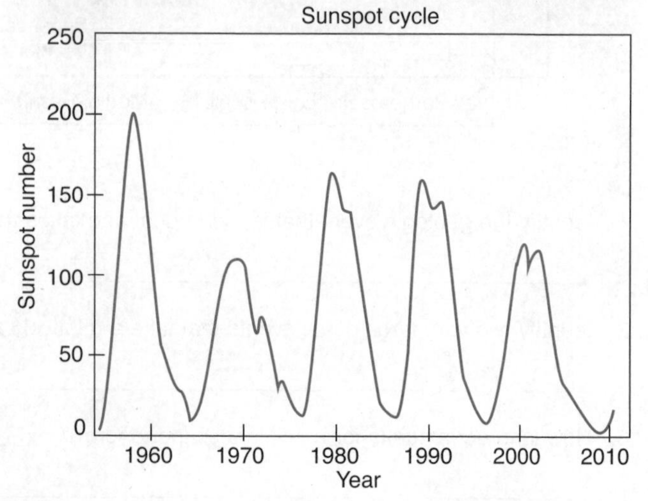

**ISBN: 978-1-927309-37-7**

## El Niño Southern Oscillation

▶ Interactions between atmospheric and the oceanic circulation are at the core of most global climate patterns. The El Niño-Southern Oscillation cycle (ENSO) is the most prominent of these global oscillations, causing weather patterns involving increased rain in specific places but not in others. It is one of the many causes of drought.

### Normal climatic conditions

In non-El Niño conditions (right), a low pressure system over Australia draws the southeast trade winds across the eastern Pacific from a high pressure system over South America. These winds drive the warm South Equatorial Current towards Australia's coast. Off the coast of South America, upwelling of cold water brings nutrients to the surface.

### El Niño effect

In an El Niño event (right), the pressure systems over Australia and South America are weakened or reversed, beginning with a rise in air pressure over the Indian Ocean, Indonesia, and Australia. Warm waters block the nutrient upwelling along the west coast of the Americas. El Niño brings drought to Indonesia and northeastern South America, while heavy rain over Peru and Chile causes the deserts to bloom.

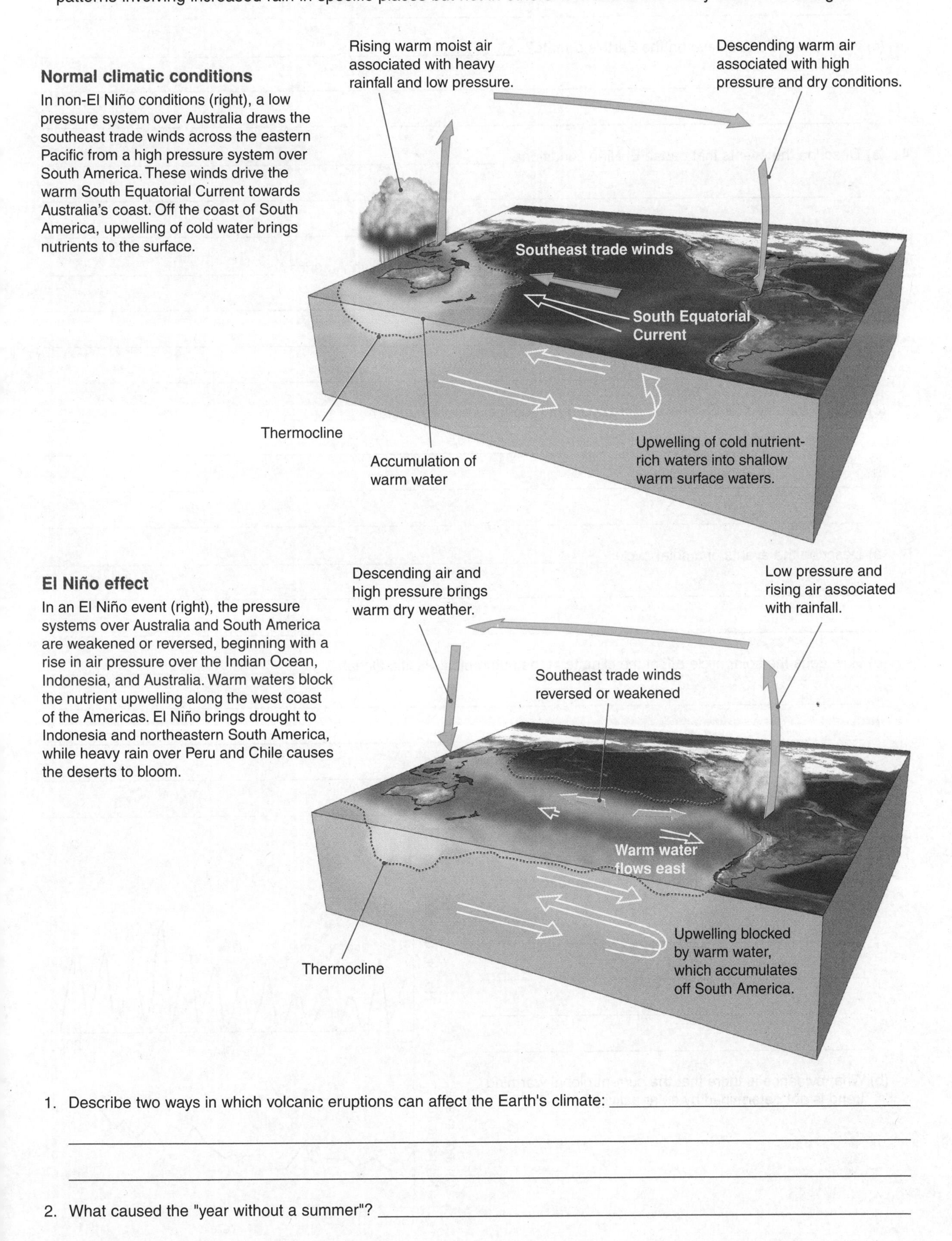


1. Describe two ways in which volcanic eruptions can affect the Earth's climate: ______________________________

______________________________

______________________________

2. What caused the "year without a summer"? ______________________________

______________________________


**ISBN: 978-1-927309-37-7**

3. (a) What year did Mt Pinatubo erupt?:

   (b) What volume of ash and $SO_2$ were ejected from the volcano?

   (c) What effect did this have on the Earth's climate?

4. (a) Describe the events that cause El Niño conditions:

   (b) What is the effect of El Niño on the climate of the western coast of South America?

   (c) What is the effect of El Niño on the climate of Indonesia and Australia?

5. (a) Describe the events of a solar cycle:

   (b) How does the solar cycle affect the climate of the subtropical Pacific Ocean?

6. Study the graphs shown right:

   (a) What evidence is there that sea temperature is affected by solar activity?

   (b) What evidence is there that the current global warming trend is not determined by solar activity?

No. of sunspots

1860 Year 2000

Change in sea temperature

1860 Year 2000


**ISBN: 978-1-927309-37-7**

# 70 Long-Term Changes to the Earth

**Key Idea**: Long term changes to the Earth include ice ages and glaciations, changes to land forms, and the effects of continental drift.

Some changes to the Earth happen on such vast time scales that they are not perceivable to humans. Changes may take place over thousands to million of years. Some changes appear to be cyclical, such as the advance and retreat of ice ages and glaciations, while others are continuous, such as continental drift.

## Ice ages and glaciations

The Earth has gone through five ice ages, i.e. long periods of time when large ice sheets covered large parts of the globe. The latest ice age began about 2.7 million years ago and is still ongoing. Within an ice age, there are periods of warmer climate conditions called interglacials, such as the present. The cooler periods of time are called glacials. These tend to last longer than the interglacials. The current interglacial began about 12,000 years ago. Ice cores (right) confirm the advance and recession of glacials.

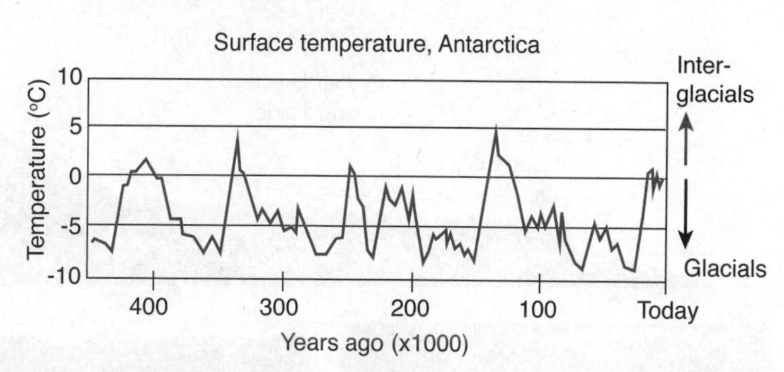

Within ice ages, there can be several periods of warming and cooling. What causes these changes is not fully understood but they may be due to changes in carbon dioxide and methane in the atmosphere, changes in solar output, and changes in the Earth's orbit.

The glaciers that spread across much of the Northern Hemisphere during the last glaciation began to retreat about 20,000 years ago. As they melted, the flood of fresh water into the oceans shut down ocean currents, which caused warming in the Southern Hemisphere, releasing $CO_2$ from the seas and warming the planet.

In North America during the last glaciation, ice covered all of Canada and extended south as far as the Upper Midwest, Idaho, Montana, and Washington. Evidence of this glaciation includes the grooves and U-shaped valleys formed as the glaciers advanced and retreated.

## Continental drift

- Movement of the Earth's tectonic plates drives continental drift (the movement of the continents). Over millions of years, the continents have split and converged many times and this has had major effects on the Earth's climate. Currently, Antarctica is surrounded by the Southern Ocean through which flows the Antarctic Circumpolar Current. The Antarctic Circumpolar Current did not exist until the Antarctic continent separated fully from South America. It prevented warm waters from the Atlantic, the Pacific, and the Indian Oceans reaching the Antarctic. As a result, Antarctica rapidly developed a huge ice sheet that remains in place today.
- Until about 5 million years ago North and South America were not connected. The gap between them (called the Central American Seaway) allowed tropical waters to flow between the Atlantic and the Pacific Oceans. The formation of the Central American isthmus blocked this flow and may have contributed to the beginning of the ice age 2.7 million years ago.
- The collision of India with Europe raised up the Himalayas, changing the way air currents moved about the globe. The Himalayas affect the seasonal rains in India by causing warm, moist air to rise. The cooling air then loses its moisture as precipitation as it moves south, causing the monsoons. The rain shadow effect of the Himalayas produces the large deserts of the Gobi and the Taklamakan.

40 million years ago the land bridge between Antarctica and South America prevented polar circulation of the ocean.

The appearance of the Central American isthmus prevented the flow of water between the Atlantic and the Pacific Oceans through the Central American Seaway.


ISBN: 978-1-927309-37-7

### Mountain building and rain shadows

- Mountains affect climate by deflecting air currents to higher levels of the atmosphere. As the air rises, it cools and moisture in it condenses to fall as rain, producing a wet, cool climate. Having lost its moisture the air passes over the mountain and descends as dry air, picking up moisture and producing a warm, dry climate. This effect is called a rain shadow and it occurs anywhere there are tall mountain ranges that block air flow.
- Mountain building via plate tectonics (orogeny) has occurred in many parts of the world near plate boundaries. Mountain building is a long term process, and can last tens of millions of years, thus producing long term climatic changes on large areas of land.

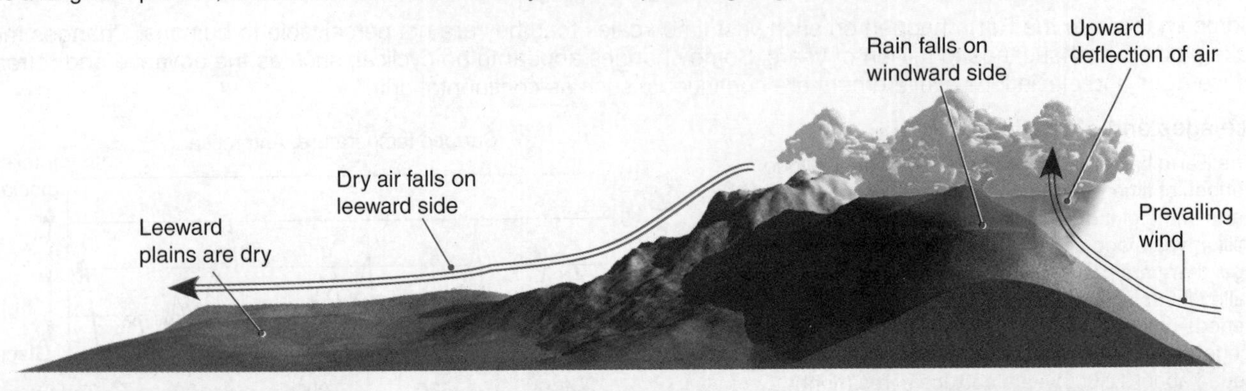

Rain shadows can cause some extreme effects. In Chile, the Andes block moist winds from the Amazon basin, producing the Atacama desert, the driest non-polar desert on Earth. The Atacama is estimated to be at least 3 million years old.

The Southern Alps in New Zealand produce one of the most extreme rain shadows in the world. On the western slopes, rainfall can reach 8900 mm a year. On the Eastern slopes, rainfall drops to just 380 mm in some areas.

Death Valley in California is in the rain shadow of the Sierra Nevada mountains. The valley has recorded the hottest temperatures on Earth (56.7°C). Death Valley formed after the last glacial period and receives just 60 mm of rain a year.

1. (a) When did the current ice age begin? ____________________

   (b) When did the last glacial period end? ____________________

   (c) What is the term for the current period of generally warm climatic conditions? ____________________

2. What are some possible reasons for the periodic occurrence of glacial periods? ____________________

3. Using examples explain how continental drift can affect ocean currents and the effect this can have on climate:

4. Use examples to explain how mountain ranges can have long term effects on local climate: ____________________


**ISBN: 978-1-927309-37-7**

# 71 Chapter Review

Summarize what you know about this topic so far under the headings provided. You can draw diagrams or mind maps, or write short notes to organize your thoughts. Use the introduction and the images and hints included to help you:

Structure of the Earth

HINT: Describe the layers of the Earth, their properties, and the techniques to investigate these.

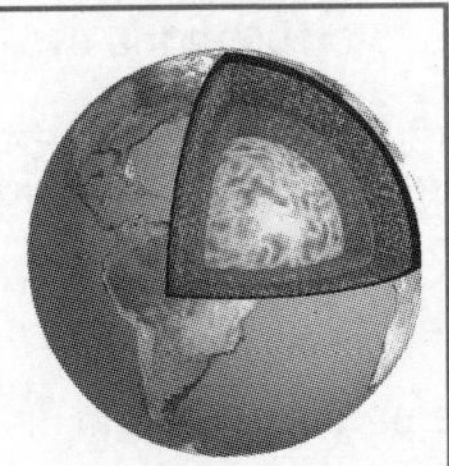

Changes to the Earth

HINT: Describe the factors that shape the Earth, how feedback mechanisms can enhance or stabilize the climate, and short and long term changes to the Earth's climate.

ISBN: 978-1-927309-37-7

# 72 KEY TERMS AND IDEAS: Did You Get It?

1. Test your vocabulary by matching each term to its definition, as identified by its preceding letter code.

|  | |  |
|---|---|---|
| albedo | A | A mechanism in which the output of a system acts to oppose changes to the input of the system. The effect is to stabilize the system and dampen fluctuations. |
| glacial period | B | A longitudinal seismic wave produced by an earthquake. Also known as a pressure wave, it is able to move through solid and liquid media and is the fastest moving seismic wave. |
| ice age | C | The proportion of light reflected off a surface, its reflectivity. |
| interglacial | D | A transverse seismic wave produced during an earthquake that is not able to move through a liquid medium. |
| negative feedback | E | The warmer periods of climate between glacial periods in which ice sheets and glaciers tend to retreat. |
| positive feedback | F | A destabilizing mechanism in which the output of the system causes an escalation in the initial response. |
| P-wave | G | A period of long term reduction in the Earth's temperature in which ice caps and glaciers are present on the Earth's surface. |
| S-wave | H | A period of time within an ice age in which glaciers and ice advance and are present on the Earth's surface in large volumes. |

2. (a) Label the layers of the Earth in the diagram below:
   (b) Beside your labels note whether the layer is solid or liquid:

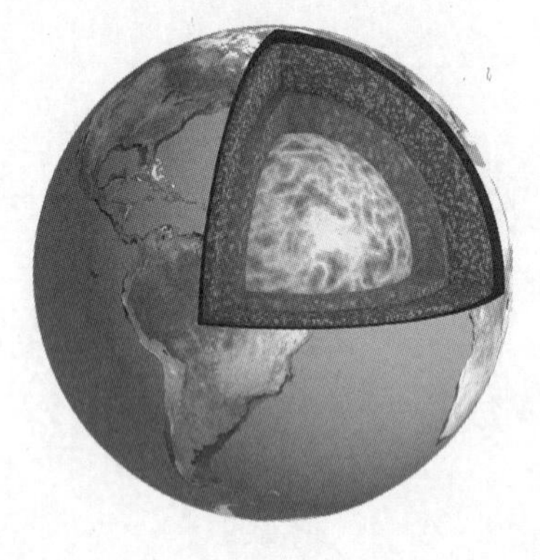

3. On the photograph below, identify, label, and describe two constructive factors and two destructive factors that are changing the landscape:

**ISBN: 978-1-927309-37-7**

# 73 Summative Assessment

1. The diagram below shows the velocity of two types of seismic waves at different depths in the Earth.

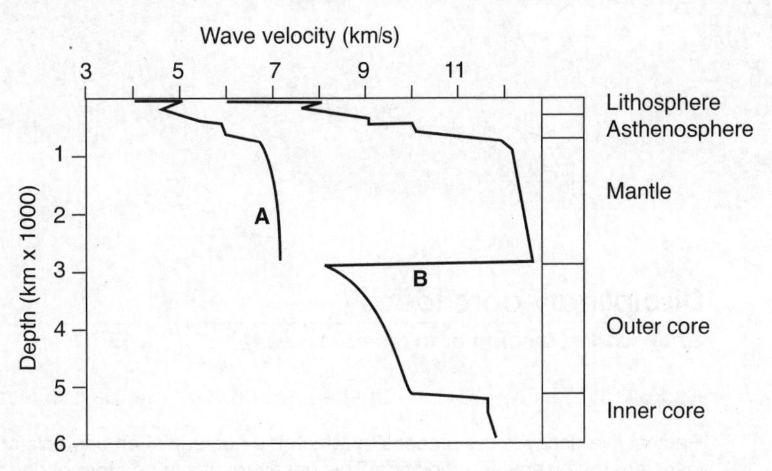


(a) Line A shows P / S waves (delete one).

(b) Line B shows P / S waves (delete one).

(c) How can you tell? ______________________________

______________________________

______________________________

______________________________

(d) Explain why the velocity of the waves increases with depth through the mantle: ______________________________

______________________________

______________________________

2. Draw a line on the graphs below to show negative feedback over time and positive feedback over time:

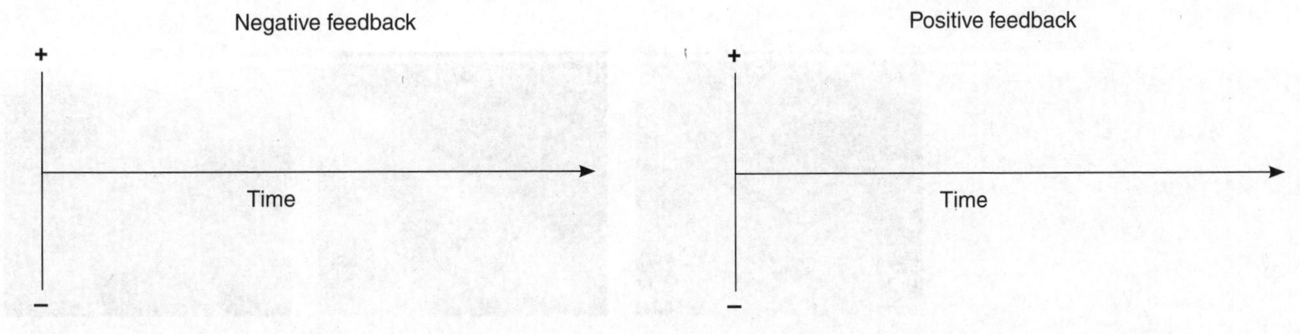


3. Study the graph below of Earth's temperature over the last 1000 years. Label the regions of apparent negative feedback and of apparent positive feedback:

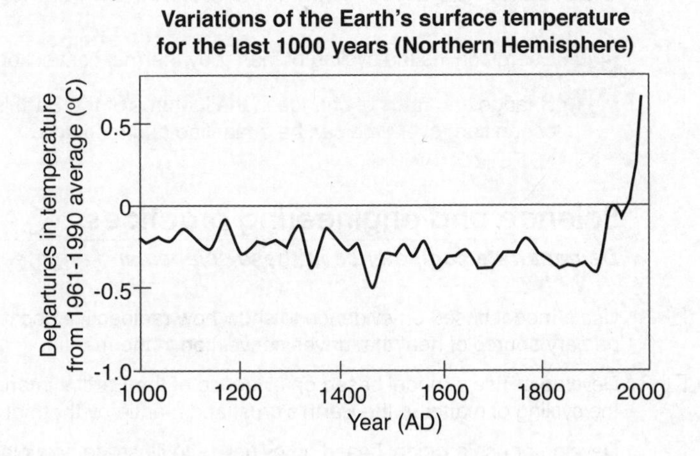



ISBN: 978-1-927309-37-7

ESS2.B

# Plate Tectonics

**Key terms**

continental drift
convection current
convergent boundary
fault
divergent boundary
mantle
plate tectonics
transform boundary
radioactive decay
subduction zone

## Disciplinary core ideas

***Show understanding of these core ideas***

Activity number

**Radioactive decay provides the heat that drives mantle convection**

- ☐ 1 Radioactive decay is the process by which the nucleus of an unstable atom loses energy by emitting radiation. This decay results in an atom of one type transforming to an atom of a different type. The decay energy is the mass difference between the parent and the daughter atom and particles. — 74
- ☐ 2 Radioactive decay provides the energy that drives mantle convection, which carries heat from Earth's interior to the surface. — 74
- ☐ 3 Describe the decay of U-238 and explain the significance of its half life being ~4.5 billion years. — 74
- ☐ 4 Plate tectonics, the scientific theory describing the large-scale motion of Earth's lithosphere, is the surface expression of the convection in the mantle. — 75

**Plate tectonics explains the movements of the Earth's surface**

- ☐ 5 Plate tectonics is a unifying theory that explains the past and current movements of the Earth's crust and helps us to understand Earth's geologic history. — 75 76 77
- ☐ 6 The lithosphere is broken up into seven large tectonic plates and many minor plates. The plate boundaries, where plates meet, may be convergent, divergent, or transform. Earthquakes, volcanism, mountain-building, and formation of oceanic trench occur along these plate boundaries. — 75 76
- ☐ 7 The amount of material lost at convergent boundaries is roughly balanced by the amount formed at divergent boundaries (seafloor spreading). — 76

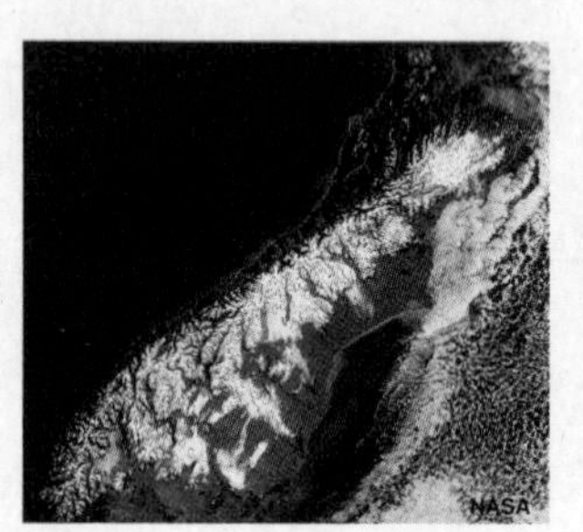

## Crosscutting concepts

***Understand how these fundamental concepts link different topics***

Activity number

- ☐ 1 **EM** Energy drives the cycling of matter by thermal convection in the mantle. — 74 75
- ☐ 2 **SC** Change and rates of change in the features of the Earth's continental and ocean floor over time can be quantified and modeled. — 75 76 77 80

## Science and engineering practices

***Demonstrate competence in these science and engineering practices***

Activity number

- ☐ 1 Use a model based on evidence to show how radioactive decay provides the primary source of heat that drives convection in the mantle. — 74
- ☐ 2 Develop or use a model based on evidence of the Earth's interior to describe the cycling of matter in the Earth's crust and mantle by thermal convection. — 75 76 79
- ☐ 3 Develop or use a model based on evidence to illustrate how plate tectonics explains the continental and ocean floor features of the Earth. — 75 76 77 79 80

# 74 The Nuclear Fires

**Key Idea**: The Earth's interior heat is caused by the decay of radioactive elements and the heat left over from the Earth's formation.

- Radioactive decay is the break down (fission) of large atoms into smaller ones. Radioactive decay occurs in many different ways but usually involves the release of highly energetic particles (e.g. alpha particles) or electromagnetic radiation (e.g. gamma waves). These high energy particles and waves collide with objects about them releasing heat and causing the temperature of the surrounding material to increase.
- The material that formed the Earth included elements such as uranium and thorium. Uranium is the heaviest naturally occurring element on Earth. It occurs in numerous isotopes (atoms with the same atomic number but different atomic masses) all of which are radioactive. U-238 is the most common and the most stable, it has a half life of about 4.5 billion years (about the age of the Earth).
- When a U-238 atom decays it produces a helium nucleus (an alpha particle) and a thorium atom. The thorium atom itself is unstable and will also decay. There is a decay chain that continues until a stable atom of lead is produced. Heat is released at each stage and this heats the interior of the planet.

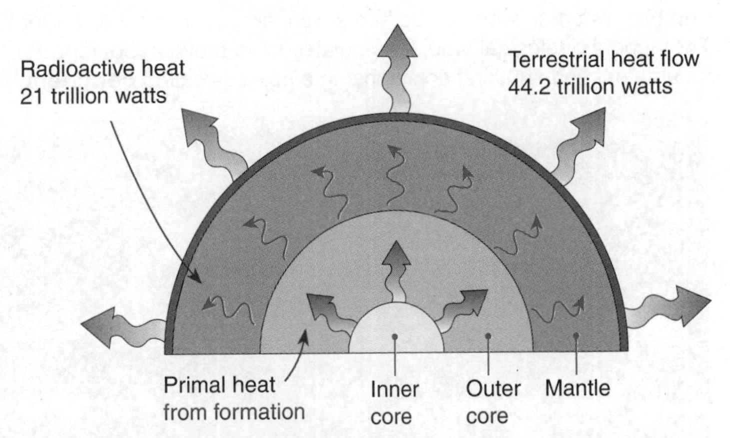


About half the heat radiated by the Earth comes from the decay of radioactive elements. Most of the rest comes from heat left over from the Earth's formation.

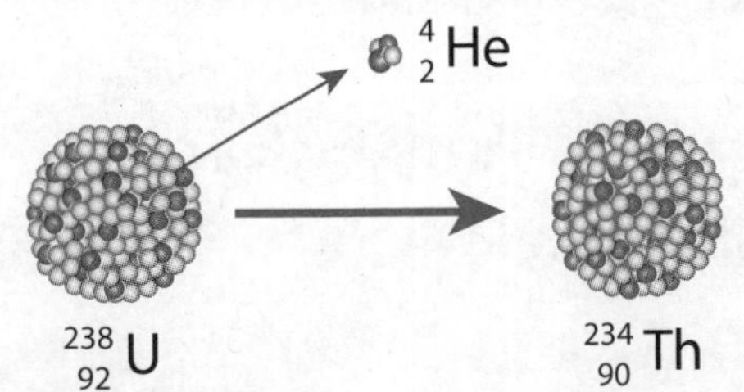


Uranium-238 has a half life of 4.5 billion years and decays to thorium-234. Thorium-234 has a half life of just 24.5 days.

Radioactive decay heats the interior of the planet. Some of this heat powers volcanic activity, helping to form volcanoes.

Earthquakes result from convection currents (produced by heat) moving Earth's interior and shifting surface layers.

The movement of Earth's mantle causes parts of the surface to collide, rising up to produce mountains, e.g. the Himalayas.

1. What is radioactive decay? ______________________________

______________________________

______________________________

2. (a) How much heat is radiated by the Earth? ______________________________

(b) How much of this is produced from radioactive decay? ______________________________

3. What forms or is emitted when a uranium-238 atom decays? ______________________________

4. Explain the importance of radioactive decay to geological processes such as plate tectonics:

______________________________

______________________________

______________________________

______________________________


ISBN: 978-1-927309-37-7

# 75 Plate Tectonics

**Key Idea**: Plate tectonics is the theory describing the large scale movement of the Earth's lithosphere and how that the movement is related to convection currents in the mantle.

The Earth's crust is broken up into seven large, continent-sized **tectonic plates** and about a dozen smaller plates. Throughout geological time, these plates have moved about the Earth's surface, opening and closing oceans, building mountains, and shuffling continents in a process called **continental drift**.

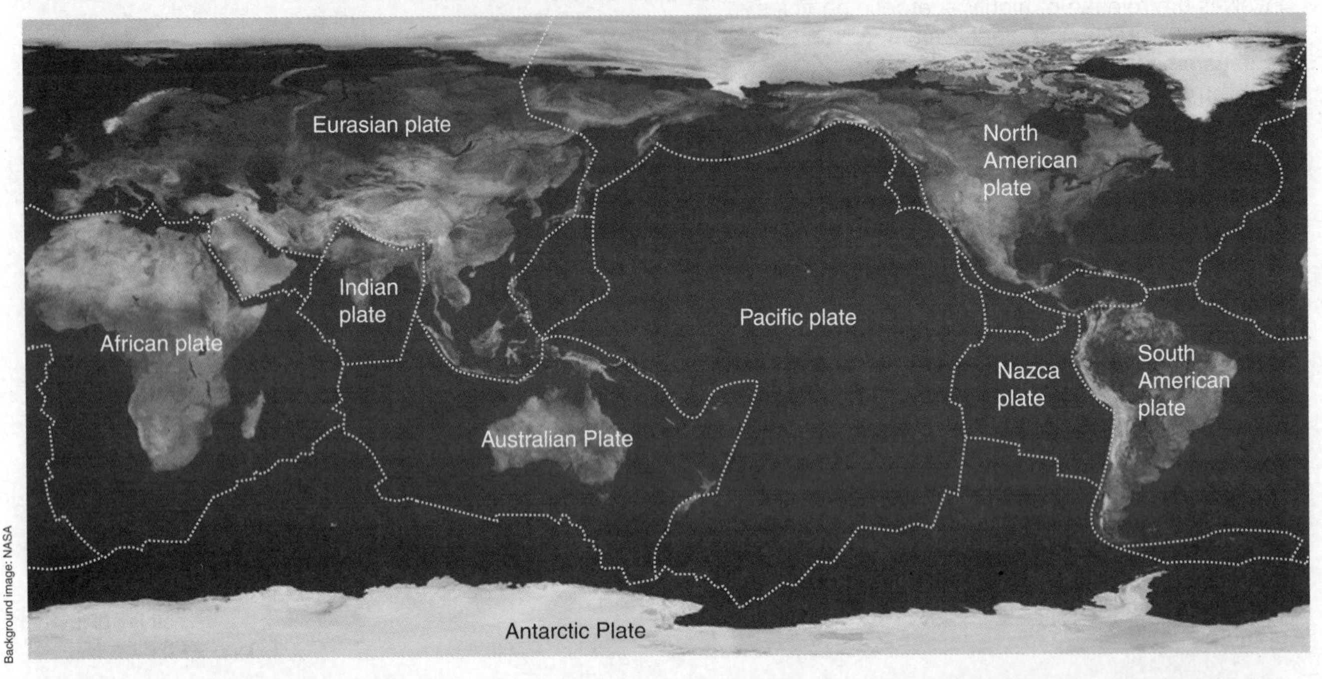


The evidence for past plate movements has come from several sources: mapping of plate boundaries, the discovery of sea floor spreading, measurement of the direction and rate of plate movement, and geological evidence such as the distribution of ancient mountain chains, unusual deposits, and fossils. The size of the plates is constantly changing, with some expanding and some getting smaller. The extent of the tectonic plates is shown in the diagram above. The Pacific plate is by far the largest, measuring 103 million km$^2$.

### The mechanism of plate movement

- The relatively cool **lithosphere** overlies the hotter, plastic and more fluid **asthenosphere**. Heat from the mantle drives two kinds of asthenospheric movement: **convection** and **mantle plumes**. Plate motion is partly driven by the weight of cold, dense plates sinking into the mantle at trenches. This heavier, cooler material sinking under the influence of gravity displaces heated material, which rises as mantle plumes.
- The movements of the tectonic plates puts the brittle rock of the crust under strain, creating **faults** where rocks fracture and slip past each other. Earthquakes are caused by energy release during rapid slippage along faults. Consequently, the Earth's major earthquake (and volcanic) zones occur along plate boundaries.

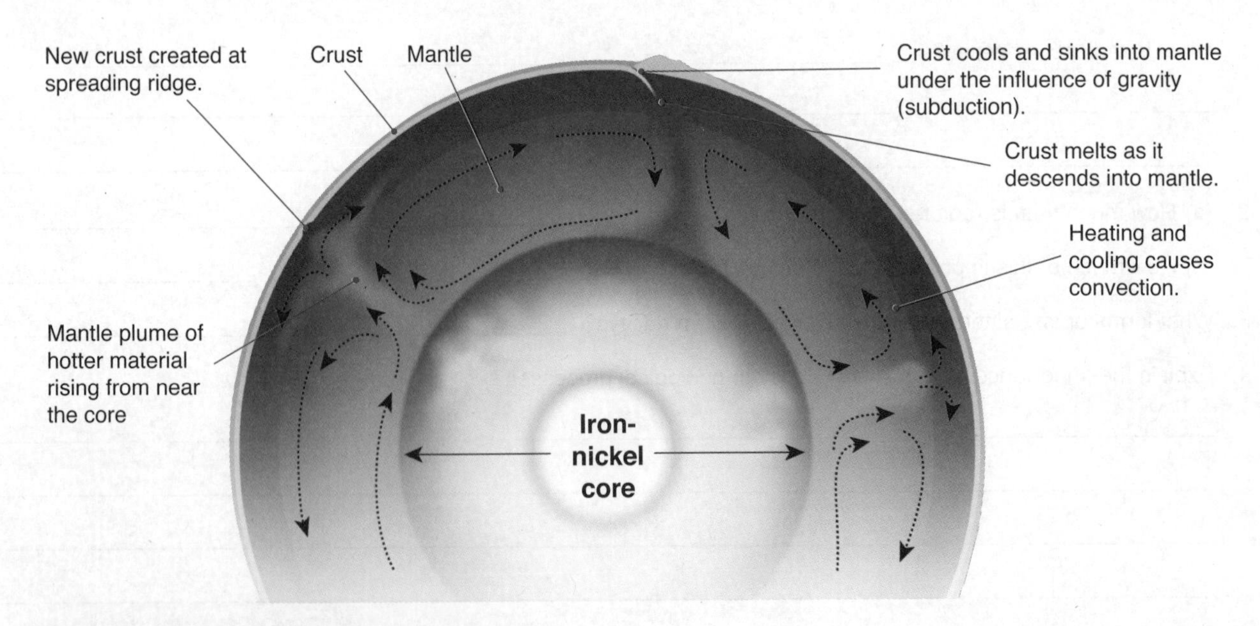

**ISBN: 978-1-927309-37-7**

## Subduction and volcanoes

Much of the Pacific plate boundaries form subduction zones (where one tectonic plate moves under another and into the mantle). This produces a region of extreme seismic and volcanic activity. Along with parts of the Nazca plate, it forms the Pacific Ring of Fire. Around three quarters of the world's active and dormant volcanoes are found around the edge of the ring and nearly 90% of all earthquake activity is located there.

**Subduction zones around the Pacific rim**

**Volcanoes around the Pacific rim**

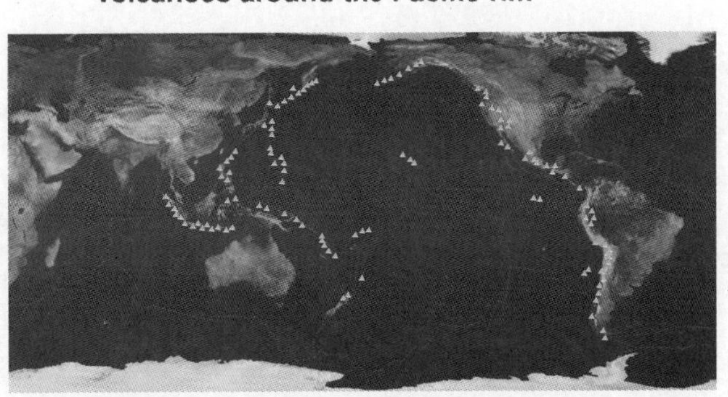

Background images: NASA

1. Name the seven major tectonic plates: ______________________

______________________

______________________

2. Study the images of the Hawaiian island chain below. These islands formed as the Pacific plate moved over a "hot spot" in the mantle. From the information, determine which direction the Pacific plate is moving at this point and indicate this direction by drawing an arrow on the left diagram.

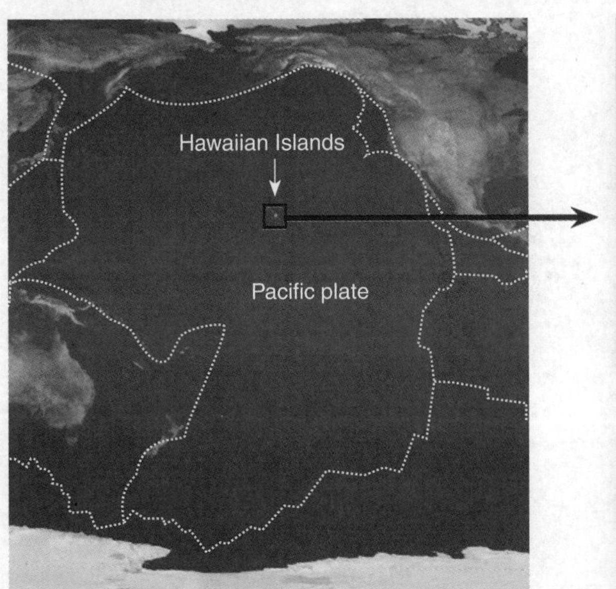

3. Identify three types of evidence for the movement of tectonic plates: ______________________

______________________

______________________

4. (a) Describe the mechanism of plate movement: ______________________

______________________

______________________

(b) How does this account for continental drift? ______________________

______________________

5. Study the images at the top of the page. Describe the relationship between the position of subduction zones and volcanoes in the Pacific:

______________________

______________________


**ISBN: 978-1-927309-37-7**

# 76 Plate Boundaries

**Key Idea**: When tectonic plates meet they form either convergent, divergent, or transform boundaries. Movement at these boundaries is caused by convection currents in the mantle.

Plate boundaries are marked by well-defined zones of seismic and volcanic activity. Plate growth occurs at **divergent boundaries** along sea floor spreading ridges (e.g. the Mid-Atlantic Ridge and the Red Sea) whereas plate attrition occurs at **convergent boundaries** marked by deep ocean trenches and subduction zones. Divergent and convergent zones make up approximately 80% of plate boundaries. The remaining 20% are called **transform boundaries**, where two plates slide past one another with no significant change in the size of either plate.

USGS

The San Andreas fault is a transform boundary running for over 1000 km through California.

**Island arcs** form from a chain of volcanoes parallel to the edge of a subduction zone

Convergent plate boundary

Trench

Island arc

Stratovolcano

Island chain

Shield volcano

Transform fault

Divergent plate boundary

Ocean spreading ridge

Lithosphere

Asthenosphere

Hot spot, e.g. Hawaiian islands

1. Describe what is happening at each of the following plate boundaries and identify an example in each case:

(a) Convergent plate boundary: ______________________

(b) Divergent plate boundary: ______________________

(c) Transform plate boundary: ______________________




**ISBN: 978-1-927309-37-7**

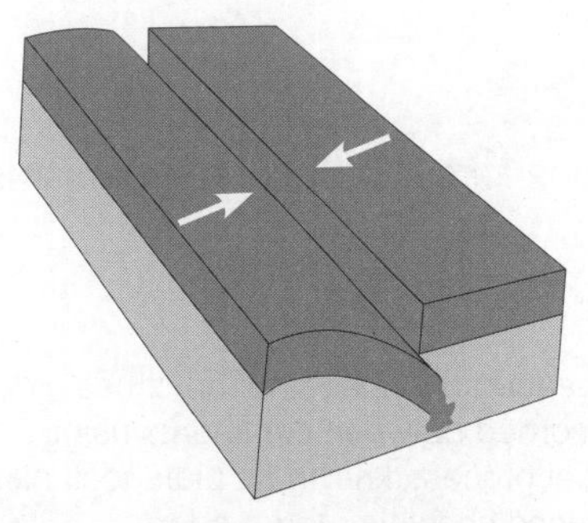

Plate boundaries moving towards each other are called **convergent plate boundaries**. Where oceanic crust and continental crust meet, the oceanic crust will subduct under the continental crust, creating a subduction zone. Volcanoes normally form along the continental border of a subduction zone. When continental crusts collide, huge mountain ranges such as the Himalayas can form.

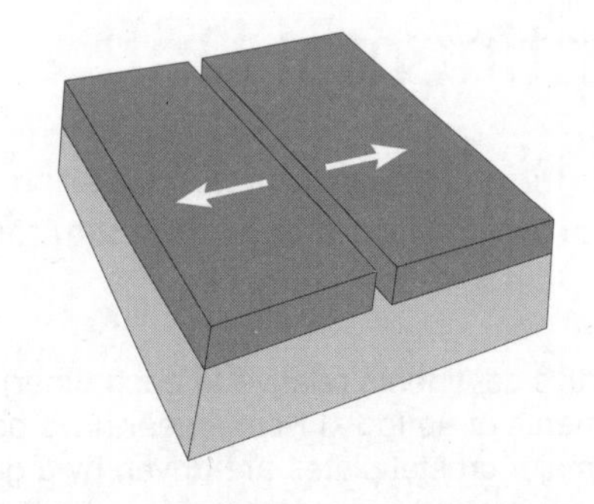

**Divergent plate boundaries** form where the tectonic plates are moving away from each other. These are commonly found along the mid ocean ridges, but occasionally are seen on land, as in the Great Rift Valley and Iceland. Divergent boundaries are also known as constructive boundaries as they produce new crust from the upwelling of magma.

**Transform boundaries** are formed when the tectonic plates are moving past each other. They are therefore neither constructive nor destructive. Examples include the San Andreas fault in California and the Alpine Fault in New Zealand.

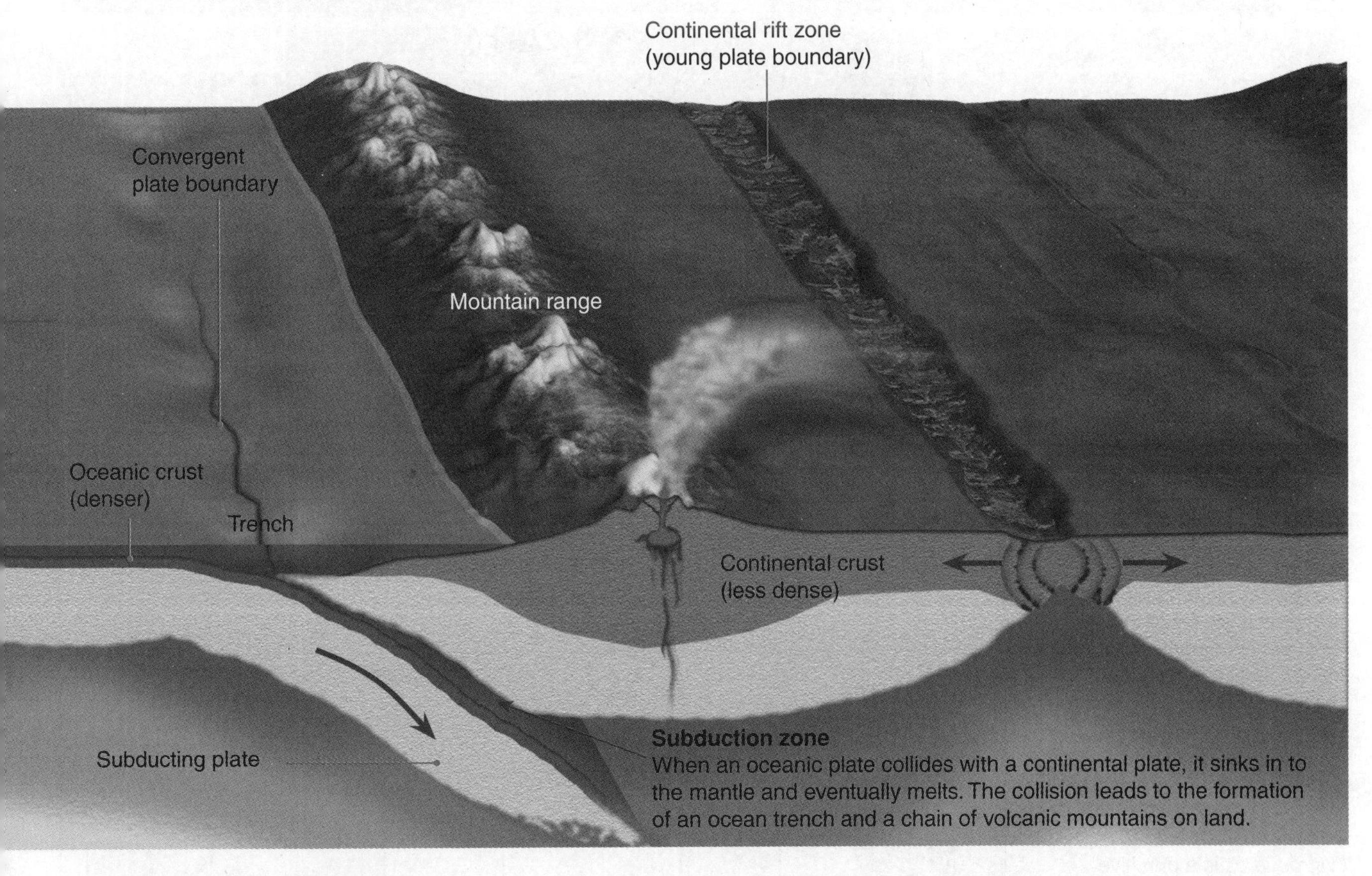


2. Identify the type of plate boundary at which each of the following occurs:

    (a) Mountain building: ______________________ (c) Creation of new ocean floor: ______________________

    (b) Subduction: ______________________ (d) Island arc: ______________________

3. (a) Explain why the oceanic crust subducts under the continental crust in a subduction zone: ______________________

    ______________________

    ______________________

    (b) What cause volcanoes to form along the continental plate boundary of a subduction zone? ______________________

    ______________________

    ______________________

    ______________________


**ISBN: 978-1-927309-37-7**

# 77 Modeling Continental Drift

**Key Idea**: Continental drift results from plate tectonics. Evidence includes fossils of related organisms and similar deposits being found on different continents.

**Continental drift** (the movement of the Earth's continents relative to each other) is a measurable phenomenon; it has continued throughout Earth's history. Movements of up to 2-11 cm a year have been recorded between continents using GPS. The movements of the Earth's seven major crustal plates are driven by a geological process known as plate tectonics. Some continents are drifting apart while others are moving together. Many lines of evidence show that the modern continents were once joined together as 'supercontinents'. One supercontinent, **Gondwana**, was made up of the southern continents some 200 mya.

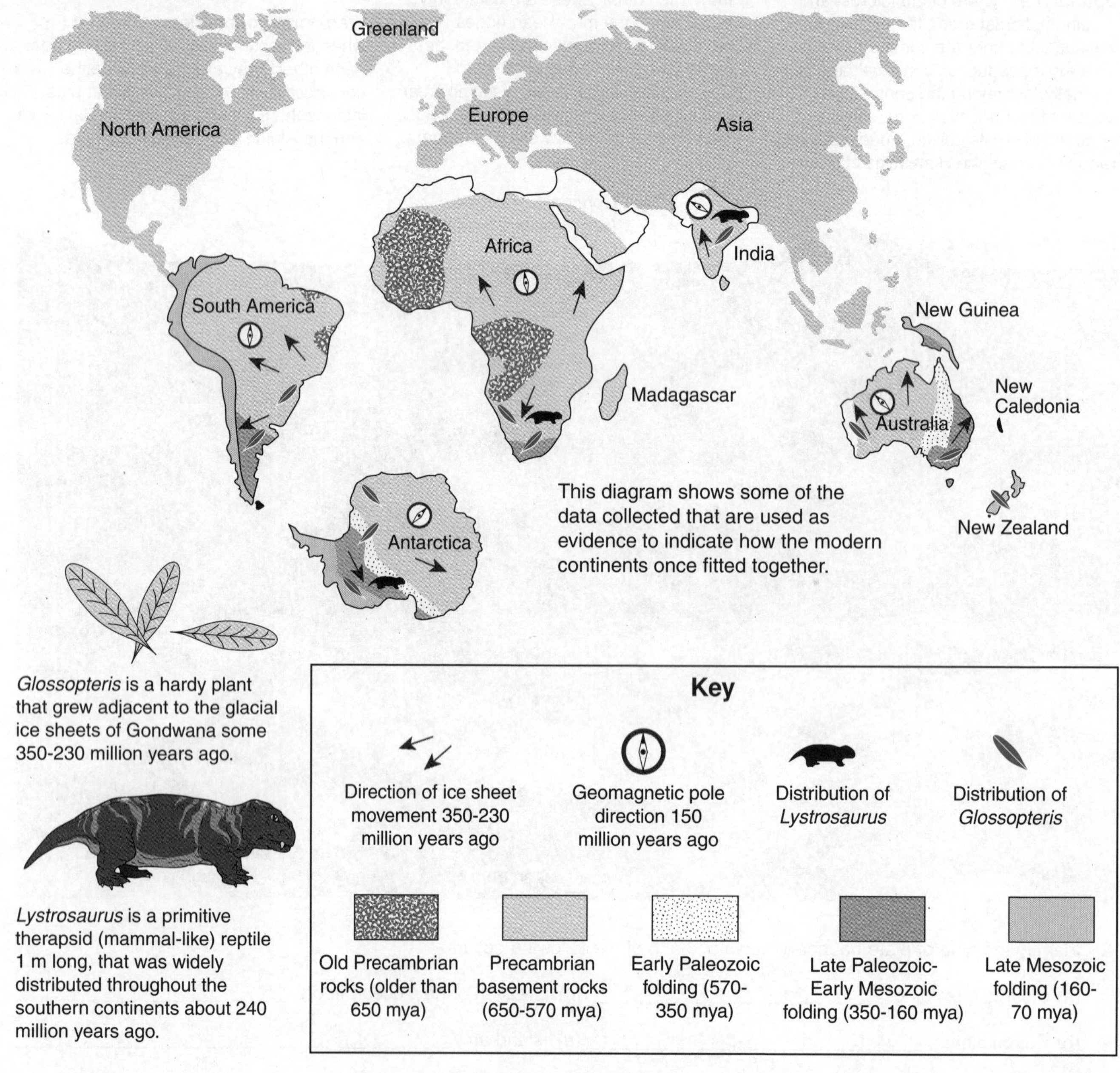


This diagram shows some of the data collected that are used as evidence to indicate how the modern continents once fitted together.

*Glossopteris* is a hardy plant that grew adjacent to the glacial ice sheets of Gondwana some 350-230 million years ago.

*Lystrosaurus* is a primitive therapsid (mammal-like) reptile 1 m long, that was widely distributed throughout the southern continents about 240 million years ago.

1. Name the modern landmasses (continents and large islands) that made up the supercontinent of Gondwana:

2. Cut out the southern continents on page 129 and arrange them to recreate the supercontinent of Gondwana. Take care to cut the shapes out close to the coastlines. When arranging them into the space showing the outline of Gondwana on page 131, take into account the following information:
   (a) The location of ancient rocks and periods of mountain folding during different geological ages.
   (b) The direction of ancient ice sheet movements.
   (c) The geomagnetic orientation of old rocks (the way that magnetic crystals are lined up in ancient rock gives an indication of the direction the magnetic pole was at the time the rock was formed).
   (d) The distribution of fossils of ancient species such as *Lystrosaurus* and *Glossopteris*.

**ISBN: 978-1-927309-37-7**

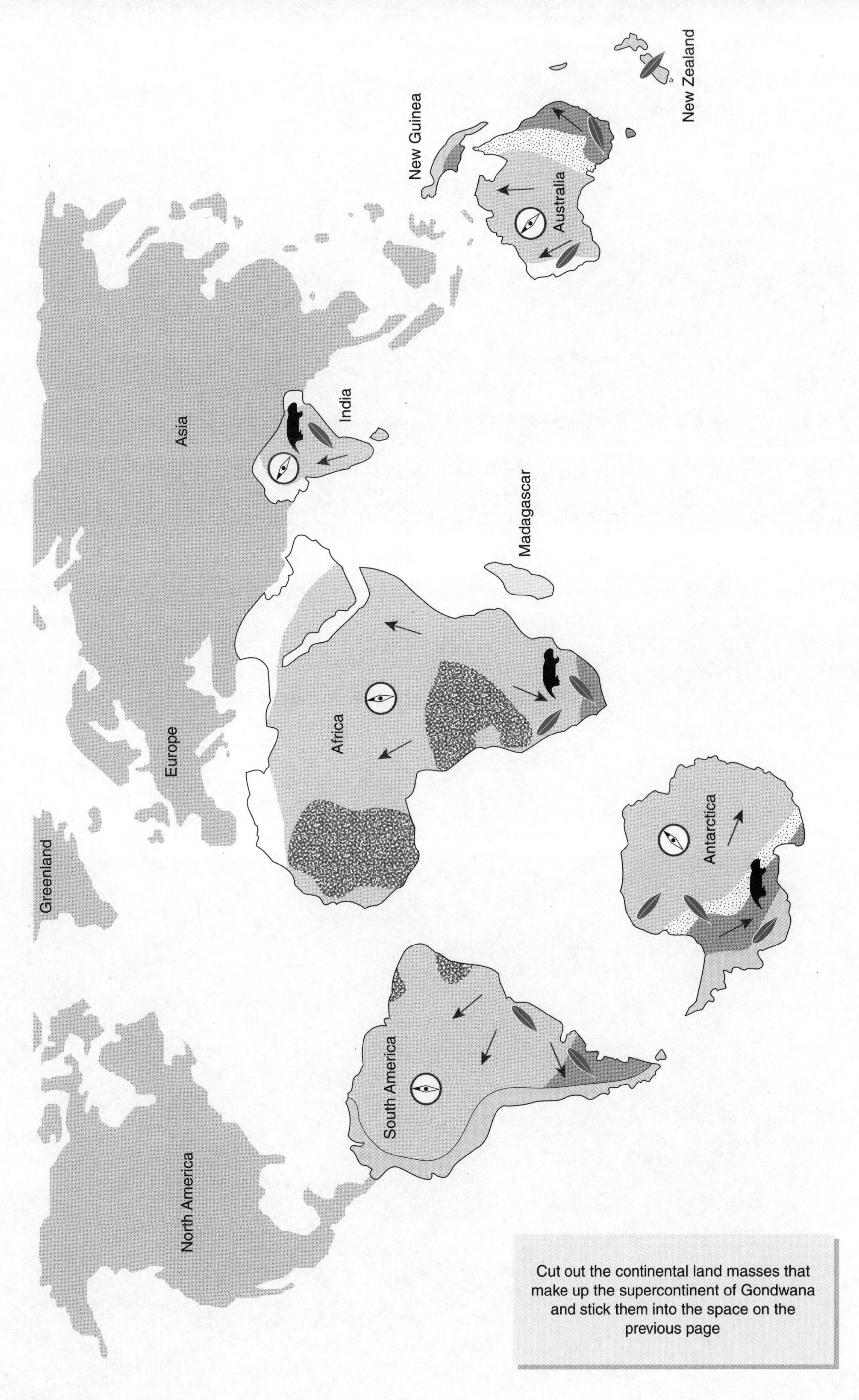


Cut out the continental land masses that make up the supercontinent of Gondwana and stick them into the space on the previous page


**ISBN: 978-1-927309-37-7**

This page has been deliberately left blank


**ISBN: 978-1-927309-37-7**

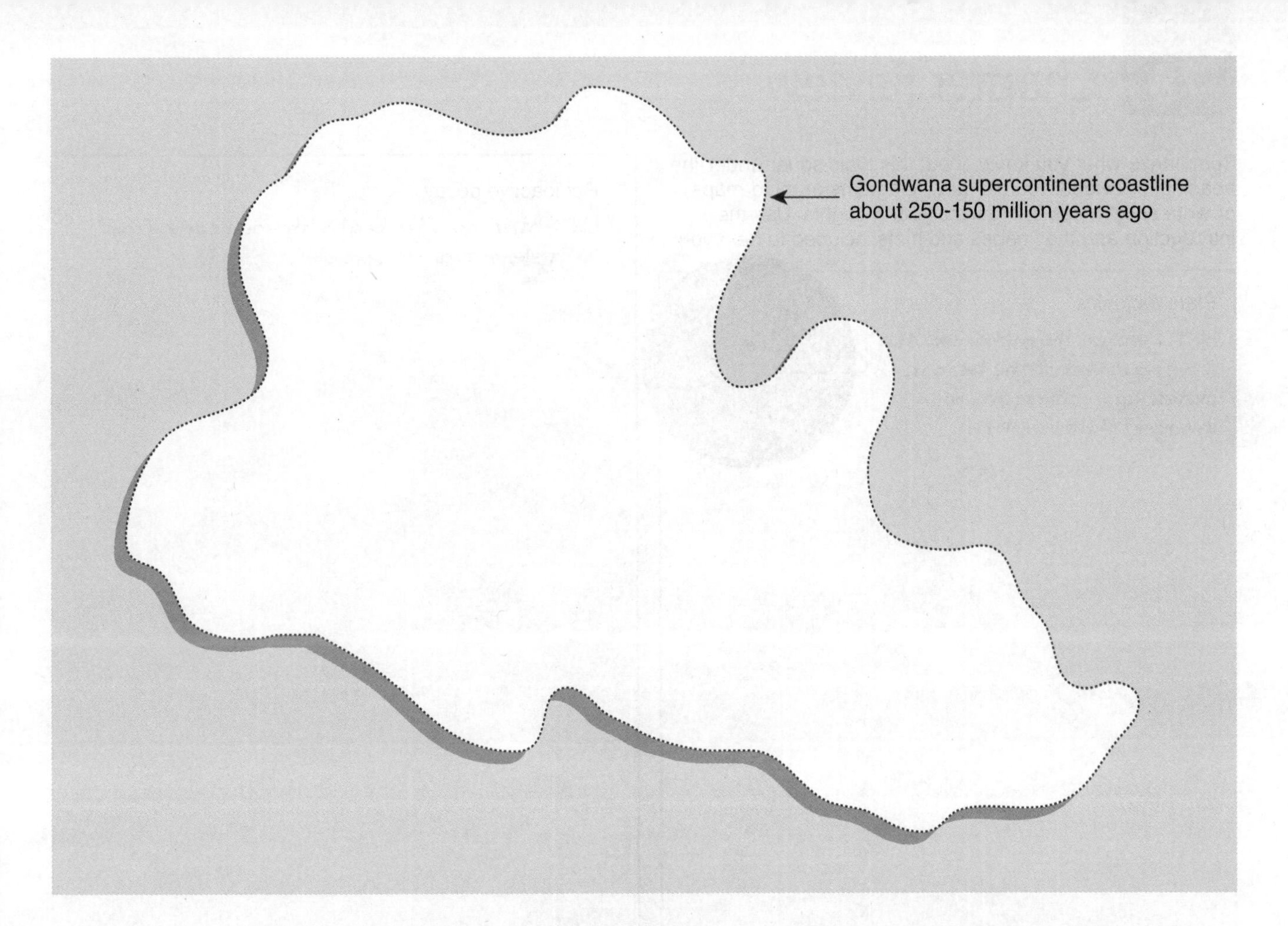


3. Once you have positioned the modern continents into the pattern of the supercontinent, mark on the diagram:
   (a) The likely position of the South Pole 350-230 million years ago (as indicated by the movement of the ice sheets).
   (b) The likely position of the geomagnetic South Pole 150 million years ago (as indicated by ancient geomagnetism).

4. State what general deduction you can make about the position of the polar regions with respect to land masses:

5. Fossils of *Lystrosaurus* are known from Antarctica, South Africa, India and Western China. With the modern continents in their present position, *Lystrosaurus* could have walked across dry land to get to China, Africa and India. It was not possible for it to walk to Antarctica, however. Explain the distribution of this ancient species in terms of continental drift:

6. The Atlantic Ocean is currently opening up at the rate of 2 cm per year. At this rate in the past, calculate how long it would have taken to reach its current extent, with the distance from Africa to South America being 2300 km (assume the rate of spreading has been constant):

7. How do the different lines of evidence allow scientists to recreate the positions of the Earth's continents through time?


**ISBN: 978-1-927309-37-7**

# 78 Chapter Review

Summarize what you know about this topic so far under the headings provided. You can draw diagrams or mind maps, or write short notes to organize your thoughts. Use the introduction and the images and hints included to help you:

Plate tectonics

HINT: Describe the mechanisms that drive movement of the tectonic plates. Draw convergent and divergent plate boundaries.

Radioactive decay

HINT: Where does the heat in the Earth comes from? What elements are responsible?

Continental drift

HINT: Describe the evidence for continental drift. How do we know the continents were once joined together?


**ISBN: 978-1-927309-37-7**

# 79 KEY TERMS AND IDEAS: Did You Get It?

1. (a) Use the following list to label the diagram below: *continental crust, mid ocean ridge, subduction zone, convergent boundary, divergent boundary, hot spot volcano, island chain, oceanic crust*

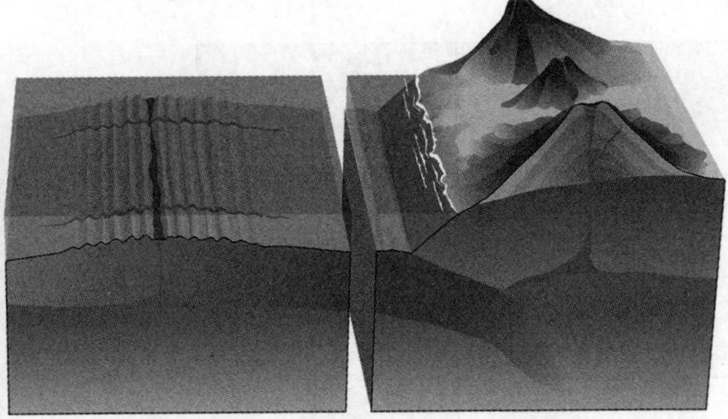

(b) Draw arrows on the diagram to show the direction of the crust and convection currents:

2. (a) Use the information below to produce a graph of the age of the volcanoes in the Hawaiian island chain compared to their distance from the Kilauea volcano on Hawai'i (Big island).

| Name | Distance from Kilauea (km) | Age (millions of years) |
|---|---|---|
| Kilauea | 0 | 0 |
| Mauna Kea | 54 | 0.375 |
| Kohala | 100 | 0.43 |
| East Maui | 182 | 0.75 |
| West Maui | 220 | 1.32 |
| East Molokai | 256 | 1.76 |
| West Molokai | 280 | 1.9 |
| Koolau | 339 | 2.6 |
| Waianae | 374 | 3.7 |
| Kauai | 519 | 5.1 |
| Nihoa | 780 | 7.2 |
| Necker | 1058 | 10.3 |

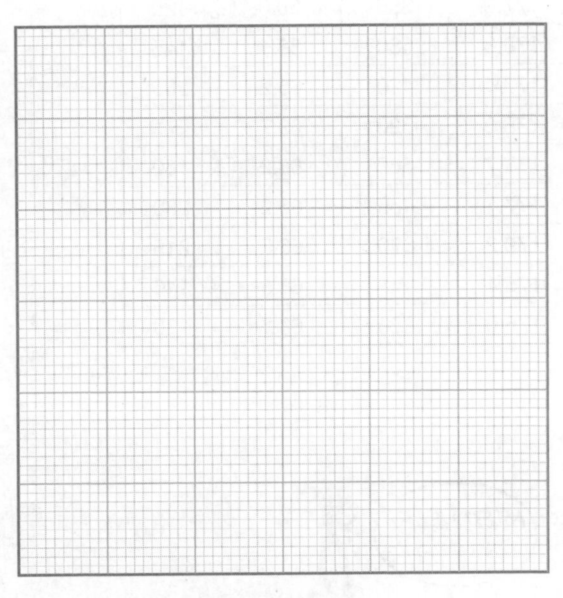

(b) Use the data to calculate the rate of the movement of the Pacific plate over the Hawaiian hot spot:

3. (a) Explain why convergent plate boundaries are also destructive plate boundaries:

(b) Explain why divergent plate boundaries are also constructive plate boundaries:


**ISBN: 978-1-927309-37-7**

# 80 Summative Assessment

1. Earthquakes normally occur along plate boundaries. Measuring the depth of these earthquakes can give an idea of the shape of the boundary and how the plate are interacting. The data below shows earthquake depths for the Tonga Trench in the Pacific Ocean and along the coast of Chile.

(a) Plot a scatter graph of the data on the grids provided:

| Tonga trench | | Chile coast | |
|---|---|---|---|
| Longitude (°W) | Depth (km) | Longitude (°W) | Depth (km) |
| 176.2 | 270 | 67.5 | 180 |
| 175.8 | 115 | 68.3 | 130 |
| 175.7 | 260 | 62.3 | 480 |
| 175.4 | 250 | 62.0 | 600 |
| 176.0 | 160 | 69.8 | 30 |
| 173.9 | 60 | 69.8 | 55 |
| 174.9 | 50 | 67.7 | 120 |
| 179.2 | 650 | 67.9 | 140 |
| 173.8 | 50 | 69.2 | 35 |
| 177.0 | 350 | 68.6 | 125 |
| 178.8 | 580 | 68.1 | 145 |
| 177.4 | 420 | 65.2 | 285 |
| 178.0 | 520 | 69.7 | 50 |
| 177.7 | 560 | 68.2 | 160 |
| 177.7 | 465 | 66.2 | 230 |
| 179.2 | 670 | 66.3 | 215 |
| 175.1 | 40 | 68.5 | 140 |
| 176.0 | 220 | 68.1 | 130 |

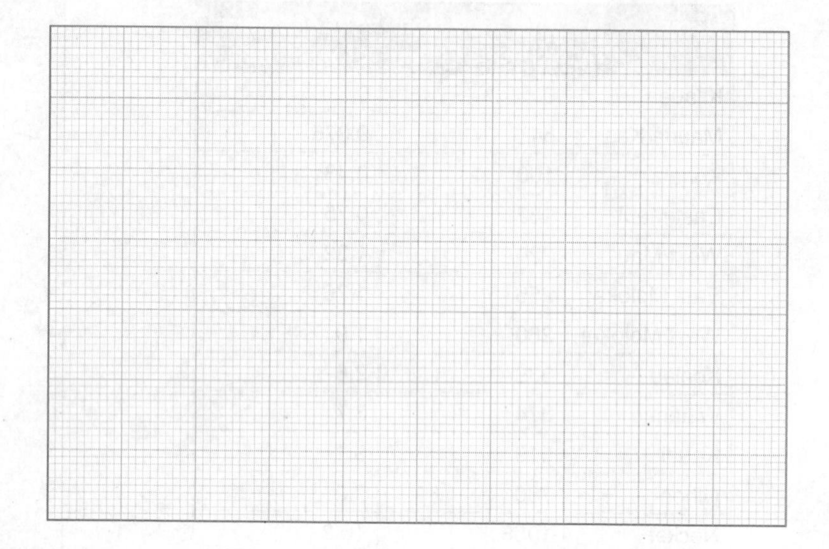

(b) Add a line of best fit through the data points:

(c) What type of plate boundary appears to be present at the locations plotted?

___

(d) Draw a diagram in the space below to show the how the layers of the Earth are moving at the Tonga Trench:

TEST

CCC

PRACTICES

ISBN: 978-1-927309-37-7

2. Draw a diagram in the space below to show the mechanisms of plate tectonics that shift the plates of the Earth's crust. You should make sure your diagram has divergent and convergent boundaries and the layers of the Earth that are significant in plate tectonics.

3. Explain the significance of radioactive decay to plate tectonics: ____________________

4. The diagram right shows the Pacific plate and the Nazca plate. The white dotted line shows the location of the plate boundaries The blue dotted line shows the location of subduction zones along those boundaries.



(a) On the diagram circle the area that would likely be a divergent plate boundary.

(b) Where would you expect to find volcanoes on this diagram?

(c) Explain why volcanoes form in the places you have indicated in (b): ____________________


**ISBN: 978-1-927309-37-7**

# ESS2.C The Roles of Water in Earth's Surface Processes

**Key terms**

adhesion
cohesion
deposition
dipole
erosion
hydrologic cycle
ice
igneous rock
magma
metamorphic rock
rock cycle
sedimentary rock
water
weathering

## Disciplinary core ideas

***Show understanding of these core ideas***

| | | Activity number |
|---|---|---|
| | **Water is central to Earth's dynamics** | |
| ☐ | 1 The unique physical and chemical properties of water are a function of its dipole nature and the strong attraction of its molecules to each other. | 81 |
| ☐ | 2 Water is the only substance on Earth that exists naturally in solid, liquid, and gaseous forms. A lot of energy is required before water changes state, so it has an exceptional capacity to absorb, store, and release large amounts of energy. This is important in regulating the Earth's climate and in the transport of energy around the globe in the hydrologic cycle. | 81 82 |
| ☐ | 3 Water transmits light, making it a suitable medium in which photosynthesis can occur, even at depth. | 81 |
| ☐ | 4 Water plays a central role in the rock cycle and is important in chemical and physical weathering processes and erosion. Water (as ice) can shape landscapes through glacial scouring and (as liquid) can dissolve and transport material. | 83 84 |
| ☐ | 5 Water is important in lowering the viscosities and melting points of rocks. This is important in creating magma (below ground molten rock) and lava (above ground molten rock) which bring mantle material to the surface. | 86 |
| ☐ | 6 The cohesive and adhesive properties of water are important in binding soil particles together. Soils low in water are vulnerable to erosion (transport away). However, if soils become saturated they can be more vulnerable to slipping. | 87 |

NASA

## Crosscutting concepts

***Understand how these fundamental concepts link different topics***

| | | Activity number |
|---|---|---|
| ☐ | 1 SF The functions and properties of water and its effects on Earth's materials can be inferred from its molecular substructure. | 81-87 |
| ☐ | 2 EM Energy drives the cycling of water within and between the Earth's systems. | 82 83 |

## Science and engineering practices

***Demonstrate competence in these science and engineering practices***

| | | Activity number |
|---|---|---|
| ☐ | 1 Plan and conduct an investigation to produce data to serve as evidence for the effects of water on Earth's materials and surface processes. | 85 86 |
| ☐ | 2 Use a model to describe the behavior of water in its various states. | 81 82 |
| ☐ | 3 Use a model to illustrate how water moves between Earth's systems: the geosphere, hydrosphere, atmosphere, and biosphere). | 82 |
| ☐ | 4 Develop and use a model to illustrate the processes involved in the rock cycle. | 83-86 |
| ☐ | 5 Use quantitative data to support a claim about the effect of soil moisture on wind-induced rate of erosion. | 87 90 |

# 81 The Properties of Water

**Key Idea**: Water's unique physical and chemical properties make it a central chemical in many biological and geological systems.

- Water has a simple chemical formula, containing just two hydrogen atoms and one oxygen atom ($H_2O$) held together by covalent bonds. The electrons in these bonds are not shared equally between the atoms. On average, the electrons spend a greater amount of time near oxygen than near hydrogen. This gives the oxygen a slightly negative ($\delta^-$) charge and the hydrogens a slightly positive ($\delta^+$) charge (called a **dipole**).
- This difference in charge produces a strongly polar molecule. The positive hydrogen end of the water molecule is attracted to the negative oxygen end of other water molecules (hydrogen bonding). The attraction is strong and it is this feature that gives water many of its unique properties, including its high surface tension and its chemical behavior as a powerful solvent.

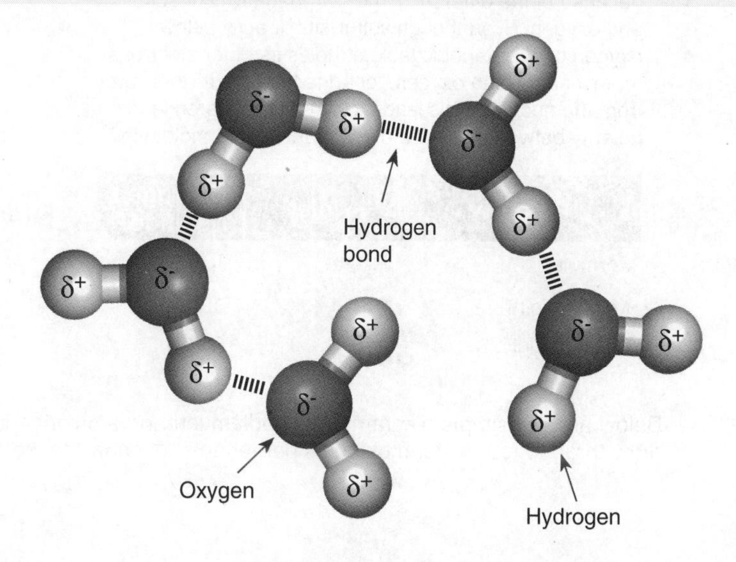


**Important properties of water**

Water has the highest heat capacity of all liquids. It takes a lot of energy to raise its temperature. Water has a high boiling point because a lot of energy is need to break the hydrogen bonds between water molecules and make water boil.

Water molecules are **cohesive**, i.e. they stick together. This is due to hydrogen bonding. Cohesion allows water to form drops and is responsible for surface tension. Water molecules also **adhere** to other substances.

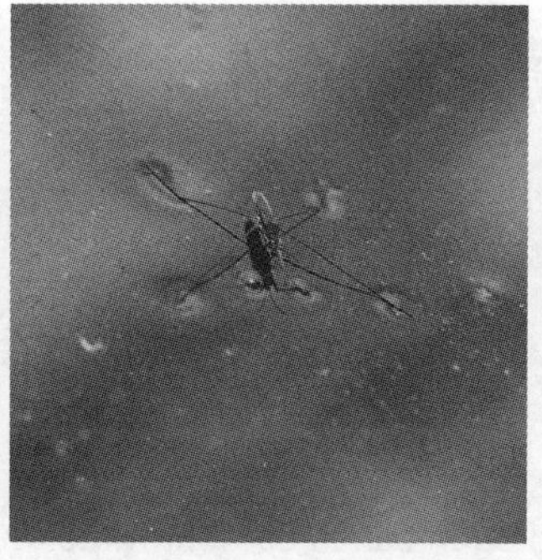

Surface tension is the ability of a fluid's surface to deform without breaking. Water can deform more than most fluids before its surface tension is broken. Thus a water skater can float on the surface because it doesn't break the surface.

Water's polarity makes it an extremely good solvent. Ionic substances such as salts (e.g. sodium chloride, NaCl) dissolve relatively easily in water, one of the reasons the sea is so salty (carrying 35 g/L of salt).

One of the most important properties of water is that its solid state (ice, right) is less dense that its liquid state. This means that ice floats on water. When freezing, water molecules align into a crystal structure that increases its volume by about 9% compared to liquid water at the same temperature. Water is colorless and transparent.

Water plays an important part in plate tectonics and the melting of rock. Water in the rock disrupts the chemical bonds and lowers its melting point. It also lubricates subduction zones, helping the crust to sink into the mantle.

**ISBN: 978-1-927309-37-7**

## Comparing water and hydrogen sulfide

Hydrogen sulfide ($H_2S$) has the same molecular shape as water (right) but has a melting point of -82°C and a boiling point of -60°C, far lower than water. This is because of the difference in electronegativity of sulfur and oxygen. Even though sulfur sits directly below oxygen on the periodic table, it holds its outer electrons more weakly than oxygen because they are further away from the nucleus. This leads to a greater difference in polarity between a $H_2O$ molecule and a $H_2S$ molecule.

| Property | Water | Hydrogen sulfide |
|---|---|---|
| Formula | $H_2O$ | $H_2S$ |
| Melting point | 0°C | -82°C |
| Boiling point | 100°C | -60°C |

Water

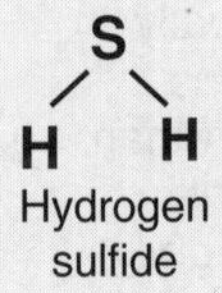

Hydrogen sulfide

*The bonds that water molecules form with each other require a lot of energy to break. This is why water has a much higher boiling point than hydrogen sulfide*

1. Below are two simple diagrams of a sodium ion and a chloride ion. Note the charge they carry. In the space around them draw three to four water molecules per ion to show how the water molecules interact with the them:

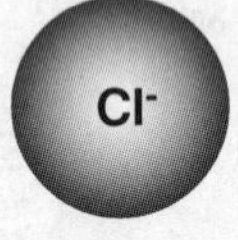


2. Explain the formation of hydrogen bonds between water molecules ________________

3. Explain why aquatic environments have relatively stable temperatures: ________________

4. Why does ice float on water? ________________

5. Explain why water has a high surface tension: ________________

6. Why does hydrogen sulfide have a much lower melting point and boiling point than water? ________________


**ISBN: 978-1-927309-37-7**

# 82 The Hydrologic Cycle

**Key Idea**: The hydrologic cycle is the cycling of water from the oceans to the land and back.

## Earth's water

- About 97% of Earth's water is stored in the oceans, which contain more than 1.3 billion $km^3$ of water. Less than 1% of Earth's water is freely available fresh water (in lakes, rivers, and streams).
- Water evaporates from the oceans and lakes into the atmosphere and falls as precipitation (e.g. rain, snow, or hail). Precipitation falling on the land is transported back to the oceans by rivers and streams or is returned to the atmosphere by evaporation or transpiration (evaporation from plant surfaces).
- Water can cycle very quickly if it remains near the Earth's surface, but it can also remain locked away for hundreds or even thousands of years, e.g. if trapped in deep ice layers at the poles or in groundwater storage (aquifers).
- Humans intervene in the water cycle by using water for their own needs. Withdrawing water from rivers and lakes for irrigation changes evaporation patterns, lowers lake levels, and reduces river flows.

Water is the only substance on Earth that is found naturally as a solid, liquid, or gas. It has an unexpectedly high boiling point compared to other similar molecules and requires a lot of energy to change state. This means it acts as a buffer against extreme temperature fluctuations in the environment.

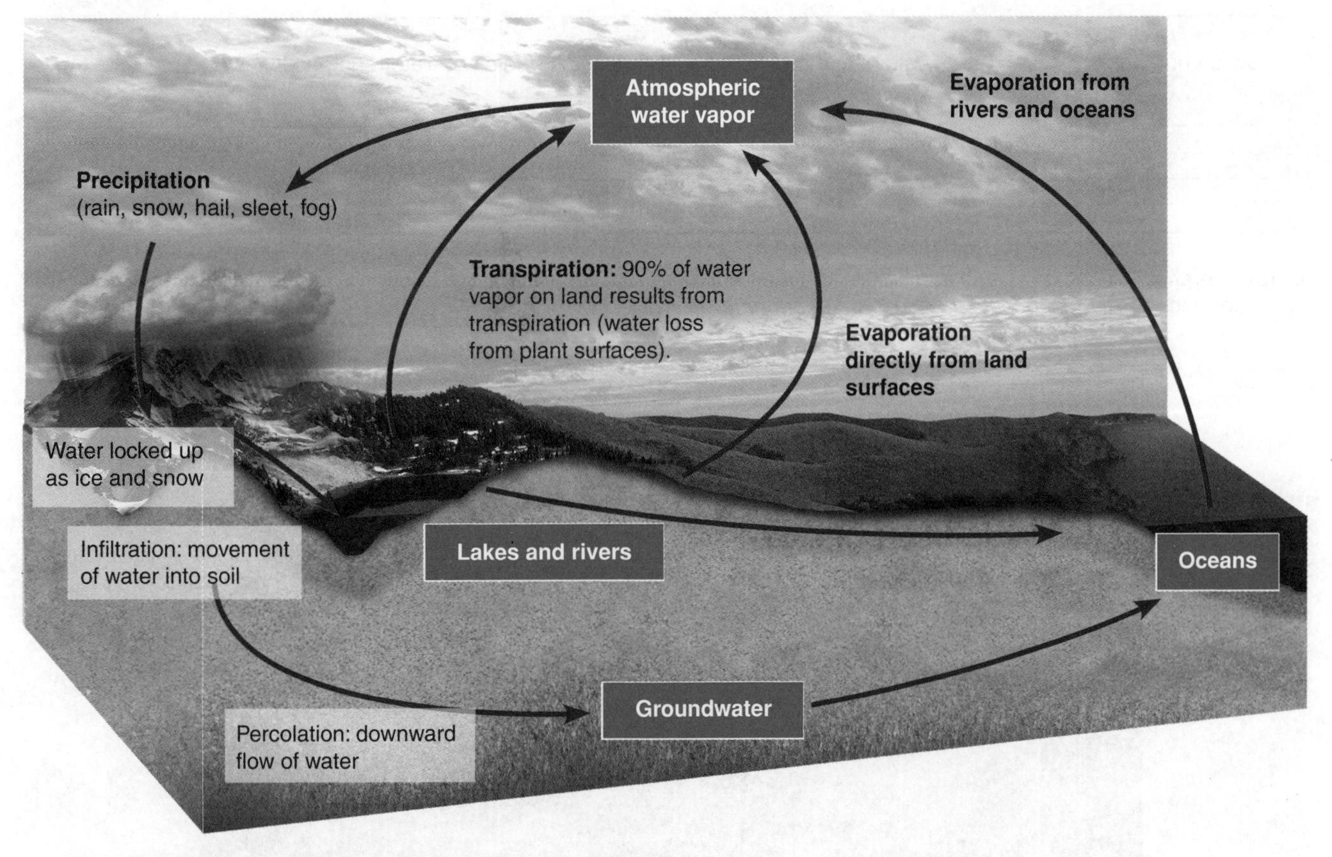


1. What is the main storage reservoir for water on Earth? ______________________

2. Describe the feature of water that allows it cycle as described above: ______________________

3. Identify the two processes by which water moves from the land or oceans to the atmosphere: ______________________


ISBN: 978-1-927309-37-7

The water cycle is important in the transport of energy about the globe. Energy (from the Sun) is absorbed by the oceans. Water evaporates from the oceans, cooling them. This energy is released again when water vapor condenses.

Water can be held inside the Earth itself. The largest reservoir of water on Earth is in fact in the mantle, bound with minerals. Some estimates put the amount of water in the mantle at ten times the volume in the Earth's oceans.

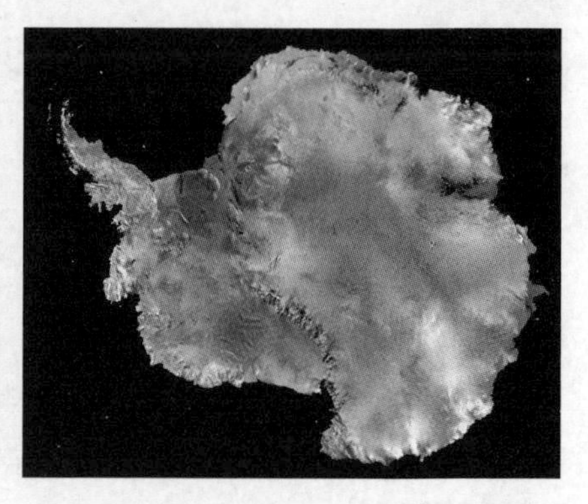

Water can be locked up in ice for tens of thousands of years. The ice in the ice sheet of Greenland is about 100,000 years old, while ice cores from Antarctica have dated some ice to at least 800,000 years old.

4. Explain how the hydrologic cycle helps to move energy around the globe: ______

5. How do humans intervene in the water cycle and how might this affect bodies of water such as lakes?

6. (a) The photograph below shows a set up for modeling the water cycle in the classroom. Use the following labels to label the model: *Clouds, oceans, rain, evaporation, land*

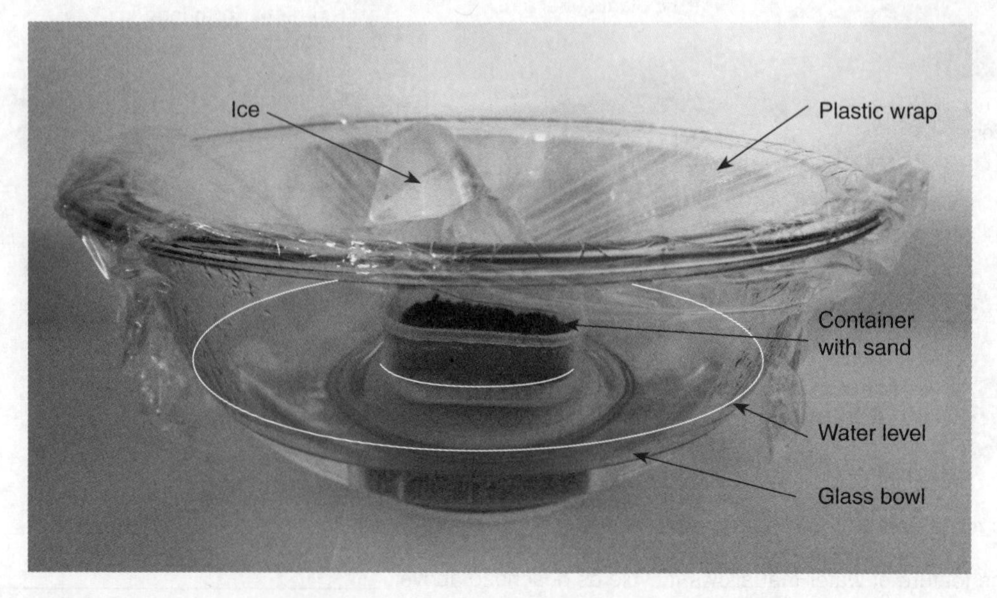


(b) i. What represents the clouds in the model? ______

ii. Explain how the "clouds" model this part of the actual water cycle: ______


**ISBN: 978-1-927309-37-7**

# 83 The Rock Cycle

**Key Idea**: The continual erosion, burial, melting, and reforming of the Earth's rocks forms a continuous cycle.

- The Earth's many rock types can be grouped as **igneous**, **metamorphic**, and **sedimentary rocks**.
- These rocks form in a continuous cycle. Erosion of surface rocks produces sediments. Burial of these transforms them into sedimentary rocks. Heat and pressure within the Earth can then transform pre-existing rocks to form metamorphic rocks such as slate and schist. Contact with magma may melt the rock which may then form as volcanic extrusions or plutonic intrusions (rocks formed underground e.g. cooling magma).
- When rocks are exposed at the surface, they are then subjected to the physical, chemical, and biological processes collectively known as **weathering**. This cycle of rock formation, exposure, weathering, erosion, and deposition is known as the **rock cycle**.

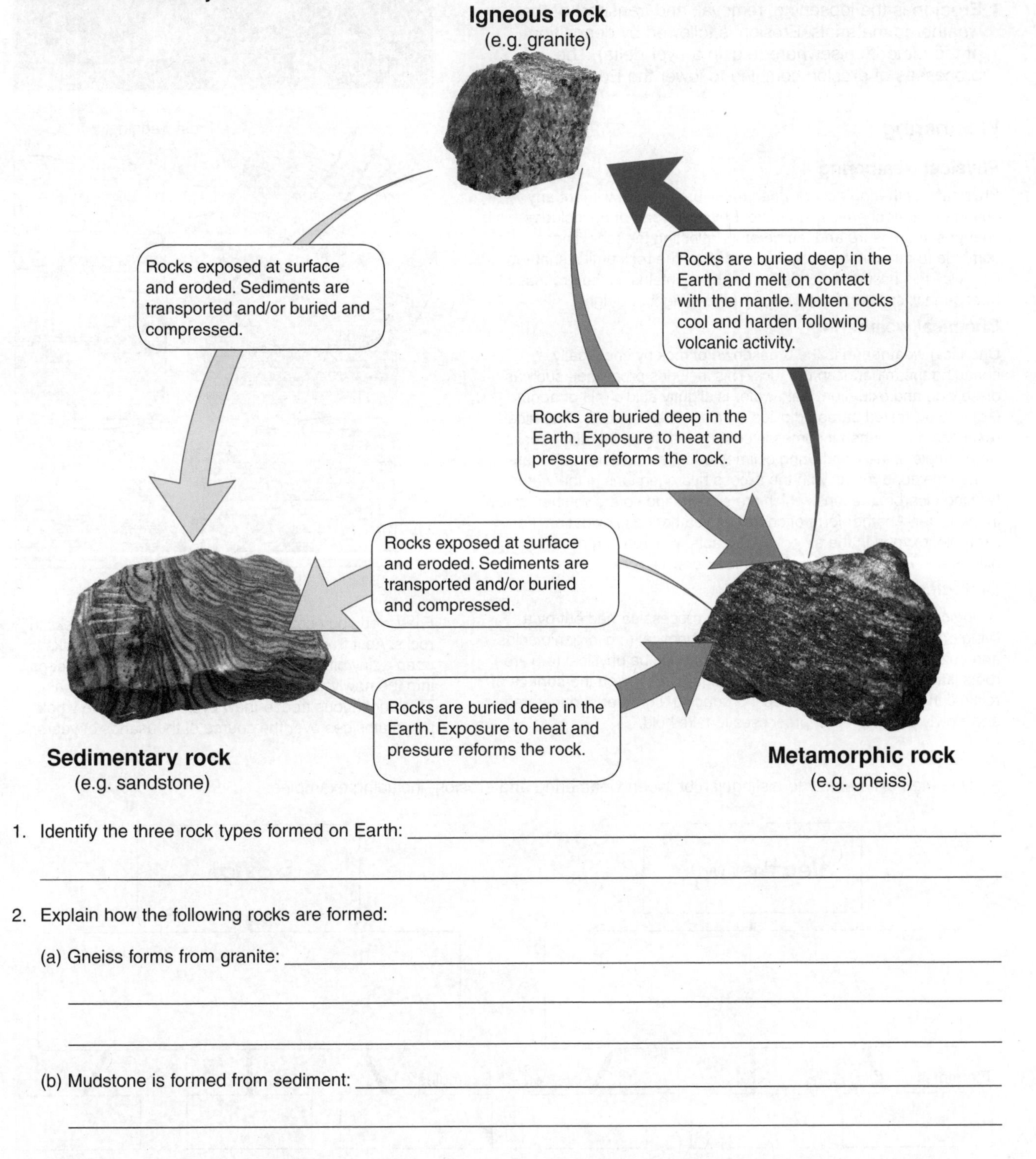


1. Identify the three rock types formed on Earth: ______

2. Explain how the following rocks are formed:

(a) Gneiss forms from granite: ______

(b) Mudstone is formed from sediment: ______


**ISBN: 978-1-927309-37-7**

# 84 Weathering and Erosion

**Key Idea**: Weathering is the physical, chemical, and biological breakdown of rock. Erosion is the transport of the weathered material to other places.

- **Weathering** and **erosion** are important processes in shaping the Earth's surface. They usually work closely together and are often confused as the same thing. It is important to remember weathering and erosion are quite different and separate processes.
- **Weathering** is the chemical, physical, and biological process of breaking rocks and minerals down into smaller pieces, e.g. dissolving limestone by rain.
- **Erosion** is the loosening, removal, and transport of the weathered materials. Erosion is followed by deposition of the material elsewhere (e.g. in a river delta). The processes of erosion combine to lower the Earth's surface.

Frost wedging

Till Niermann

## Weathering

### Physical weathering

Physical weathering occurs when rocks break apart without any change to their chemical structure. Physical weathering includes changes in pressure and temperature affecting the rock. These combine to put constant physical stress on the rock until it shatters. One way this occurs, especially in high mountains, is the process of freeze-thaw, causing frost wedging (above right and right).

### Chemical weathering

Chemical weathering is the breakdown of rock by chemically changing the minerals in the rock. This includes processes such as dissolving and oxidation. Rain water is slightly acidic (pH of about 6 due to dissolved carbon dioxide forming carbonic acid). Chemical reactions occur when it comes in contact with the minerals in rocks. An example is the weathering of limestone. The calcium carbonate in the limestone reacts with the excess hydrogen ions in the water forming bicarbonate ions, which are soluble and so are washed away in the water. Another form of chemical weathering is oxidation. For example, oxygen in the air or water reacts with iron in rocks forming oxides and rust, which can slowly break down a rock.

### Biological weathering

Biological weathering is any weathering process carried out by a living organism. It can therefore also be chemical (e.g. organic acids and enzymes produces by organisms) or it can be physical (e.g. tree roots lifting pavements). Lichens and algae growing on the surface of rocks can slowly etch the surface, producing a greater surface area and slowly allowing other processes to take hold.

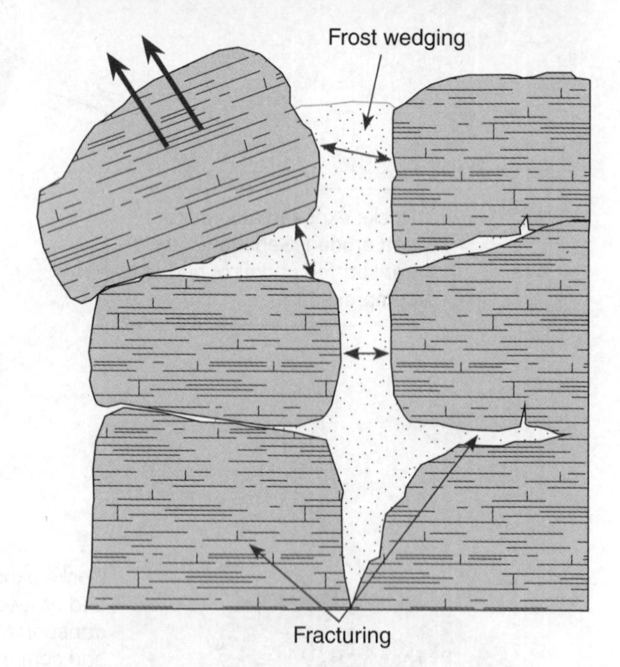


Frost wedging occurs when water seeps into cracks in rocks. As it freezes it expands and forces the cracks open a tiny bit more. When the ice thaws, water seeps into the newly widened cracks, ready to freeze again. This continuous freeze-thaw cycle can crumble whole mountainsides over the course of thousands of years.

1. Use the boxes below to distinguish between weathering and erosion, including examples:

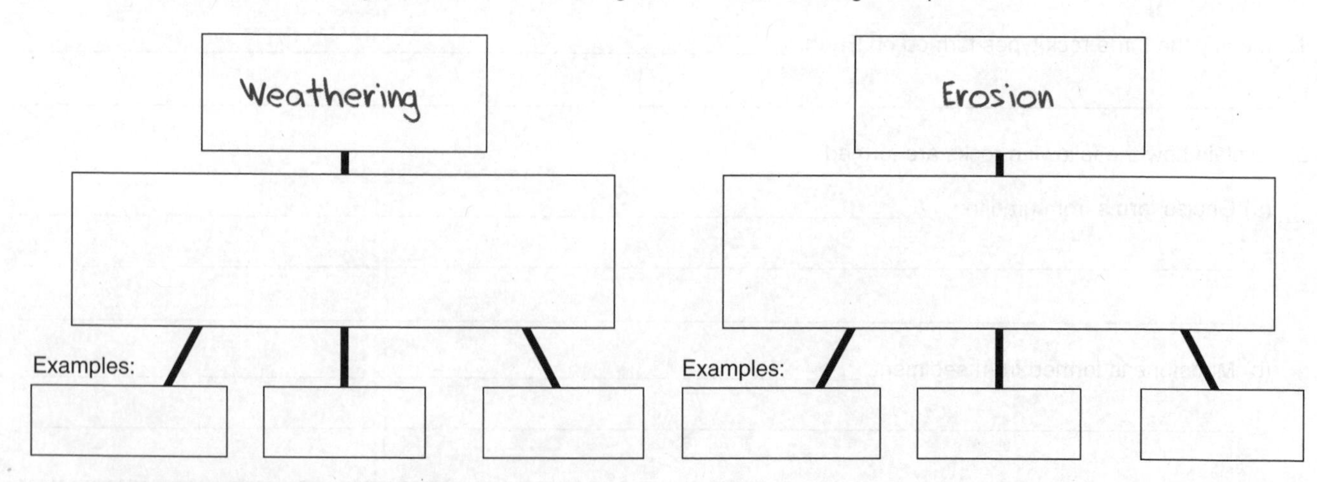

ISBN: 978-1-927309-37-7

## Erosion

### Water

Water is a major contributor to erosion. The force of water movement can be small, moving small pebbles or fine sediment, or it can be massive, with flood waters moving huge boulders down river beds. The continual flow of water over a stony river bed or the continual action of waves can shape the rocks into rounded smooth shapes. The transport of sediment by rivers plays a major role in layering sediment and debris according to size and weight. In a delta, where water flow slows, heavier debris drops out of the water flow first with the finer sediment being carried further out to sea.

### Wind

Much like water, wind has the power to carry sediment and shape rocks. In deserts, where there is a lack of vegetation to cover the dry ground, sand is picked up by the wind and transported across continents and hurled at cliffs and mountains, physically weathering their rocks and helping to carve them into the most extraordinary shapes.

### Glaciers

Glaciers form in areas where snow and ice accumulate year round and flow slowly downhill. Currently about 10% of the Earth's surface is covered by glaciers. At the height of the last glaciation this figure was about 30%. Glaciers have enormous erosive power, and are capable of moving massive boulders and scouring out huge U shaped valleys in the process. The rock frozen in the ice acts as a most effective abrasive as it slowly moves down the valley, removing soil and earth as it goes.

2. Explain how frost wedging breaks down large rocks: ________________

3. Describe two ways in which chemical weathering breaks down rocks:

   (a) ________________

   (b) ________________

4. Explain how biological weathering can incorporate both physical and chemical weathering: ________________

5. Describe how each of the following take place:

   (a) Water erosion: ________________

   (b) Wind erosion: ________________

   (c) Glacial erosion: ________________

6. Express the chemical weathering of limestone by water as a chemical equation:


**ISBN: 978-1-927309-37-7**

# 85 Modeling Erosion

**Key Idea**: The erosive action of water can be modeled using a stream tray and different mixes of sand and shingle.

Stream trays are a simple way of modeling and observing how rivers develop, erode the land, and deposit sediment. Any long tray can be used as long as an outlet for the water is drilled at the lower end.

- Set up the tray by placing it on slight angle with the outlet at the lower end. Place gravel or sand in the tray and work the gravel so that it becomes thinner near the lower end.
- The simplest set up is to make a gravel "mountain" near the upper end of the tray to initially block water flow, forming a "lake".

Observe how the lake overflows and forms a channel (below). The variation of sand and stone in the gravel and shape of the mountain range influences how the channel will form. Try this several times with different shaped mountain ranges. Using different materials can simulate the effect of a river meeting different types of rock. Softer layers of gravel and sand could be topped with a harder layer of clay, and vice versa.

Another simulation is to premake a river meander and observe how it changes over time as water moves at different speeds around the bends, depositing and eroding material at different places. Other investigations include adding vegetation to observe its effect on erosion and river channel formation.

River meanders occur because of deposition on the inner bank and erosion on the outer bank. Over time, river channels can meander vast distances and either meet up or close up, forming oxbow lakes (below). This usually occurs in the shallower end-stages of a river's course.

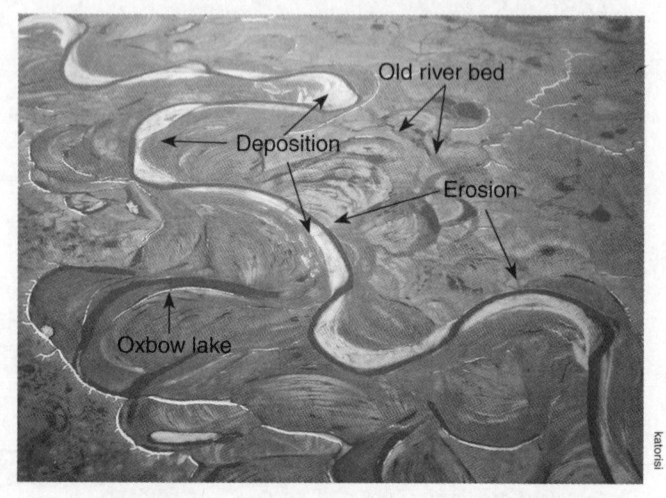


1. (a) Before you do the modeling, predict the following: the effect of flow velocity on erosion, where the most erosion will occur in a meandering stream, the effect of layering materials of different hardness. Record your predictions on the sheet with the results of your modeling (see b). Were your predictions supported?
   (b) On a separate sheet, draw the river flow through the different river models you have made. Include notes on how the model changes over time and how features such as large rocks and vegetation affect the river's shape.

2. Study the notes on this page and your river models and use them to explain the pattern of deposition and erosion in a meandering river:

ISBN: 978-1-927309-37-7

# 86 Water's Role in the Melting of Rocks

**Key Idea**: Water lowers the melting point of the rocks in the lithosphere and asthenosphere, allowing them to melt at the temperatures encountered there.

- The lithosphere, asthenosphere and the mantle are solid, they are not liquid or molten. Yet volcanoes located above a hot spot or a subduction zone spew out molten lava (molten rock above ground) and magma (molten rock below ground) oozes out of fissures along mid-ocean ridges.
- This implies special conditions must be being encountered for magma to form. Indeed there are three conditions that cause the local melting of rocks and the formation of magma chambers:
  1. **Heat** (the most obvious but not the most important cause).
  2. **Decreased pressure**: The pressure on hot material decreases as it rises towards the crust, allowing particles more room to move about. Decreased pressure is responsible for magma forming at mid ocean ridges.
  3. **Addition of water:** Water disrupts the bonds in rocks and lowers their melting point. This can be modeled using ice and salt (NaCl). In this model ice acts as the rock in the mantle and salt as the water held inside the rock. The addition of water is responsible for magma forming at subduction zones.

## Sodium chloride and water solution

- Several solutions were made using fresh water and sodium chloride salt to produce concentrations of 0 g/L, 50 g/L, 100 g/L, 150 g/L, 200 g/L, and 250 g/L.
- These were poured into identical beakers and placed into a freezer at -50°C. The temperature of each solution was measured to record its freezing (and thus melting) point.
- The results are shown below:

| Solution concentration (g/L) | Freezing/melting point (°C) |
|---|---|
| 0 | 0 |
| 50 | -3 |
| 100 | -6.5 |
| 150 | -10.9 |
| 200 | -16.5 |
| 250 | -24.5 |

NEED HELP? See Activities 4 & 7

## Water and plate tectonics

- As a tectonic plate descends in a subduction zone it drags down water-laden sediment and rocks. The rocks are heated and squeezed and at a depth of about 100 km the water is driven out and begins to rise through the rock as vapor.
- As it rises, the vapor encounters hotter rocks above that are close to their melting point. The water vapor enters these rocks, lowering their melting point and producing magma.

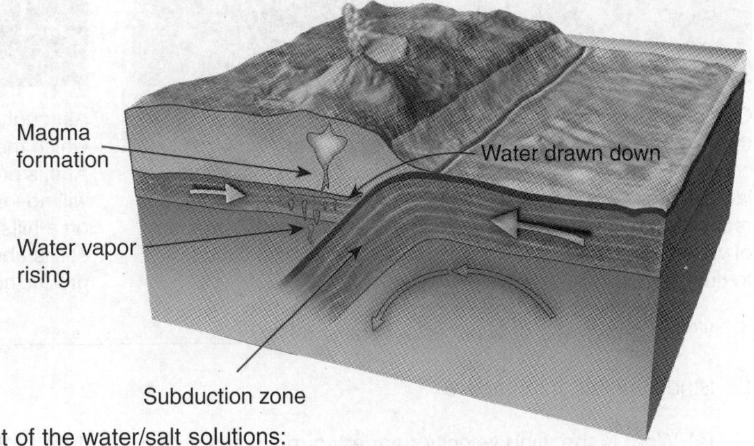


1. Use the tabulated data to graph the melting point of the water/salt solutions:
2. Describe the shape of the graph: ____________________
3. Explain why water lowers the melting point of rocks: ____________________

____________________

____________________


ISBN: 978-1-927309-37-7


PRACTICES  PRACTICES  CCC SF DATA

# 87 Moisture Content and Soil Erosion

**Key Idea**: The cohesive and adhesive properties of water bind soil particles together. This binding can influence the rate of erosion in soils.

## Effect of soil moisture on wind erosion

Water plays a major role in binding soil together. Dry soils are easily blown away by the wind. In deserts, this effect can be seen in the movement of sand dunes. The dune slowly creeps in the direction of the wind flow as dry sand is blown up a dune and tumbles over the front. When water is at low to medium concentrations, its cohesive and adhesive properties bind soil particles together into clumps, increasing soil stability and making it harder for the wind to move the soil particles.

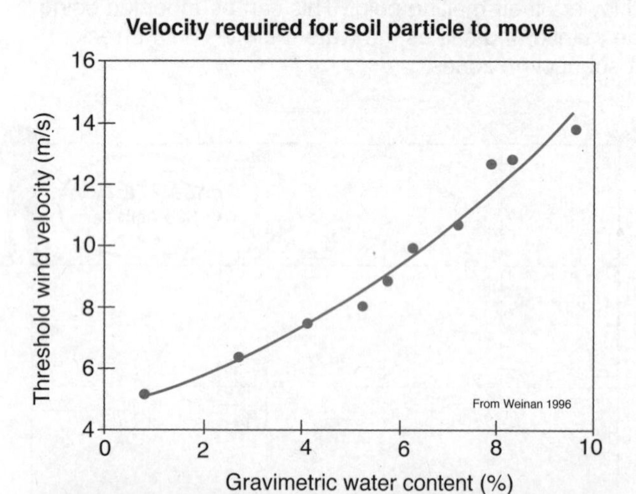


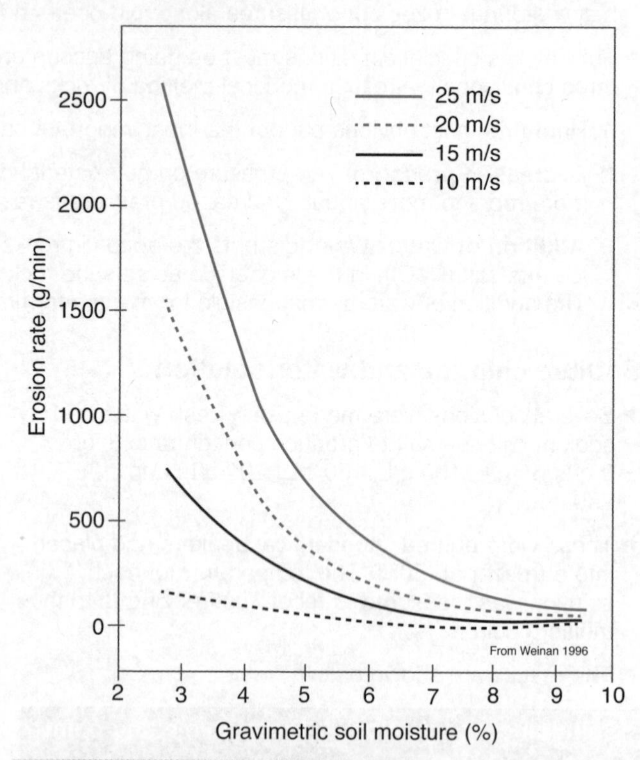

Cropping, plowing, and planting in many farms occurs in the summer months when soil moisture is low (but growing conditions are favorable). Valuable top soil can be lost if the land is struck by high winds at this time. Soil moisture may be reduced due to drought, poor irrigation management, or a lack of vegetation cover. The lack of moisture in soil also contributes to dust storms.

Although water helps bind soil particles, there is a point at which the soil becomes saturated and cannot hold more water. At this point (or even before, depending on the soil type) the soil will no longer be able to maintain its structure. If this happens on a hillside then a landslide may occur. Landslides also occur if soil is above an impermeable layer that water cannot penetrate, producing a lubricated area that the soil will slide over.

1. Which properties of water affect soil stability? ______________________

2. Study the left graph above:

   (a) What is the wind velocity needed to move soil with a moisture of 3%? ______________________

   (b) What is the wind velocity needed to move soil with a moisture of 8%? ______________________

3. How many times more erosion occurs in a soil with 3% moisture than a soil with 7% moisture is a 15 m/s wind?

   ______________________

4. At what soil moisture does wind speed make little difference to erosion rate? ______________________

DATA

**ISBN: 978-1-927309-37-7**

# 88 Chapter Review

Summarize what you know about this topic so far under the headings provided. You can draw diagrams or mind maps, or write short notes to organize your thoughts. Use the introduction and the images and hints included to help you:

Properties of water

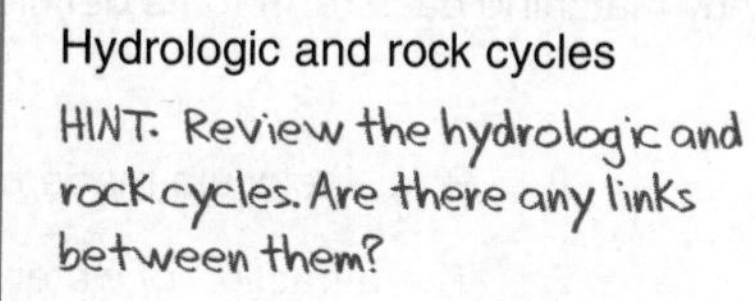

HINT: Describe the properties of water including its structure and polarity.

Hydrologic and rock cycles

HINT: Review the hydrologic and rock cycles. Are there any links between them?

Water and erosion

HINT: What is water's role in erosion?


**ISBN: 978-1-927309-37-7**

# 89 KEY TERMS AND IDEAS: Did You Get It?

1. Test your vocabulary by matching each term to its definition, as identified by its preceding letter code.

|  |  |  |
|---|---|---|
| adhesion | A | Polar molecule made of two hydrogen atoms and one oxygen atom. |
| cohesion | B | The attraction of water molecules to themselves, caused by their polarity. |
| erosion | C | The chemical, physical, and biological process of break rocks and other minerals down into smaller pieces. |
| hydrologic cycle | D | The solid form of water. |
| ice | E | The attraction of water to other kinds of molecules or to surfaces. |
| water | F | The loosening, removal, and transport of weathered material. |
| weathering | G | The cycling of water (in its various states) through the environment. |

2. On the diagram below add the δ+ and δ- signs to the oxygen and hydrogen atoms in the water molecules:

3. Identify the type of weathering occurring in the following photos.

(a) ______________ (b) ______________ (c) ______________

4. The diagram below shows the current shape of a short part of a river's course. Draw over the top of the diagram what you might expect the river to look like at a time in the future:

5. When the mineral peridotite is in the presence of water its melting point is about 800°C. When water is absent it melts at about 1500°C. Explain what causes this change in melting point and the effect of this on volcanic activity and magma/lava formation:

______________

______________

______________


**ISBN: 978-1-927309-37-7**

# 90 Summative Assessment

1. Wind is an agent of erosion. How much it erodes soil depends on the moisture content of the soil. The data below shows the effect of wind velocity on the erosion of soil:

| Wind velocity (m/s) | Soils lost ($kg/m^2/min$) | |
|---|---|---|
| | 2.67% moisture | 5.20% moisture |
| 2 | 0.2 | 0.1 |
| 4 | 0.8 | 0.2 |
| 6 | 1.3 | 0.35 |
| 8 | 2.2 | 0.6 |
| 10 | 3.6 | 0.8 |

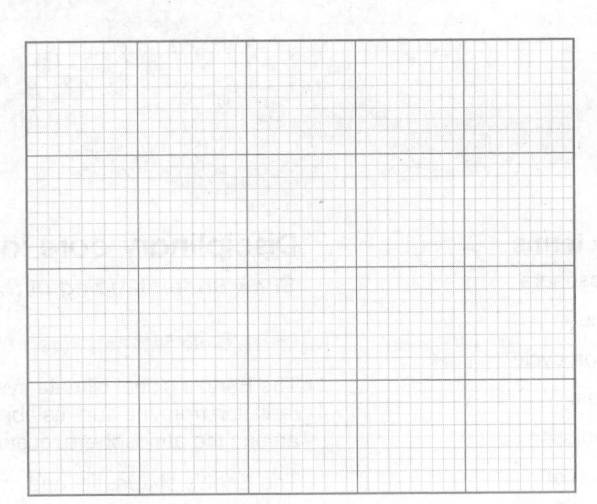

(a) Graph the data on the grid provided:

(b) Describe the shape of lines: ______________________________

(c) What is the effect of soil moisture of erosion? ______________________________

(d) What is the effect of wind velocity on erosion? ______________________________

2. Some students decided to carry out an investigation into frost wedging. They set up the experiment as shown below:

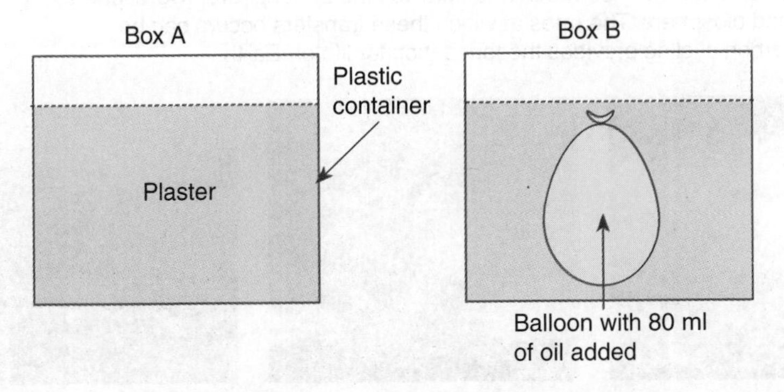


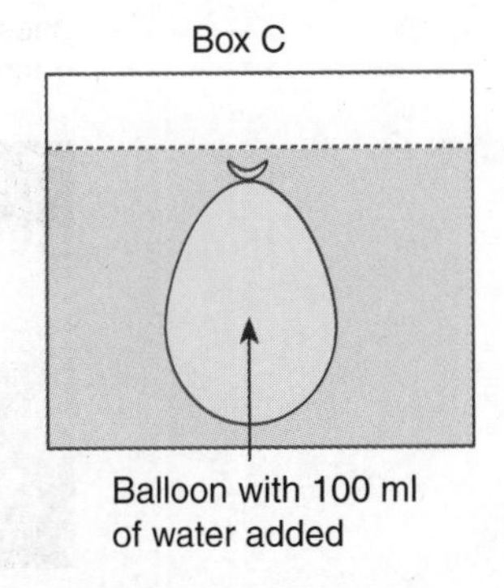


After the boxes were set up and the plaster was set, they were placed in a freezer over night.

(a) Which box(es) acted as the control for the test? ______________________________

(b) When they inspected the boxes the next day the students found the plaster in box C had split. Explain why:

______________________________

(c) Identify an error in the test carried out by the students. How would it be corrected to make this a fair test?

______________________________


ISBN: 978-1-927309-37-7

# ESS2.D ESS2.E Weather, Climate, and Biogeology

## Key terms

atmosphere
biome
carbon cycle
climate
coevolution
coral reef
equinox
global warming
Great Oxygenation Event (GOE)
greenhouse effect
greenhouse gas
season
snowball Earth
soil
solstice
tricellular model
troposphere

## Disciplinary core ideas

***Show understanding of these core ideas***

| | | Activity number |
|---|---|---|
| | **Earth's climate is driven by the Sun** | |
| ☐ | 1 The Earth's global climate systems are founded on the continual electromagnetic radiation from the sun, its absorption, reflection, storage, and redistribution among the atmosphere, ocean, and land, and its re-radiation into space. | 91 |
| ☐ | 2 The Earth's shape, tilt, and rotation influence the distribution of energy about the globe, create the seasons, and affect atmospheric circulation. | 91-94 |
| ☐ | 3 Atmospheric circulation transports warmth from the equatorial areas to high latitudes and returning cooler air to the tropics. The tricellular model of atmospheric circulation describes how atmospheric cells circulate at specific regions in the atmosphere. The circulation of air in these cells produce belts of prevailing winds around the globe. | |
| | **Organisms caused change in the early atmosphere** | |
| ☐ | 4 The concentration of gases in the Earth's atmosphere fluctuates on small scales over the short term and has changed on a large scale over the long term. Changes in gas concentrations over Earth's are related to changes in climate. | 95 |
| ☐ | 5 The increase in concentration of oxygen on Earth was the result of organisms that captured carbon dioxide and released oxygen. The presence of oxygen in the atmosphere triggered a snowball Earth event that lasted 300 million years and was important in the evolution of eukaryotes and multicellular life. | 96 97 98 |
| ☐ | 6 The current oxygen concentration in the atmosphere is ~21%. Atmospheric oxygen peaked at about 35% 300 million years ago in the Carboniferous. | 95 |
| ☐ | 7 Carbon cycles at different rates within and between the atmosphere, hydrosphere, geosphere, and biosphere. The rates at which these transfers occurs can be quantified. Carbon cycling provides the foundation for life on Earth. | 99 |

PlaneMad cc 3.0

| | | Activity number |
|---|---|---|
| | **Human activity has affected climate** | |
| ☐ | 8 The greenhouse effect is the natural process by which heart is retained in the atmosphere by greenhouses gases such as methane and carbon dioxide. Human activities have resulted in an increase in atmospheric carbon dioxide concentration. This is causing an increase in average global temperatures or global warming and this is affecting Earth's climate systems. | 100 |
| | **Current models predict that global temperatures will continue to rise** | |
| ☐ | 9 Current climate models predict that global temperatures will continue to rise, although regional climate changes will be highly variable. | 100 |
| ☐ | 10 Predicted outcomes of climate models depend on the volumes of greenhouse gases added annually and how these are absorbed by the oceans and biosphere. | 100 |
| | **The Earth's surface and the life that exists on it coevolve** | |
| ☐ | 11 Feedback between the biosphere and Earth's other systems is responsible for the continual coevolution of life and the Earth's surface. The presence of organisms influences how landforms develop by influencing sediment deposition or water movements. The development of these landforms influences the evolution of life. | 96 101 102 |

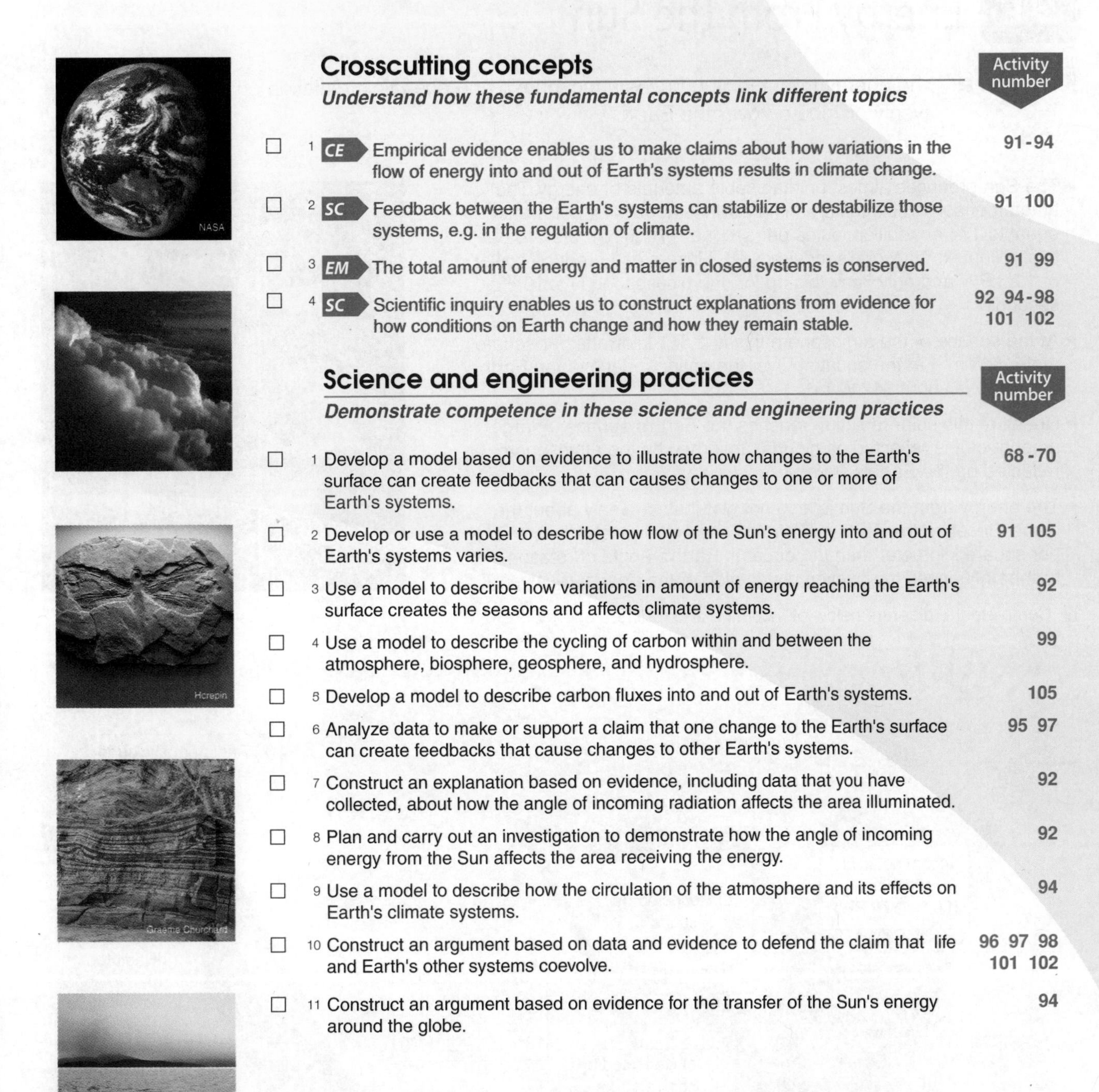


## Crosscutting concepts

***Understand how these fundamental concepts link different topics***

| | | Activity number |
|---|---|---|
| ☐ | 1 CE Empirical evidence enables us to make claims about how variations in the flow of energy into and out of Earth's systems results in climate change. | 91-94 |
| ☐ | 2 SC Feedback between the Earth's systems can stabilize or destabilize those systems, e.g. in the regulation of climate. | 91 100 |
| ☐ | 3 EM The total amount of energy and matter in closed systems is conserved. | 91 99 |
| ☐ | 4 SC Scientific inquiry enables us to construct explanations from evidence for how conditions on Earth change and how they remain stable. | 92 94-98 101 102 |

## Science and engineering practices

***Demonstrate competence in these science and engineering practices***

| | | Activity number |
|---|---|---|
| ☐ | 1 Develop a model based on evidence to illustrate how changes to the Earth's surface can create feedbacks that can causes changes to one or more of Earth's systems. | 68-70 |
| ☐ | 2 Develop or use a model to describe how flow of the Sun's energy into and out of Earth's systems varies. | 91 105 |
| ☐ | 3 Use a model to describe how variations in amount of energy reaching the Earth's surface creates the seasons and affects climate systems. | 92 |
| ☐ | 4 Use a model to describe the cycling of carbon within and between the atmosphere, biosphere, geosphere, and hydrosphere. | 99 |
| ☐ | 5 Develop a model to describe carbon fluxes into and out of Earth's systems. | 105 |
| ☐ | 6 Analyze data to make or support a claim that one change to the Earth's surface can create feedbacks that cause changes to other Earth's systems. | 95 97 |
| ☐ | 7 Construct an explanation based on evidence, including data that you have collected, about how the angle of incoming radiation affects the area illuminated. | 92 |
| ☐ | 8 Plan and carry out an investigation to demonstrate how the angle of incoming energy from the Sun affects the area receiving the energy. | 92 |
| ☐ | 9 Use a model to describe how the circulation of the atmosphere and its effects on Earth's climate systems. | 94 |
| ☐ | 10 Construct an argument based on data and evidence to defend the claim that life and Earth's other systems coevolve. | 96 97 98 101 102 |
| ☐ | 11 Construct an argument based on evidence for the transfer of the Sun's energy around the globe. | 94 |

# 91 Energy From the Sun

**Key Idea**: The energy from the Sun drives atmospheric and oceanic circulation on Earth. It is not evenly distributed over the Earth.

- The Sun produces almost unimaginable amounts of energy. The amount of solar radiation reaching Earth is about 174 petawatts, equal to 174 quadrillion joules per second ($174 \times 10^{15}$ J/s). To put this in context the world's most powerful lasers can produce power of 1.25 PW and only keep this up for one picosecond ($1 \times 10^{-12}$ seconds).
- At the surface of the atmosphere this is 1.361 kilowatts per square meter (kW $m^{-2}$) at the equator. Over the entire surface of the Earth the figure is about 342 W/$m^2$.
- Not all of this solar radiation reaches the Earth's surface. A large amount of it is reflected off clouds, absorbed by the atmosphere, or reflected off the Earth's surface.
- The energy from the Sun is also not distributed evenly about the globe. Because the Earth is spherical the poles receive less energy per square kilometer than the equator. Earth's angle of rotation further influences the uneven distribution of the energy received.

1. Complete the diagram below of incoming and radiated energy.

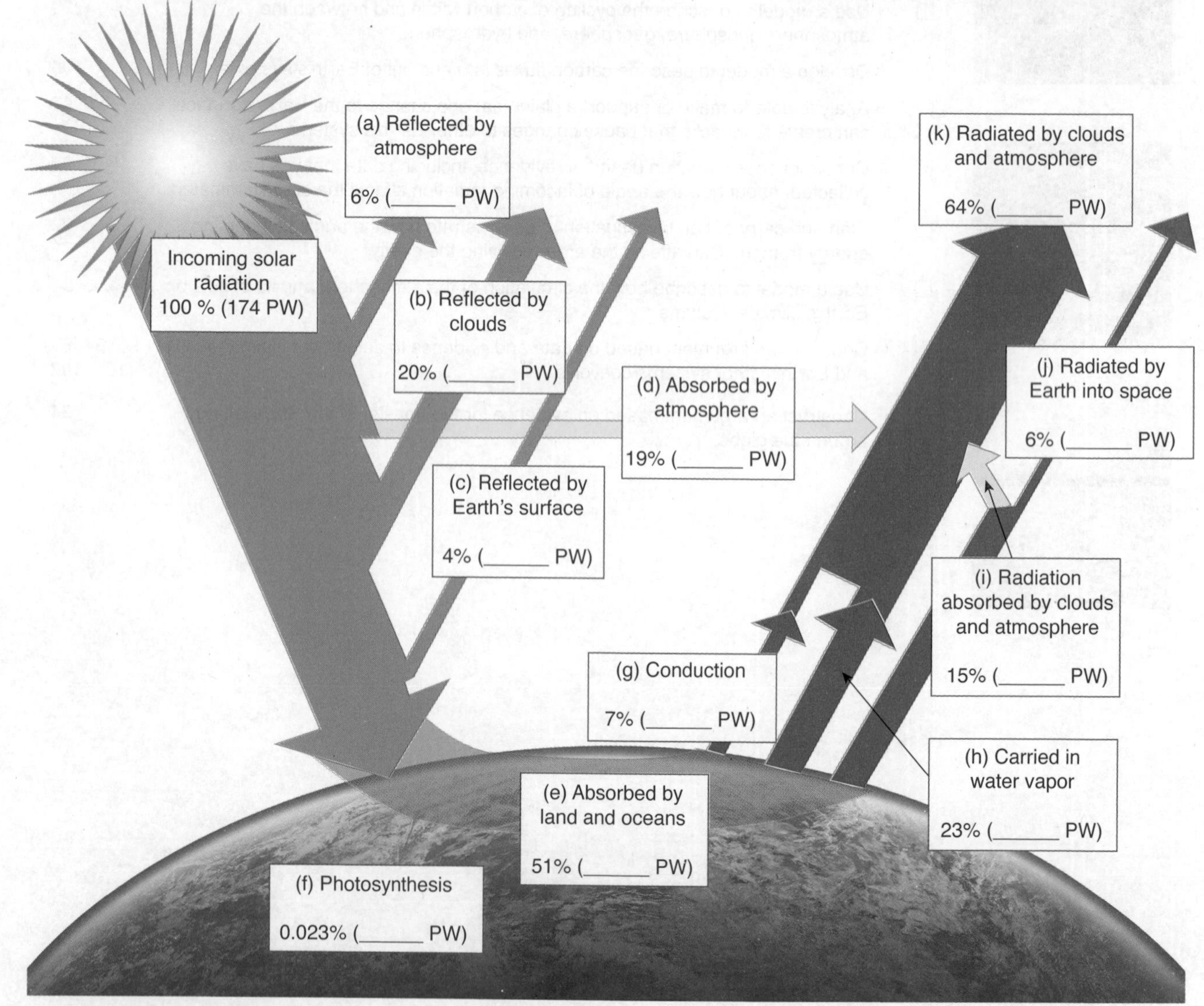


- About 51% of the incoming solar radiation reaches the Earth's surface. Some (0.023%) of this is used by photosynthesis in plants to build organic molecules. The rest drives atmospheric winds and ocean circulation and eventually radiated back into space.

ISBN: 978-1-927309-37-7

- The curvature of the Earth changes the angle at which solar radiation hits the planet's surface. At the equator, the Sun is directly overhead and solar radiation hits the Earth perpendicular to the surface (at noon). The Tropic of Cancer and the Tropic of Capricorn are lines of latitude on the Earth that denote the limit of the Sun being overhead at noon (depending on the season).
- Outside the tropics, solar radiation continues to hit the Earth more and more obliquely towards the poles until it hits virtually parallel with the surface. This makes an enormous difference to the amount of useful solar radiation received. At the equator, almost 100% of the light to reach the surface is available, while at the poles almost none is. Moreover, light at the poles has to travel through more of the atmosphere and so even less light actually reaches the surface.

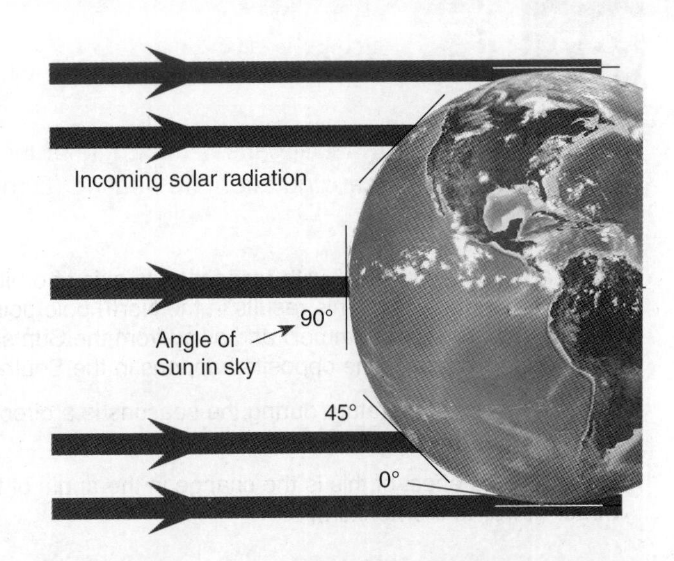

Below the Arctic and Antarctic circles, the Earth receives only about 40% of the solar energy that is received at the equator.

The tropics receive the full amount of sunlight and energy available. This causes heating, which carries water into the air, creating a hot, wet climate.

Differential heating between the tropics and poles drives air currents from the tropics towards the poles. This is because as air rises at the equator and falls at the poles.

2. (a) Describe the relationship between the curvature of the Earth and the amount of solar energy received at the surface:

(b) Explain how this affects the climate at the following points:

i: The equator:

ii: The poles:

3. The amount of solar energy received at any point on the Earth can be calculated using the equation:

$$A' = A \times \text{cosine } x$$

Where **A** = the amount of solar radiation at the equator, **A'** = the amount of solar radiation at latitude x.
**x** = the latitude of the point on the Earth.

For each of the following lines of latitude, calculate the amount of solar energy received, assuming 700 $W m^{-2}$ of energy is received at the equator and no tilt to the Earth (i.e. at the spring and fall equinoxes). The first one is completed for you:

(a) Tropic of Cancer (23.5° North): $A' = 700 \times \cos 23.5 = 700 \times 0.917 = 641.9 \ W/m^2$

(b) 45° North:

(c) Arctic circle (66° North):

(d) North pole (90° North):


**ISBN: 978-1-927309-37-7**

# 92 Seasons

**Key Idea**: The Earth's tilt causes the angle of the Sun in the sky at noon to change over the course of a year, causing alternate heating (summer) and cooling (winter).

- The Earth is tilted at 23.4° with respect to its axis of orbit around the Sun. The angle remains the same as it travels around the Sun. This results in the North pole pointed towards the Sun during the months of June, July, and August, (the northern summer) and away from the Sun six months later during December, January, and February (the northern winter). The opposite happens in the Southern Hemisphere (below).
- The change in temperature during the seasons is a direct result of this change in whether the pole points towards the Sun or away from it.
- The observed effect of this is the change in the angle of the Sun at noon during the summer (more overhead) and winter (lower to the horizon).

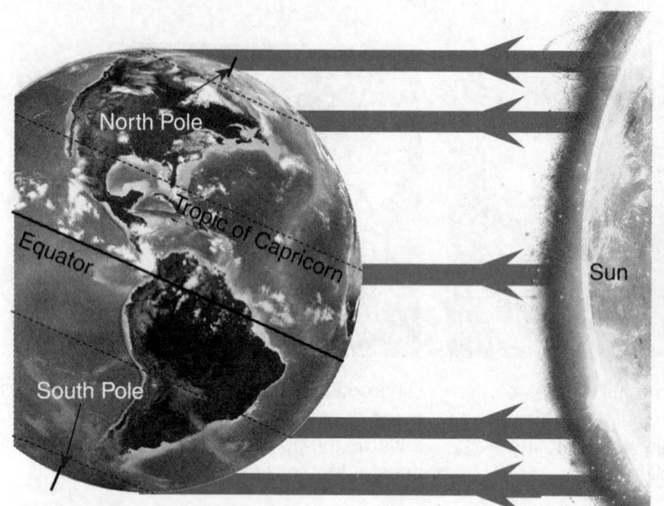


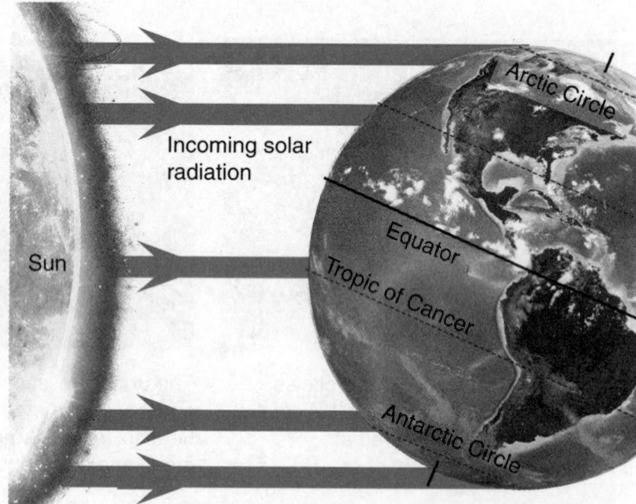


**Northern Hemisphere: Summer**
**Southern Hemisphere: Winter**

**Northern Hemisphere: Winter**
**Southern Hemisphere: Summer**

- The energy received from the Sun is more or less constant, so why is it so much colder during the winter? The answer is that the energy from the Sun is spread over a much larger area during the winter because the angle of sunlight hitting the ground so much less in winter.
- For example a shadow at noon in the summer is short. The shadow represents the area of ground that the sunlight would have hit. In winter the shadow is much longer and therefore the area of ground the sunlight would have hit is greater. Thus the energy per square meter is less, resulting in a lower overall temperature.

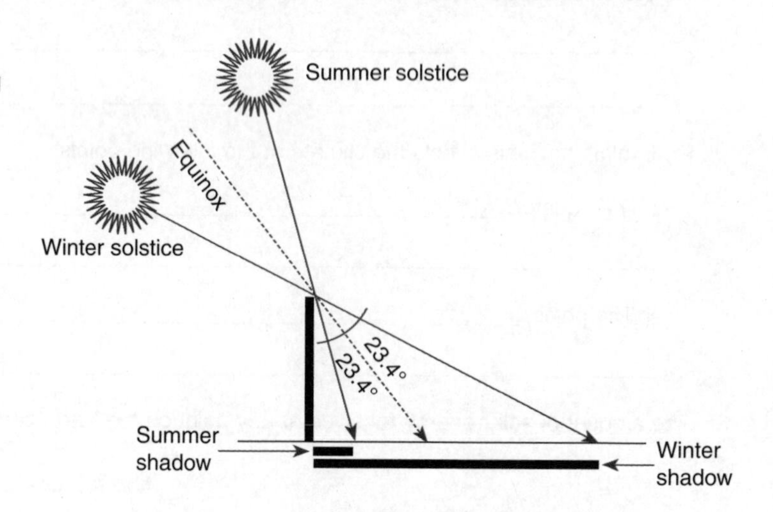


1. What causes the change in seasons on Earth? ______________________

2. What effect does the change of seasons have on the angle of the Sun above the horizon at noon?

______________________

3. Compare the length of a shadow at noon in summer and winter: ______________________




**ISBN: 978-1-927309-37-7**

4. (a) Using a stand and protractor to measure angle, set up a flashlight pointing straight down (90° angle) at a piece of grid paper from a distance of 30 cm (a darkened room works best) as shown in the diagram below:

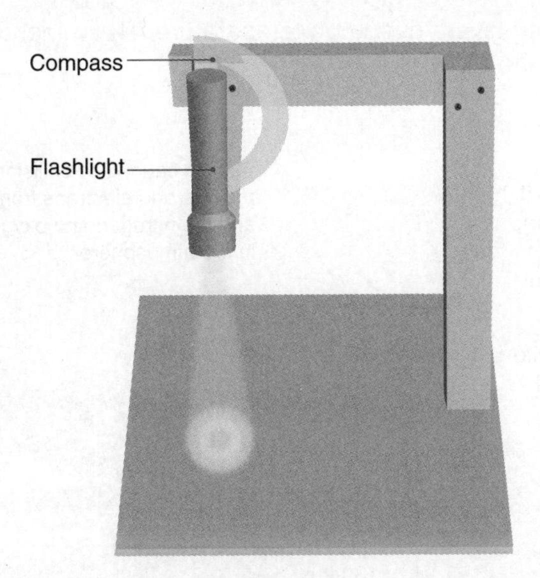


(b) Draw around the illuminated part of the grid paper and then use the grid paper to help find the area illuminated.

Area illuminated (90°): ______________________________

(c) Tilt the clamp stand to a 66° angle, measuring 30 cm from the center of the flashlight front along the angle to the center of the illuminated area on the grid paper. Again, draw around and calculate the area of grid paper illuminated.

Area illuminated (66°): ______________________________

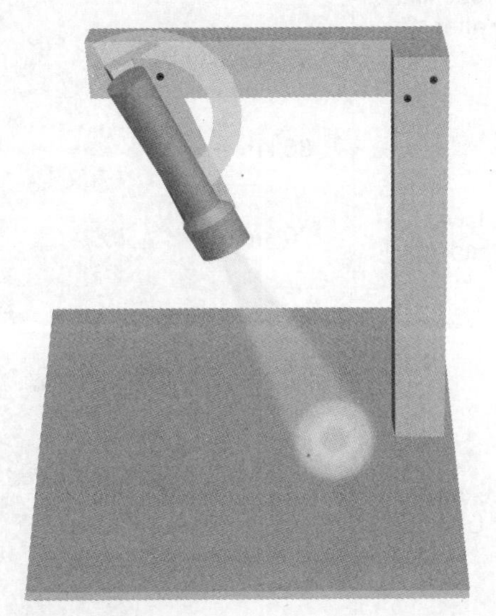

(d) Repeat this procedure for 45° and 33°.

Area illuminated (45°): ______________ Area illuminated (33°): ______________

(e) How does the area of graph paper illuminated change with the angle of the torch? ______________________________

______________________________

______________________________

(f) Calculate the amount of light energy (watts) received per $cm^2$ on the grid paper. You could use a light meter to measure the light output of the flashlight if you don't know it or (assuming its output is constant) start with a general light output of 3 W. (Hint if the light bulb is a 3 W bulb, each square cm of paper is receiving how many watts?)

______________________________

______________________________

Attach all your grids to this page as a record.


**ISBN: 978-1-927309-37-7**

# 93 The Earth's Atmosphere

**Key Idea**: The atmosphere is divided into layers based on temperature. The atmosphere helps to carry energy from the Sun around the globe.

- The Earth's atmosphere is a layer of gases surrounding the globe and retained by gravity. It contains roughly 78% nitrogen, 20.95% oxygen, 0.93% argon, 0.038% carbon dioxide, trace amounts of other gases, and a variable amount (average around 1%) of water vapor.
- This mixture of gases, known as air, protects life on Earth by absorbing ultraviolet radiation and reducing temperature extremes between day and night. The atmosphere consists of layers around the Earth, each one defined by the way temperature changes within its limits.
- The outermost troposphere thins slowly, fading into space with no boundary. The air of the atmosphere moves in response to heating from the Sun and, globally, the atmospheric circulation transports warmth from equatorial areas to high latitudes and returning cooler air to the tropics.

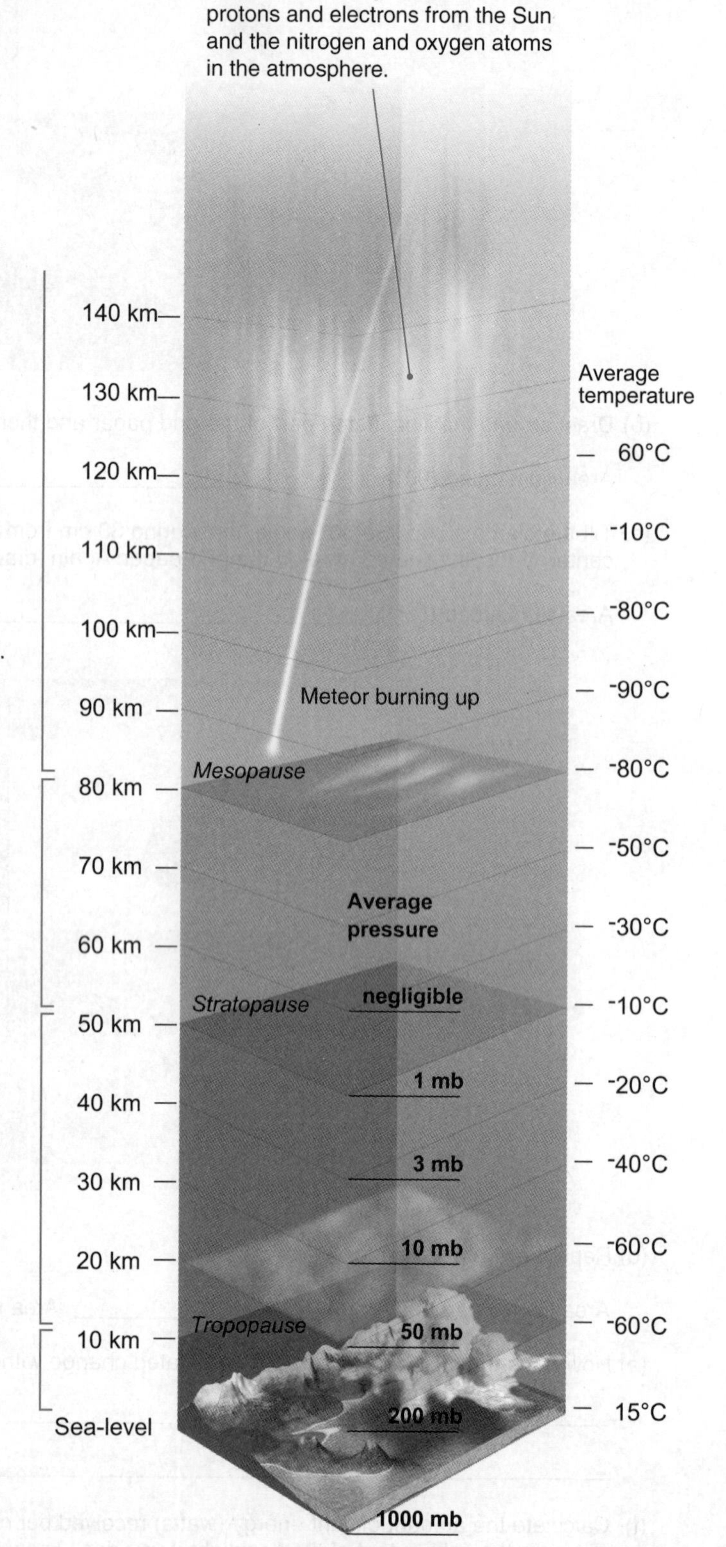


**Thermosphere**

This layer extends as high as 1000 km. Temperature increases rapidly after about 88 km.

**Mesosphere**

Temperature is constant in the lower mesosphere, but decreases steadily with height above 56 km.

**Stratosphere**

Temperature is stable to 20 km, then increases due to absorption of UV by the thin layer of ozone.

**Troposphere**

Air mixes vertically and horizontally. All weather occurs in this layer.

1. Describe two important roles of the atmosphere: ____________________

____________________

2. What causes aurora? ____________________

____________________




ISBN: 978-1-927309-37-7

# 94 Atmospheric Circulation and Climate

**Key Idea**: The rotation and differential heating of the Earth has a major effect on the circulation of the planet's atmosphere and therefore climate.

## Atmospheric circulation and the tricellular model

- High temperatures over the equator and low temperatures over the poles, combined with the rotation of the Earth, produce a series of separated atmospheric cells that circulate at specific areas in the atmosphere. This model of atmospheric circulation, with three cells in each hemisphere, is known as the **tricellular model**.

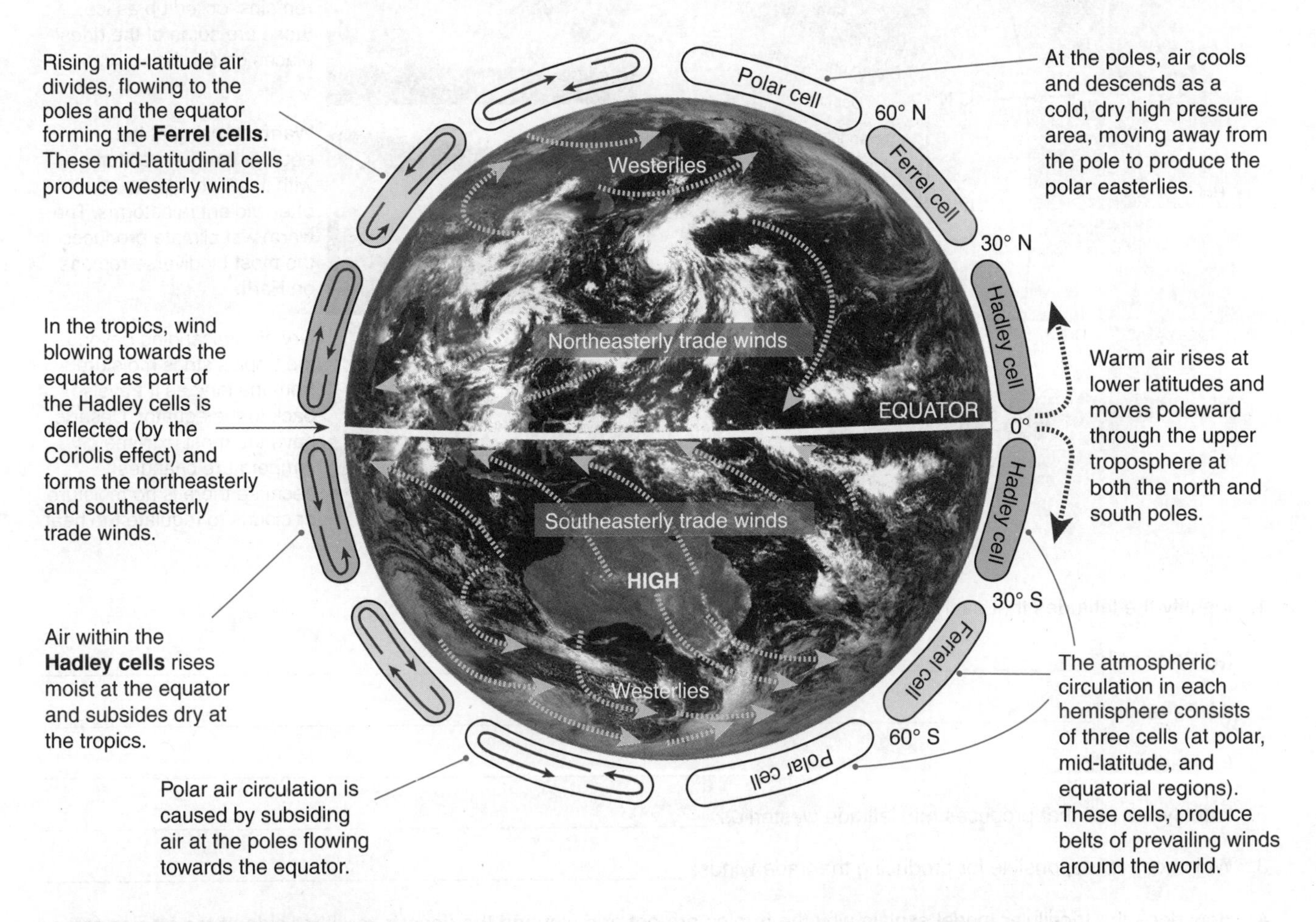


- The energy gained from solar radiation becomes progressively less from the equator to the poles. Heat gained at the tropics is transferred to cooler regions by atmospheric circulation, producing a more even spread of temperatures over the globe than would otherwise occur if there was no atmosphere. Similarly heat gained by the oceans also transfers heat about the globe. The ice caps of the poles reflect so much of the sunlight they receive that they produce a permanently cold climate.
- If there was no heat flow, the poles would be about 25°C cooler and the equator about 14° C warmer.

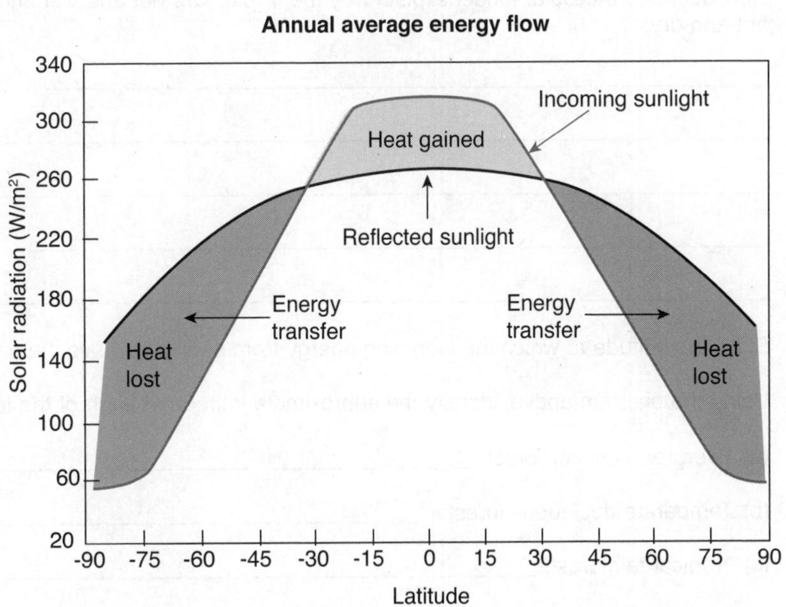



**ISBN: 978-1-927309-37-7**


PRACTICES

PRACTICES

CCC

SC
CCC

CE
WEB
94
KNOW

## Atmospheric circulation influences climates and biomes

The climate is affected by the circulation of the atmosphere. Air rising at the equator produces a warm wet climate. As the air travels away from the equator its cools and descends as dry air. The division of atmospheric circulation into three separate cells in each hemisphere produces climatic conditions that are mirrored on each side of the equator. These conditions produce **biomes**, large areas with the same climatic conditions and vegetation characteristics. Tropical rainforests are one such biome and are found circling the equator, bordered by deserts, then temperate forests and finally polar deserts.

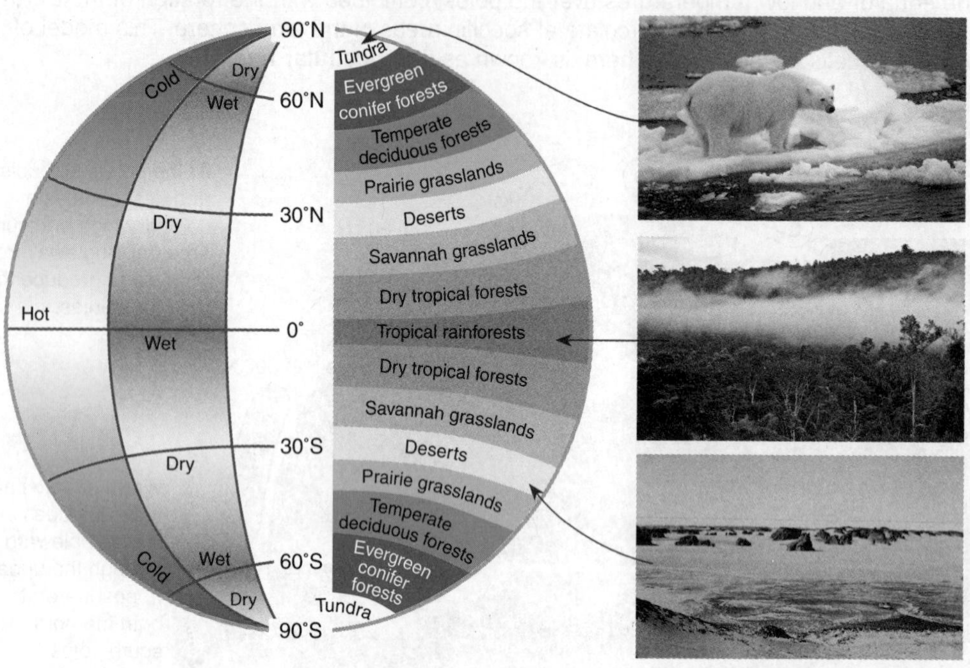


Cool dry air descending at the poles produces polar deserts. Because water remains locked up as ice, these are some of the driest places on Earth.

Warm air rising at the equator carries moisture with it, which falls as rain in often violent rainstorms. The warm wet climate produces the most biodiverse regions on Earth.

Dry air descending beyond the tropics strips moisture from the land as it travels back to the equator. Deserts have the most extreme temperature changes because there is no moisture or clouds to regulate the heat.

1. Identify the latitudes that each of the atmospheric cells lie between:

   (a) Polar cell: ______________________

   (b) Ferrel cell: ______________________

   (c) Hadley cell: ______________________

2. Identify the cell that produces mid latitude westerlies: ______________________

3. Which cell is responsible for producing the trade winds: ______________________

4. How does the tricellular model explain why the tropics are hot and wet and the deserts to either side of the tropics are hot and dry:

   ______________________

   ______________________

   ______________________

   ______________________

   ______________________

5. State the latitude at which the incoming energy from sunlight equals the energy lost? ______________________

6. Using the diagram above, identify the approximate latitude of each of the following biomes and describe their climate:

   (a) Evergreen conifer forest: ______________________

   (b) Temperate deciduous forest: ______________________

   (c) Tropical rainforest: ______________________

   (d) Deserts: ______________________


**ISBN: 978-1-927309-37-7**

# 95 Changes in Earth's Atmosphere

**Key Idea**: The concentrations of gases in the Earth's atmosphere has changed over time.

▶ The concentration of gases in the Earth's atmosphere fluctuate on small scales over the short term and on large scales over the long term. Earth's early atmosphere was very different to today's. Even after a nitrogen-oxygen atmosphere was established, the concentrations of gases such as oxygen and carbon dioxide ($CO_2$) changed greatly causing major changes to the Earth's climate.

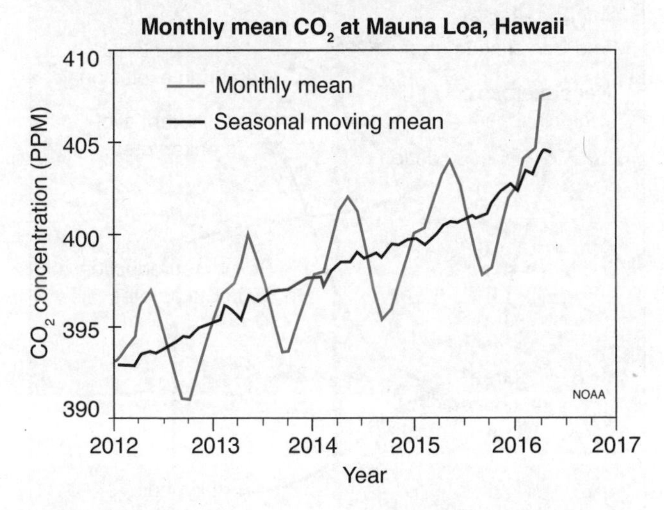


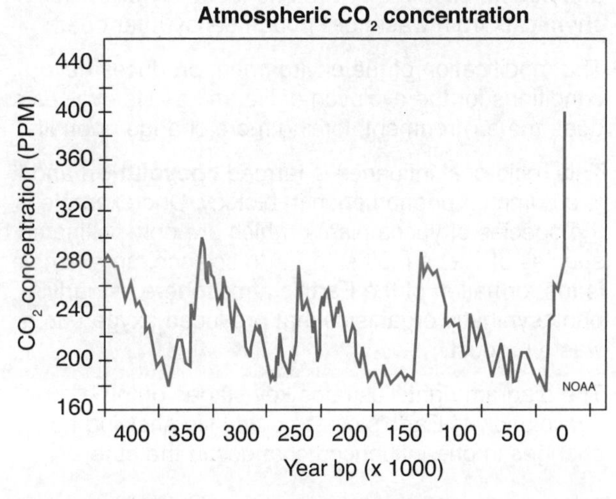


### Seasonal $CO_2$

The vast majority of the land on Earth is situated in the Northern Hemisphere. Across North America and northern Eurasia there are vast tracts of conifer and deciduous forest. During winter, plants in these forests carry out little or no significant photosynthesis. However the rest of the world keeps on respiring and decomposing, so $CO_2$ levels rise over the winter season. During spring, deciduous trees and other plants begin photosynthesizing again, removing $CO_2$ from the atmosphere. Because the winds from the Northern and Southern Hemispheres do not readily mix, this effect is not cancelled out by forests in the Southern Hemisphere (which is also mostly covered in oceans).

### Long term $CO_2$

The concentration of $CO_2$ has cycled relatively consistently over the last several hundred thousand years. The rise and fall of the atmosphere's $CO_2$ concentration correlates with the rise and fall of the Earth's surface temperature over at least the last 400,000 years. Although $CO_2$ is not the direct cause of the change in temperature (which is actually caused by changes in Earth's orbit) it does contribute to a positive feedback effect that helps warm the planet. The steep rise in $CO_2$ concentration over the last 60 years has been attributed to the burning of fossil fuels. Climate scientists fear that an atmospheric $CO_2$ concentration "point of no return" may soon be reached.

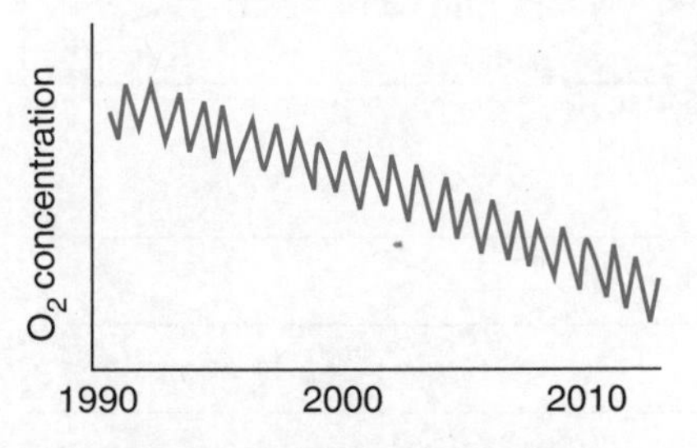


The **oxygen concentration** in the atmosphere is currently 21%. It rises and falls seasonally in a similar way to the $CO_2$ concentration. Measurements show that overall concentration has been reducing since the 1950s.

Atmospheric oxygen peaked at about 35% 300 million years ago during the Carboniferous period. During this time, huge insects, such as the giant dragonfly *Meganeura*, could exist because the high oxygen concentration allowed oxygen to easily diffuse into their bodies

The Earth has had at least three atmospheres. The earliest was probably similar to the atmospheres of the solar system's giant gas planets or the moon Titan (above). Oxygen didn't exist in the atmosphere in any large amount until about 2.4 billion years ago.

1. What causes the seasonal changes in $CO_2$ concentration? __________

2. What global changes over the last 400,000 years correlate with the rise and fall of $CO_2$ concentrations?


ISBN: 978-1-927309-37-7

# 96 The Coevolution of Earth's Systems

**Key Idea**: The Earth and life on Earth have evolved together. Life modifies the environment and the environment shapes the evolution of life.

- Life began on Earth about 3.8 billion years ago. Life takes resources from its surroundings, modifies them, and then uses them to replicate itself. In doing this, life modifies the environment, either by modifying chemicals and structures in the environment directly or by adding chemicals from waste or through decay after death.
- The modification of the environment produces new conditions for the evolution of life and as life evolves so does the environment, forcing more change upon life.
- This reciprocal influence is termed **coevolution** and it is a common phenomenon in biology. One example is the species of yucca plants, which are only pollinated by species of yucca moths. In Earth science, an example is the formation of the Earth's atmosphere by early photosynthetic organisms that produced oxygen as a waste product.
- The diagram right illustrates key stages of the coevolution of Earth's systems and life showing how changes in one influenced changes in the other.

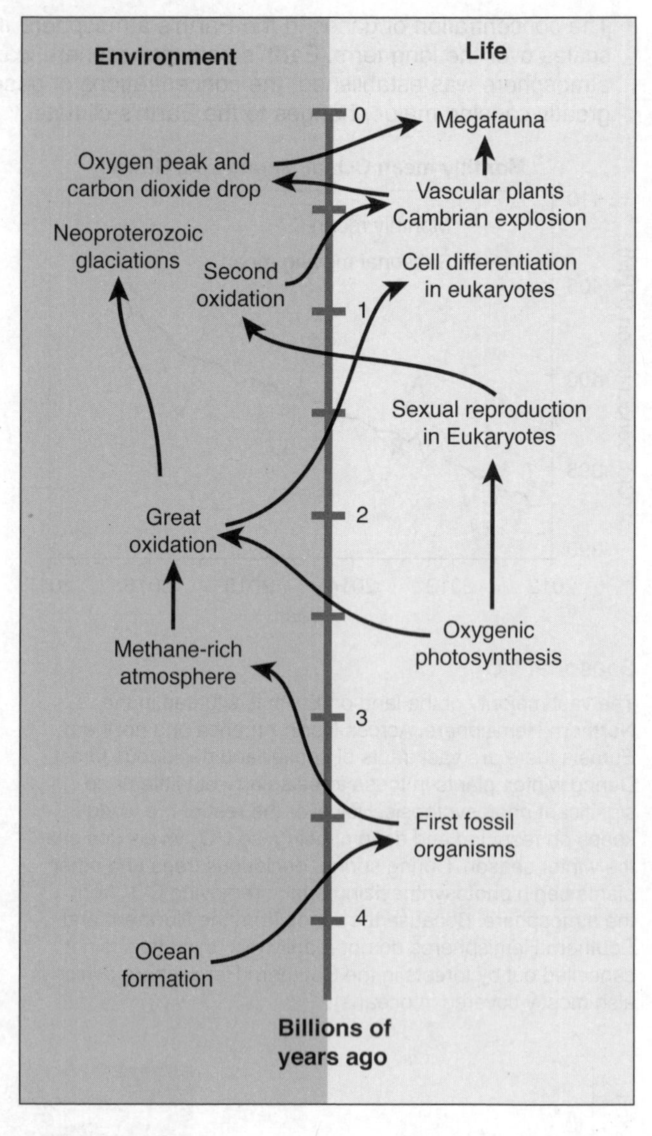

Mangroves have evolved to inhabit estuarine environments. They modify the estuary by collecting sediment and thus the environment evolves into a terrestrial instead of a marine one.

1. What is meant by coevolution?

2. Describe how life causes the environment to change:

3. How does the environment allow for life to evolve?

4. Describe a specific example of life modifying the environment:

ISBN: 978-1-927309-37-7

# 97 The Great Oxygenation Event

**Key Idea**: Oxygen produced as the waste product of oxygenic photosynthesis fundamentally changed the Earth's atmosphere and led to a snowball Earth.

- The rise of photosynthetic life produced changes to the Earth that in turn affected life itself. Photosynthesis requires electrons in order to fix $CO_2$ as organic compounds. Early photosynthetic prokaryotes probably used $H_2S$ as a source of electrons, producing sulfur compounds as waste (as do sulfur bacteria today). The evolution of an oxygen-producing photosynthetic process filled the atmosphere with oxygen and provided a new way for life to extract energy from organic molecules.
- It also changed Earth's rocks, created new minerals, formed the ozone layer, and plunged the Earth into a 300 million year long ice age. The initial rise in free oxygen is called the Great Oxygenation Event (GOE).
- The rise in oxygen allowed aerobic organisms, including eukaryotes, to evolve. There is evidence to suggest that the evolution of multicellular algae (e.g. seaweeds) triggered another ice age by extracting $CO_2$ from the atmosphere faster than it was being replaced, thus reducing its greenhouse effect.

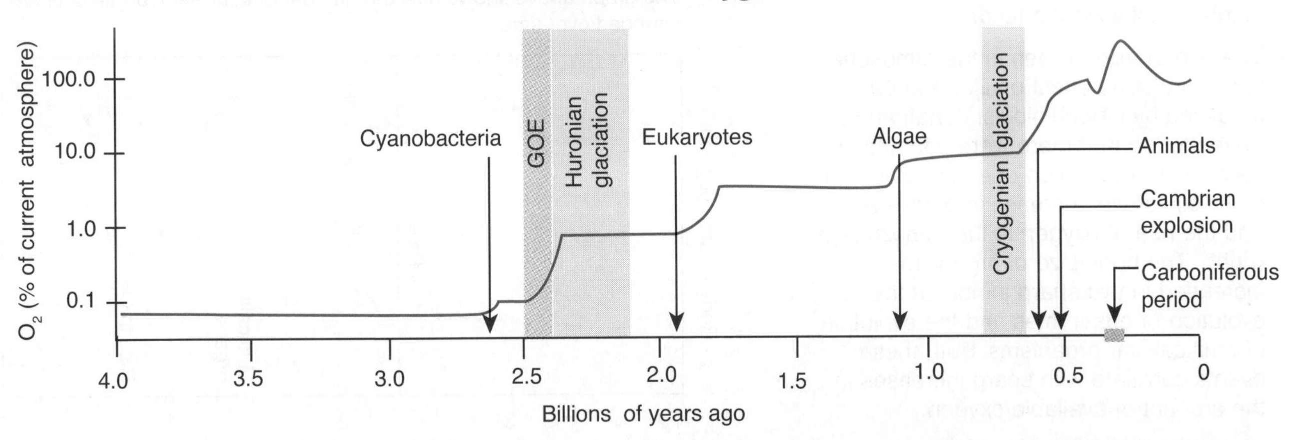


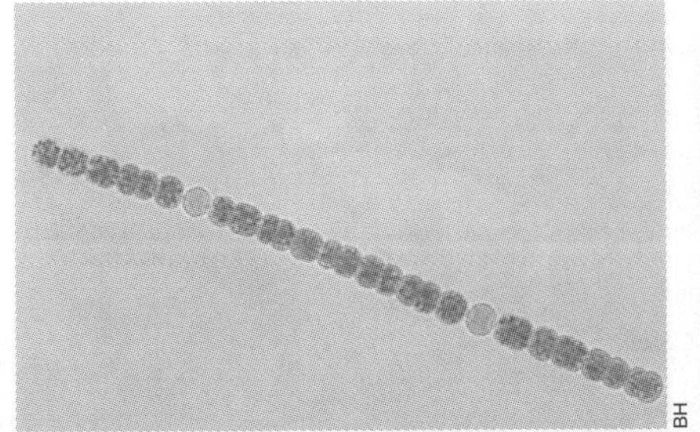


A key step in the evolution of photosynthesis was to use $H_2O$ as an electron donor for the process, releasing $O_2$ gas as a by-product. This occurred in cyanobacteria (above) more than 2.4 billion years ago.

At first this new oxygen reacted with iron in the sea forming vast tracts of banded iron formations (above). These are mostly dated between 2.4 and 1.9 billion years old. The oceans and other rocks and minerals also acted as oxygen sinks.

Oxygen also reacted with $CH_4$ in the atmosphere, forming $CO_2$, a less potent greenhouse gas. This caused a decrease in the Earth's temperature, triggering the Huronian glaciation, a snowball Earth that lasted for 300 million years.

1. Where did Earth's atmospheric oxygen originally come from? ______

2. (a) Why did the oxygen in the atmosphere not increase for millions of years after oxygenic photosynthesis evolved?

______

(b) What evidence is there for the Great Oxygenation Event? ______

3. Identify two geological events probably caused by the rise of biological oxygen production: ______


ISBN: 978-1-927309-37-7

# 98 Changes in Biodiversity

**Key Idea**: The environment and biodiversity are closely related. Changes to the Earth have affected the biodiversity of life throughout Earth's history.

- Although the biodiversity of this planet has not increased at a steady rate since the evolution of life it has nevertheless become greater over time. Biodiversity is closely related to the environment. As biodiversity has increased over time so it has affected the environment and produced new conditions for organisms to exploit. Examples include cyanobacteria producing the first oxygen atmosphere, which allowed the rise of eukaryotes, and microbes in the ground helping to produce the first soils, which helped plants to colonize the land.
- The rise of free oxygen in the atmosphere had a profound effect on life. A 2009 study led by paleobiologist Jonathan L. Payne found that there were correlations between the evolution of eukaryotes and multicellular life, increase in body size, and the rise of oxygen in the atmosphere (right). The body size of organisms increased in two sharp jumps: at the evolution of eukaryotes and the evolution of multicellular organisms. Both these events correlate with sharp increases in the amount of available oxygen.

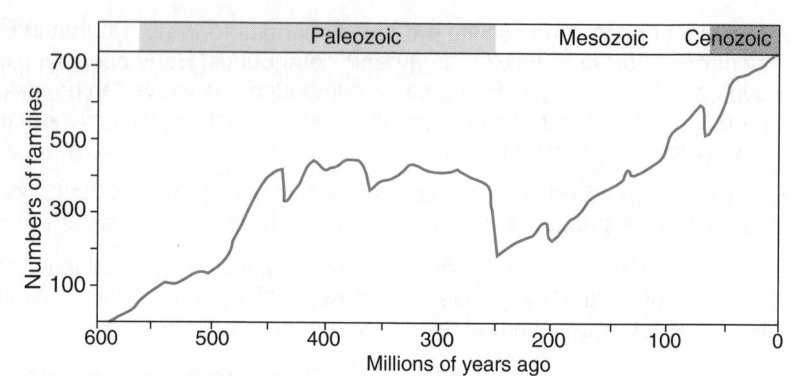


The graph above shows how the number of families of organisms has changed over time.

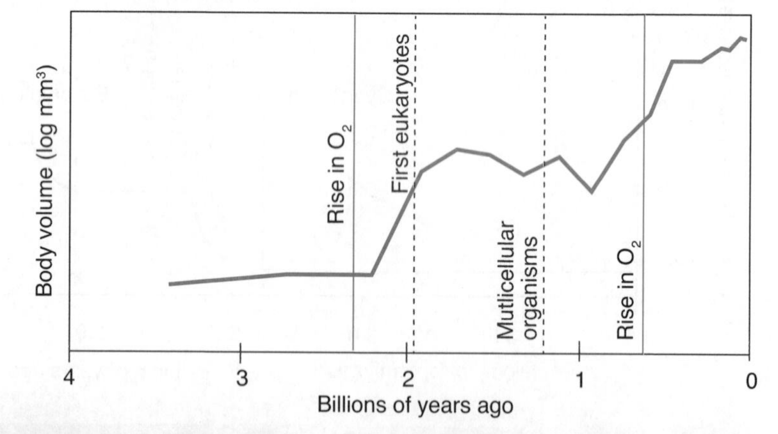

The Carboniferous period saw significant changes to the atmosphere and biodiversity. New land plants locked up large amounts of carbon and raised atmospheric $O_2$ to 35%. The plants formed huge tracts of coal and insects grew to huge sizes in the high oxygen atmosphere.

The environment can be affected by extraterrestrial sources. The impact of a giant asteroid 65 million years ago produce a nuclear winter that ended the reign of the dinosaurs. In their place, mammals diversified and became the dominant large terrestrial animal life form.

A period of global warming 53 to 47 million years ago saw a rise in the biodiversity of mammals in North America. The number of genera increased to a record of 104 and the land was covered in rainforest.

1. Give two examples to illustrate how life modifies the environment and then diversifies into the new, modified environment:

(a) ______________________________

(b) ______________________________

KNOW

**ISBN: 978-1-927309-37-7**

# 99 Carbon Cycling

**Key Idea**: All life is carbon-based. Carbon cycles between the atmosphere, biosphere, geosphere, and hydrosphere. Photosynthesis and respiration are central to these transfers.

- Carbon is the essential element of life. Its unique properties allow it to form an almost infinite number of different molecules. In living systems, the most important of these are carbohydrates, fats, nucleic acids, and proteins.
- Carbon in the atmosphere is found mainly as carbon dioxide ($CO_2$) and methane ($CH_4$). In rocks, it is most commonly found as either coal (mostly carbon) or limestone (calcium carbonate).
- The most important processes in the carbon cycle are photosynthesis and respiration. Photosynthesis removes carbon from the atmosphere and converts it into organic molecules. This organic carbon may eventually be returned to the atmosphere as $CO_2$ through respiration (the oxidation of glucose to produce usable energy for metabolism). The activity of volcanoes also releases $CO_2$ into the atmosphere, although the volumes are relatively small.
- Carbon cycles at different rates depending on where it is. On average, carbon remains in the atmosphere as $CO_2$ for about 5 years, in organisms for about 10 years, and in oceans for about 400 years. Carbon can remain in rocks (e.g. coal) for millions of years.

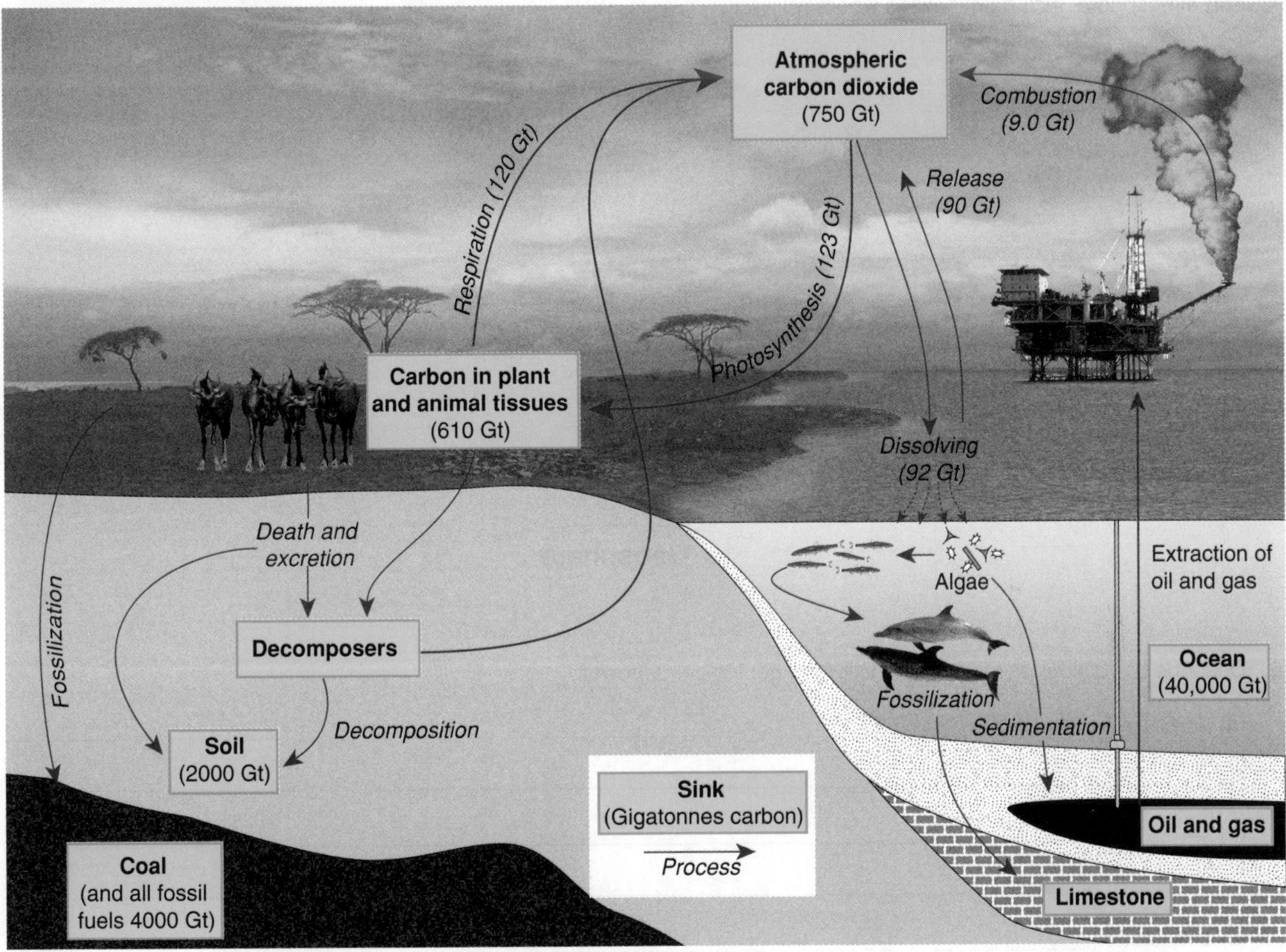


1. (a) What are the two main forms of carbon in the atmosphere? ____

   (b) In what important molecules is carbon found in living systems? ____

   (c) In what two forms is carbon found in rocks? ____

2. (a) Name two processes that remove carbon from the atmosphere: ____

   (b) Name two processes that add carbon to the atmosphere: ____

3. (a) What is the role of decomposers in the carbon cycle? ____

   ____

   (b) What is the role of fossilization in the carbon cycle? ____

   ____


ISBN: 978-1-927309-37-7

**Oil** and **natural gas** formed when dead algae and zooplankton settled to the bottom of shallow seas and lakes. These remains were buried and compressed under layers of non-porous sediment. Oil and gas, like coal, are extracted by humans.

**Limestone** is a type of sedimentary rock containing mostly calcium carbonate. It forms when the shells of mollusks and other marine organisms with $CaCO_3$ skeletons become fossilized. Weathering of exposed limestone releases carbon.

**Coal** is formed from the remains of terrestrial plant material buried in shallow swamps and subsequently compacted under sediments to form a hard black material. Coal is composed primarily of carbon and is a widely used fuel source.

4. Use the space below to complete a model to show how carbon cycles through the biosphere, geosphere, hydrosphere, and atmosphere. Include quantities where relevant:

Atmosphere

Biosphere

Hydrosphere

Geosphere

5. Describe the biological origin of the following geological deposits:

(a) Coal: ______________________________

______________________________

(b) Oil: ______________________________

______________________________

(c) Limestone: ______________________________

______________________________

6. (a) What is the effect of human activity on the amount of carbon stored in sinks? ______________________________

______________________________

______________________________

(b) Describe two global effects resulting from this activity: ______________________________

______________________________

______________________________

(c) Calculate the excess of carbon dioxide being added to the atmosphere by human activity: ______________________________

______________________________


**ISBN: 978-1-927309-37-7**

# 100 What is Global Warming?

**Key Idea**: The combustion of fossil fuels is returning ancient carbon the atmosphere, causing the atmosphere to warm.

- The Earth's atmosphere comprises a mix of gases including nitrogen, oxygen, and water vapor. Small quantities of carbon dioxide ($CO_2$), methane, and a number of other trace gases are also present. Together water, $CO_2$, and methane produce a greenhouse effect that moderates the surface temperature of the Earth.
- The term **greenhouse effect** describes the natural process by which heat is retained within the atmosphere by these greenhouse gases letting in sunlight, but trapping the heat that would normally radiate back into space. The greenhouse effect results in the Earth having a mean surface temperature of about 15°C, 33°C warmer than it would have without an atmosphere. About 75% of the natural greenhouse effect is due to water vapor.

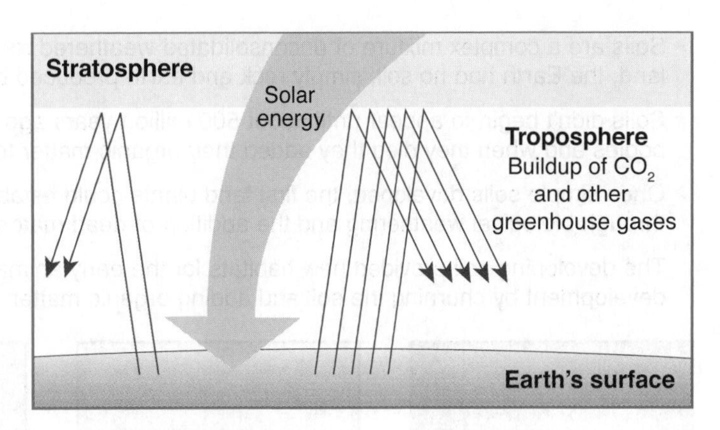


- The next most significant agent is $CO_2$. Fluctuations in the Earth's surface temperature as a result of climate shifts are normal and have occurred throughout the history of the Earth. However since the mid 20th century, the Earth's surface temperature has been increasing. This phenomenon is called global warming and the majority of researchers attribute it to the increase in atmospheric levels of $CO_2$ and other greenhouse gases emitted into the atmosphere as a result of human activity (i.e. the warming is anthropogenic).

## Potential effects of global warming

**Sea levels** are expected to rise by up to 58 cm by the year 2090. This is the result of the thermal expansion of ocean water and melting of glaciers and ice shelves. Many of North America's largest cities are near the coast. The predicted rises in sea levels could result in inundation of these cities and entry of salt water into agricultural lands.

**The ice-albedo effect** refers to the ability of ice to reflect sunlight. Cooling tends to increase ice cover, so more sunlight is reflected. Warming reduces ice cover, and more solar energy is absorbed, so more warming occurs. Ice has a stabilizing effect on global climate, reflecting nearly all the sun's energy that hits it.

The continuing effect of global warming depends upon the complex interactions of the gases in the atmosphere, solar input and ocean currents. Outcomes can be predicted using modeling programs, which depend on predictions of the amount of greenhouse gases produced by human activities. Thus a range outcomes are possible.

1. Which gases are responsible for the greenhouse effect on Earth? ______________________

______________________

2. Which of these gases is most responsible for human-caused global warming? ______________________

3. What is the difference between the greenhouse effect and global warming? ______________________

______________________

______________________

4. What are two possible effects of global warming and what are their consequences? ______________________

______________________

______________________


ISBN: 978-1-927309-37-7

# 101 Soils and Microbes

**Key Idea**: Soils are a complex mix of weathered rock and organic material. Microbes play an important role in producing soil.

- Soils are a complex mixture of unconsolidated weathered rock and organic material. Before microbes invaded the land, the Earth had no soil, simply rock and earth produced by weathering and erosion.
- Soils didn't begin to appear until about 500 million years ago. Microbes used chemicals in the rocks to build their bodies and when they died they added their organic matter to the developing soils.
- Once simple soils developed, the first land plants could establish. This resulted in further development of the soils through biological weathering and the addition of dead material to accumulate organic matter.
- The developing soil provided new habitats for the early animals that invaded the land. They too helped soil development by churning the soil and adding organic matter.

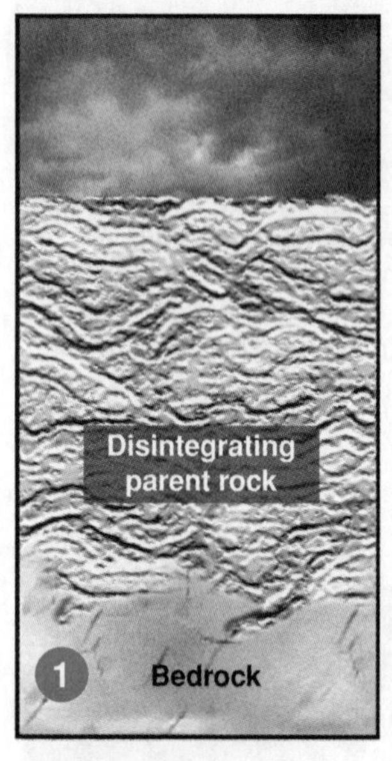

The parent rock is broken down by weathering. Some minerals are available, others are locked up in rocks.

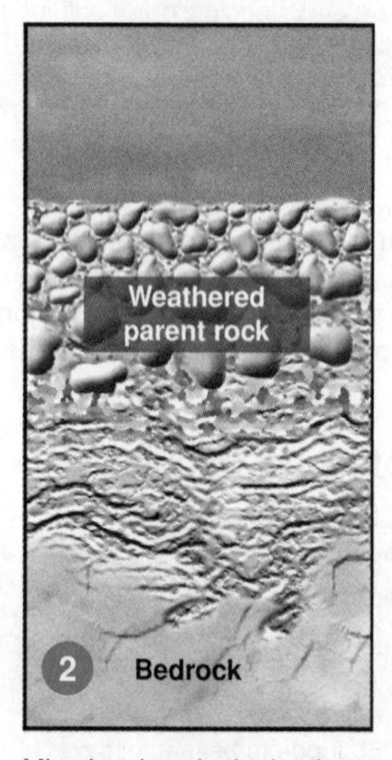

Microbes invade the land and make minerals locked in rocks available. Organic material from microbes builds up.

Early plants invade the land. They mix early soils and add organic matter. Microbes, including fungi, help plants access minerals.

The first soils have developed by 500 million years ago. The final conditions depend on the regional conditions and rock type.

1. When did the first soils appear? ______________________

2. Describe the role of weathering in soil formation: ______________________

3. Describe the role of microbes in soil formation: ______________________

4. Describe role of plants in soil development: ______________________

5. How did soils influence the evolution of plants and soil organisms? ______________________




**ISBN: 978-1-927309-37-7**

# 102 Reefs and Estuaries

**Key Idea**: Life can influence the development of landforms by accumulating sediment or influencing the movement of water and slowing erosion.

## Reefs

Coral reefs are relatively common throughout tropical oceans. They are formed by colonies of millions of tiny animals called polyps. The polyps use calcium carbonate to form the coral skeleton. Different species form different shaped corals. Over thousands of years, the coral reef can build up to cover thousands of square kilometers (e.g. the Great Barrier Reef). The first corals appeared about 500 million years ago, although modern coral reefs appeared only about 10,000 years ago.

Reefs provide a vast array of habitats into which animals can diversify

The evolution of reefs has had land-forming effects on the tropical coastlines of continents and islands, altering the seascape and the landscape. As reefs grow and eventually reach the sea surface, wave action erodes them producing deposits that form small coral islands (especially around subsiding volcanoes). These islands and reefs absorb and slow waves so that erosion on the beach behind the coral island is slowed and any sediment from the land remains where it was deposited.

Coral reefs also provided a variety of new habitats for marine life. In fact, 25% of all marine species are found on coral reefs.

## Estuaries

Estuaries are places rich in nutrients. However the mud of the estuaries is unstable and often anoxic (lacking oxygen). Any plant that could evolve to live on the unstable mud would have access to a large supply of nutrients. Mangroves are plants adapted to the estuarine environment. An important feature of mangroves is that they have roots that spread out from the central trunk and grow upwards out of the mud. This causes them to accumulate sediment and their growth stabilizes the mud. Over time, the tangle of roots allows the mud to accumulate above sea level and form land. At the same time, the seaward edge of the mud flat moves out to sea as sediment accumulates and the landward edge is taken over by land plants. Thus the mangrove forest produces land from the sea.

1. What percentage of marine species inhabit coral reefs? ______________________

2. (a) What is the effect of coral reefs on wave movement? ______________________

______________________

______________________

(b) What effect does this have on coastal development? ______________________

______________________

______________________

3. Explain how mangroves produce new land: ______________________

______________________

______________________


ISBN: 978-1-927309-37-7

# 103 Chapter Review

Summarize what you know about this topic so far under the headings provided. You can draw diagrams or mind maps, or write short notes to organize your thoughts. Use the introduction and the images and hints included to help you:

Climate and the Sun

HINT: What happens to the energy reaching Earth from the Sun? How does this energy affect the climate?

Earth's atmosphere

HINT: What is the structure of the atmosphere and how does it affect Earth's climate?

Coevolution of Earth's systems and life

HINT: Describe examples of life on Earth and Earth's systems evolving together.


**ISBN: 978-1-927309-37-7**

# 104 KEY TERMS AND IDEAS: Did You Get It?

1. Test your vocabulary by matching each term to its definition, as identified by its preceding letter code.

|  | |  |
|---|---|---|
| atmosphere | A | A diverse marine ecosystem based on the limestone skeletons of colonial marine animals called polyps. |
| biome | B | The prevailing weather conditions over long periods of time. |
| carbon cycle | C | The retention of solar energy in the Earth's atmosphere by gases that absorb heat and prevent it from escaping into space. |
| climate | D | Material that forms from the breakdown of organic matter and minerals which overlie bedrock. |
| coevolution | E | The layers of gases that surround the Earth. |
| coral reef | F | Model that describes the circulation of air cells in the atmosphere. |
| equinox | G | A major ecological area with specific climatic conditions and vegetation characteristics. |
| greenhouse effect | H | The biogeochemical cycle in which carbon is exchanged among the biosphere and inorganic reservoirs. |
| global warming | I | The point in the Earth's orbit in which the length of the day and night are the same and the Sun is directly over head at noon on the equator. |
| soil | J | The reciprocal evolution of two or more species or entities as they interact with each other over long periods of time. |
| tricellular model | K | The process of the Earth's surface steadily increasing in temperature, attributed to an increase in greenhouse gases as a result of human activity. |

2. A home owner in with a house at latitude 45°N decides to use solar energy to power her home. She has solar panels placed on the house's roof:

(a) Which direction (north, south, east, west) should the solar panels face? ______

(b) Explain why: ______

(c) The home owner finds that during summer there is enough power from the solar panels to power her house, but in winter she needs to use mains electricity as well. Explain why:

______

3. (a) How does carbon exist in the atmosphere? ______

(b) What process fixes this carbon into organic molecules? ______

(c) What process(es) return the carbon to the atmosphere? ______

4. What evidence is there that life and the environment coevolve? ______


**ISBN: 978-1-927309-37-7**

# 105 Summative Assessment

1. (a) Draw a diagram in the space below to show absorption and reflection of incoming solar radiation as it arrives from the Sun and passes through the Earth's atmosphere to the surface and is reflected back into space:

(b) Describe the greenhouse effect and explain why it is important to Earth's climate: ______________________

2. (a) What was the Great Oxygenation Event (GOE): ______________________

(b) What subsequent events in Earth's history occurred as a consequence of the GOE: ______________________

3. (a) Use the following labels to complete the diagram of the carbon cycle shown right: a*tmosphere, geosphere, oceans, respiration (R), photosynthesis (PS).*

(b) Add arrows to show deforestation (D) and combustion (C) of fossil fuels:

(c) Combustion and deforestation result in another 9 petagrams of carbon being added to the carbon cycle. Add this value to the diagram.

(d) About 3 petagrams is taken up by photosynthesis and 2 petagrams is taken up by the oceans. Add these values to the appropriate labels on the diagram.

(e) How much extra carbon is actually added to the atmosphere by deforestation and combustion?

______________________

Exchanges

Land, plants, animals, soil

Burial

Weathering

Coal, oil, gas

PRACTICES

TEST

**ISBN: 978-1-927309-37-7**

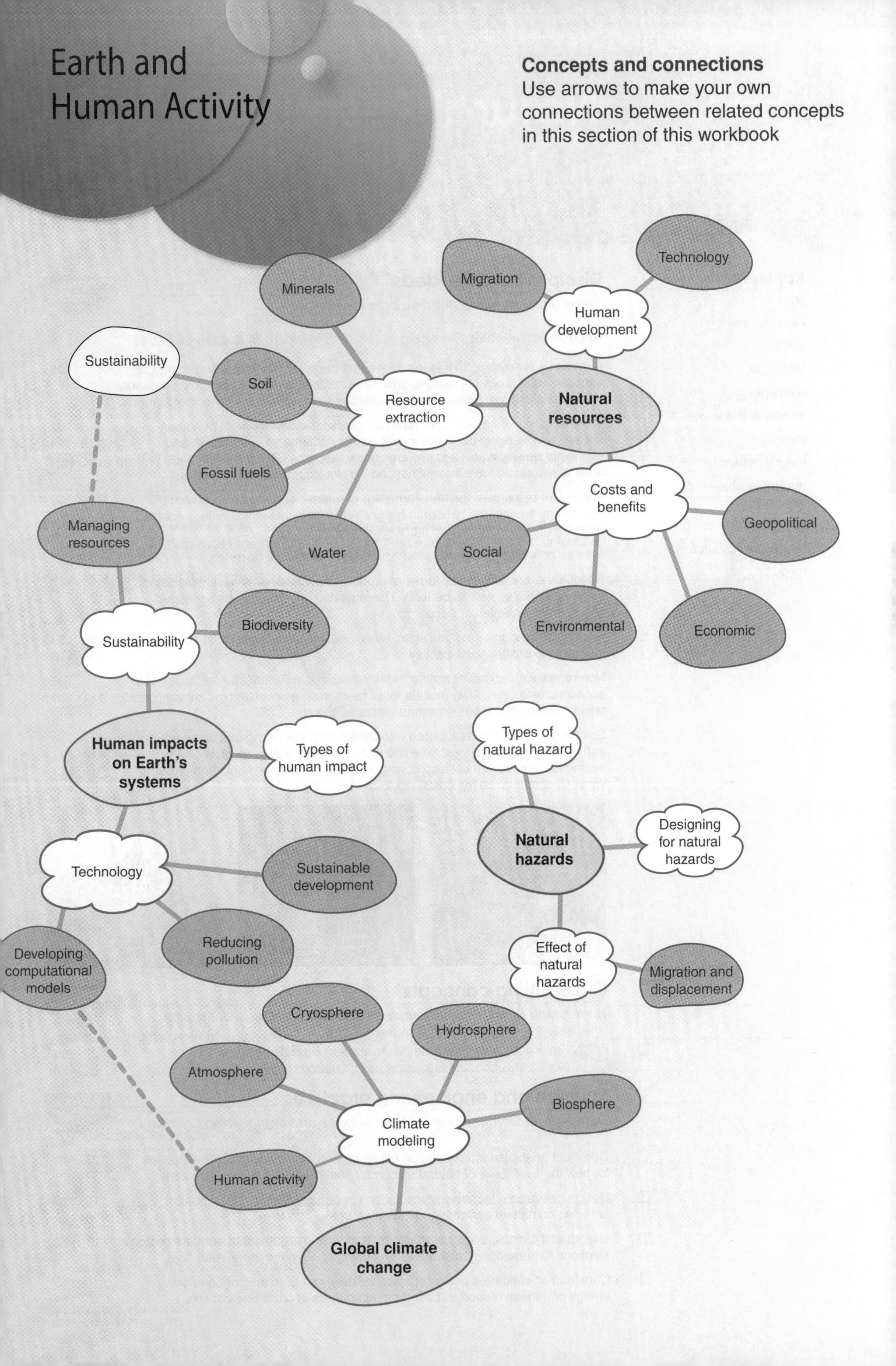

Earth and Human Activity
Concepts and connections
Use arrows to make your own connections between related concepts in this section of this workbook
Minerals
Migration
Technology
Human development
Sustainability
Soil
Resource extraction
Natural resources
Fossil fuels
Costs and benefits
Managing resources
Water
Social
Geopolitical
Sustainability
Biodiversity
Environmental
Economic
Human impacts on Earth's systems
Types of human impact
Types of natural hazard
Natural hazards
Designing for natural hazards
Technology
Sustainable development
Developing computational models
Reducing pollution
Effect of natural hazards
Migration and displacement
Cryosphere
Hydrosphere
Atmosphere
Biosphere
Climate modeling
Human activity
Global climate change

ESS3.A
ETS1.B

# Natural Resources

### Key terms

coal

electrical energy

ERoEI

fossil fuel

heat energy

mechanical energy

mineral

natural resource

non-renewable resource

oil

renewable resource

resource

## Disciplinary core ideas

*Show understanding of these core ideas*

Activity number

### Resource availability has guided the development of human societies

| | | Activity number |
|---|---|---|
| ☐ | 1 A resource is anything that is used for some purpose, e.g. materials, energy, or services. Resources are used to provide benefits and may be consumed or made unavailable in the process. Natural resources exist without the actions of humans. | 106 |
| ☐ | 2 The availability of natural resources has guided the development of human societies, influencing the location and size of settlements, commerce, and economic wealth. Water and soil are critical resources, but fossil fuel and mineral resources have also influenced the development of some nations. | 107<br>109-113 |
| ☐ | 3 The use of resources, the transformation of energy, and the development of technology are integral to human history. Advancements in technology enabled more resources to be utilized more efficiently. | 108 |

### Resource extraction and use has costs, risks, and benefits

| | | Activity number |
|---|---|---|
| ☐ | 4 Resource extraction and all forms of producing useful energy have associated costs and risks as well as benefits. These costs and risks may be economic, social, environmental, or geopolitical. | 110-117 |
| ☐ | 5 Renewable resources include solar, wind, and water but these must be carefully managed to ensure sustainability. | 106 109<br>110 |
| ☐ | 6 Non-renewable resources are not replenished at a sufficient rate for sustainable economic extraction. They include fossil fuels, such as coal and oil, conventional nuclear fission that relies on mined uranium, and soil. | 106<br>111-119 |
| ☐ | 7 Solutions to problems associated with developing, managing and using energy and mineral resources must take into account costs, safety, reliability, and aesthetics. Solutions must also consider social, cultural, and environmental impacts, and balance the costs, risks, and benefits. | 110<br>112-120<br>123 |

## Crosscutting concepts

*Understand how these fundamental concepts link different topics*

| | | Activity number |
|---|---|---|
| ☐ | 1 CE Empirical evidence enables us to support claims about how the availability of natural resources has influenced human activities. | 107-109<br>120 |

## Science and engineering practices

*Demonstrate competence in these science and engineering practices*

| | | Activity number |
|---|---|---|
| ☐ | 1 Construct an explanation based on evidence and sound scientific assumptions for how the availability of natural resources has influenced human activities. | 107-110 |
| ☐ | 2 Design or evaluate technological solutions to reduce the impacts of human activities on natural systems and their resources. | 110-113<br>116-120 |
| ☐ | 3 Use scientific ideas, empirical evidence, and logical argument to evaluate design solutions for developing, managing, and using energy or mineral resources. | 110 113<br>116-120 |
| ☐ | 4 Construct or evaluate a design solution for developing, managing, and using energy or mineral resources based on an analysis of costs and benefits. | 110<br>112-114<br>120 123 |

# 106 Humans and Resources

**Key Idea**: Resources are substances that can be used for beneficial purposes. Water is the most fundamental, but there are many other resources important to human society.

## What is a resource?

- A resource is any source or supply which can be used for some purpose. Resources can be divided into many different kinds, e.g. energy, land, and food. Resources can also be **renewable** (continuously replenished e.g. water) and **non renewable** (not replenished e.g. fossil fuels).
- Natural resources are those that are derived from the environment. Important natural resources include water, minerals, fossil fuels, and fertile soils. The availability of these resources determined where population centers became established and flourished and had a large impact on human technological and social development.

**Fossil fuels**

Human development in the last 200 years has been based on fossil fuels. Peat and coal had been burned for hundreds of years but it was not until the Industrial Revolution that fossil fuels became so important. They are so cheap, energetically concentrated, and easy to transport that they have yet to be equalled by any other kind of energy source.

**Water**

Fresh water is the most important of all resources. It is essential to all life's processes and is important in industry where it is used as a solvent and a coolant. Water covers nearly 75% of the Earth, but only a tiny fraction of this is fresh. Of that, an even smaller amount is available on the surface.

**Minerals**

Human technology is built on minerals. Minerals are solids made up of atoms arranged in a fixed composition, e.g. lime, quartz, and magnetite. Minerals are important both in their natural form (e.g. diamond coated saws) and as ores (e.g. magnetite as an ore for iron). Metals extracted from ores have been important as a basis for currency (e.g. gold) and in electrical and mechanical products (e.g. iron and copper).

**Fertile soils**

Soil fertility refers to the soil's ability to supply the nutrients for plant growth. A fertile soil is rich in nutrients and the soil microbes that make the nutrients accessible to plants. The availability of fertile soil was an important factor in the development of agricultural societies and the first towns and cities.

1. What is a resource? ______________________________

2. Why is water the most essential resource? ______________________________

3. (a) What two components are important contributors to soil fertility and why? ______________________________

   (b) How might soil fertility have contributed to the location of major population centers? ______________________________

4. Why are fossil fuels such an important and useful resource? ______________________________


**ISBN: 978-1-927309-37-7**

# 107 Resources and Civilization

**Key Idea**: Human civilization developed near important resources such as water, minerals, and fossil fuels.

- Human civilization and settlement is, for the most part, found around natural resources, be they recently found or ancient. Water is the most important natural resource and cities originally developed near water supplies such as rivers or natural springs. Today water is still important but other commodities such as fossil fuels and minerals (e.g. gold, iron, and copper) are important for economic wealth and growing populations.
- Soil is also an important resource. Across the Midwest of North America, parts of southern South America, Western Europe, and South East Asia fertile soils allow the growth of vast crops. Modern technology has boosted the size of these crops. The United States alone produced 58 million tonnes of wheat in 2013, the third largest global production, behind China and India.
- As human civilization has expanded and our ability to move resources about the globe has become more efficient, the largest population centers no longer need to be close to areas of major resources. The USA exports 50% of its wheat production. Similarly important minerals are often mined in remote parts of the world and shipped to smelters closer to population centers.

**Cities with greater than 1 million people**

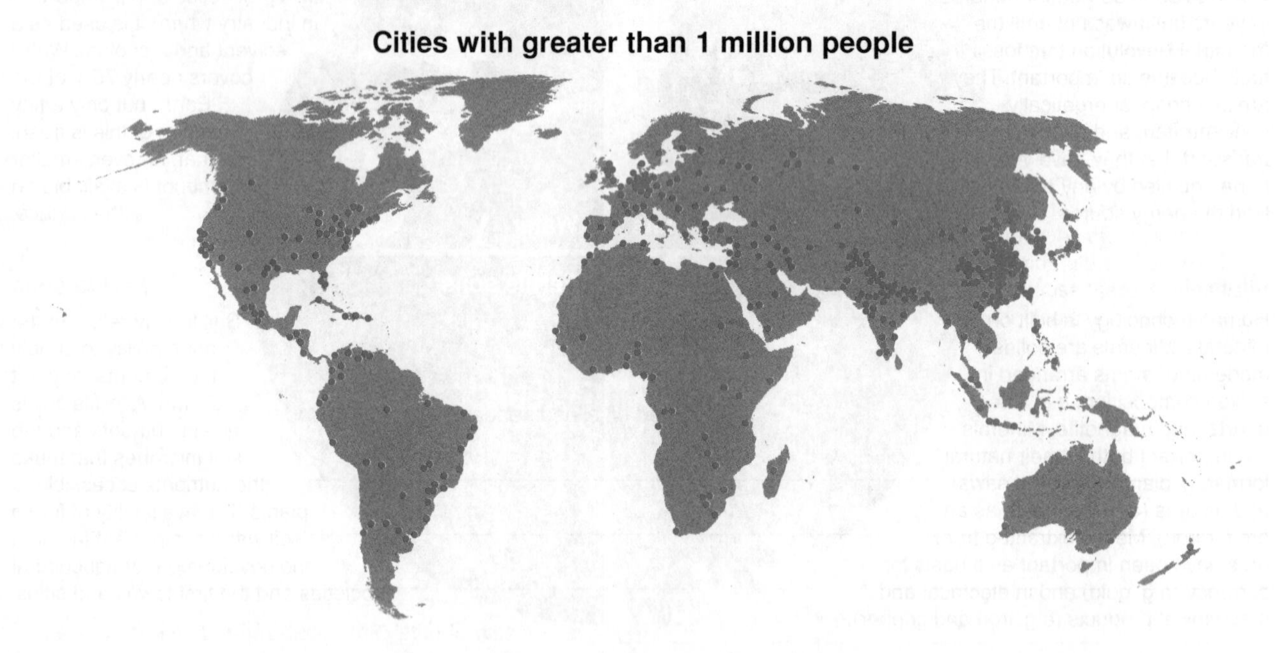

**Important mineral deposits**

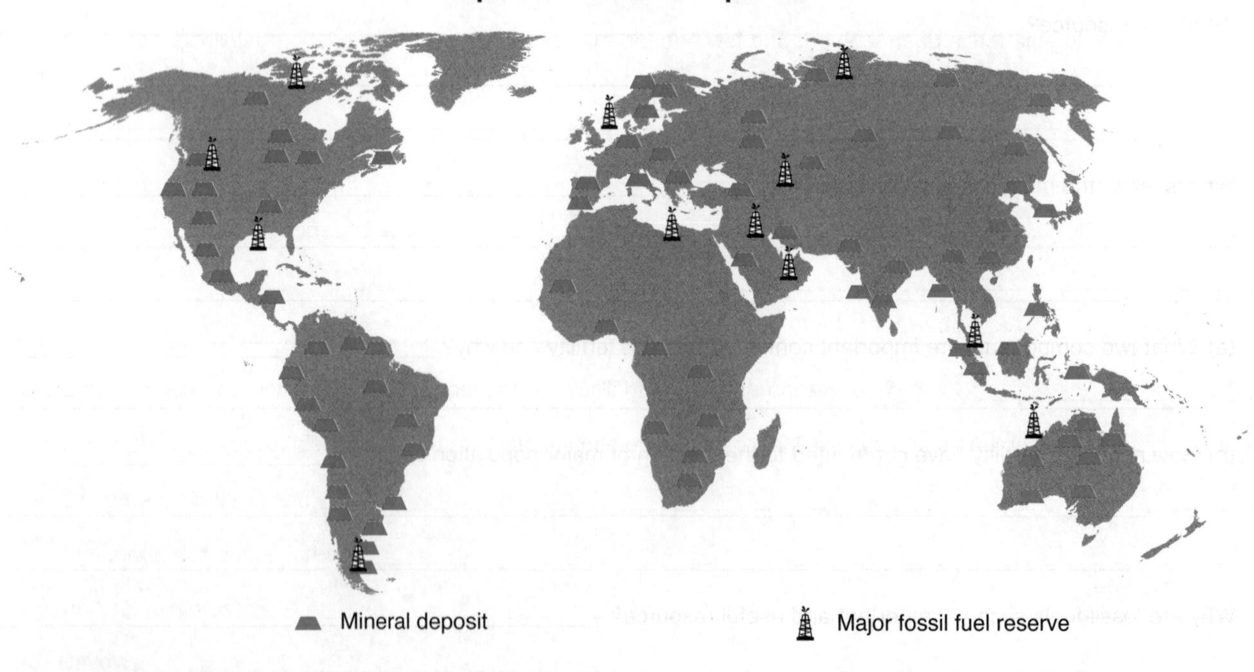

ISBN: 978-1-927309-37-7

## Soil fertility

Very high
High
Low
Very low

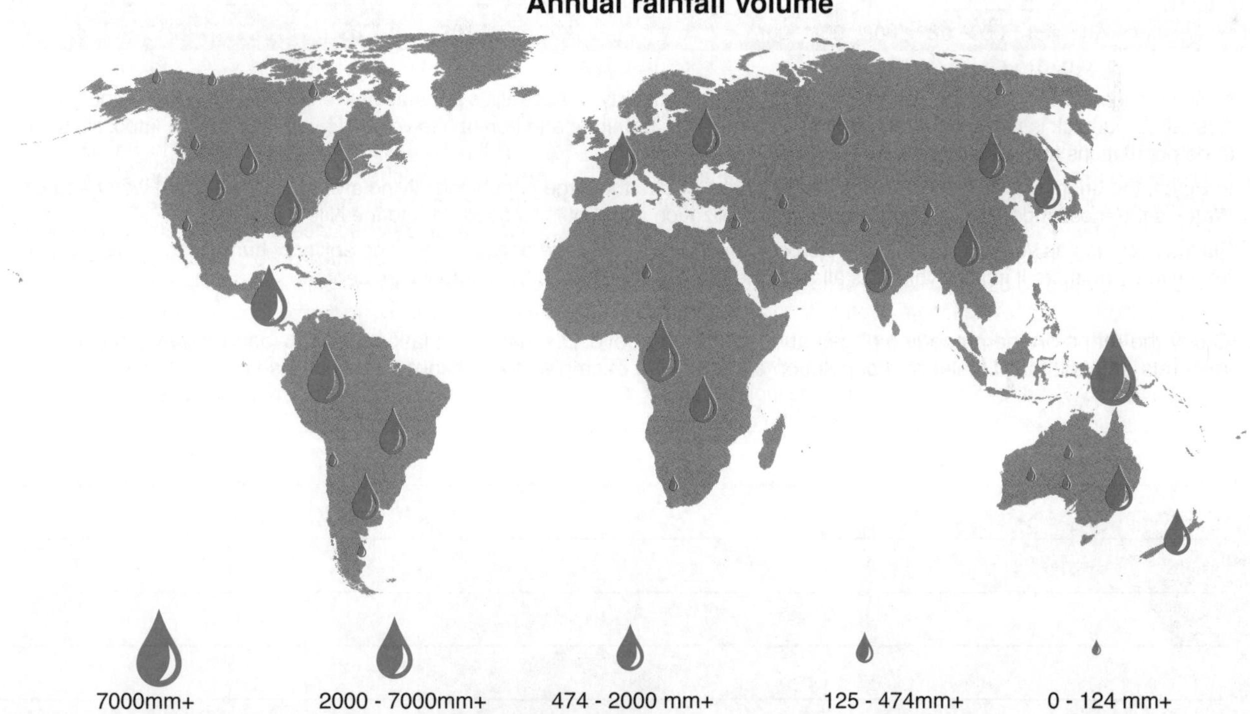


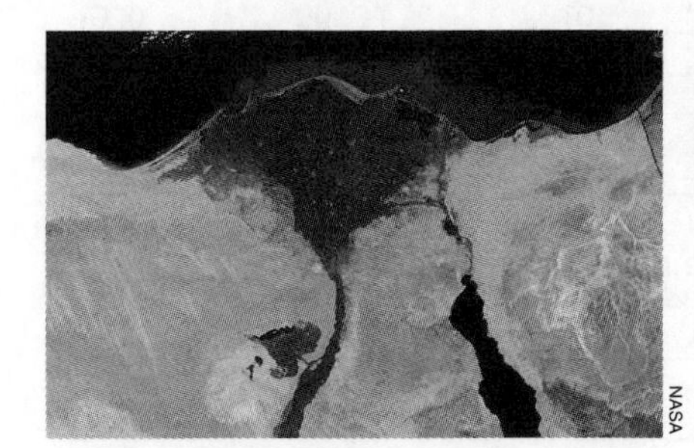


The **Nile delta** represents a classic case of human civilization developing around natural resources. The Nile provides water for drinking and irrigation, and its fertile soils allow numerous crops to be grown. However, just outside the delta lies a vast desert wasteland.

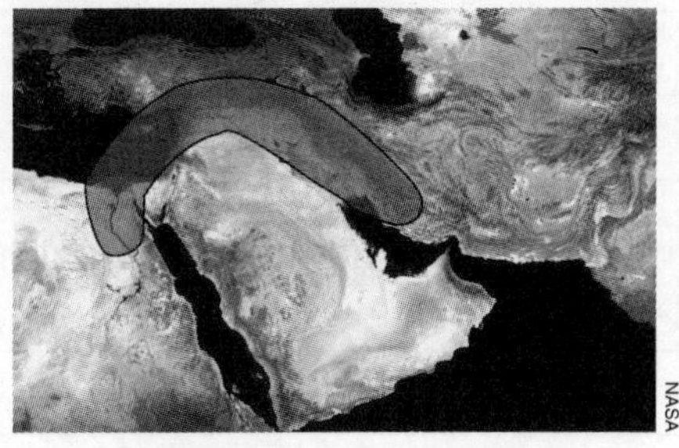


The **Fertile Crescent** is another area where human civilization developed around the resources of water and soil. The area is often referred to as one of the cradles of civilization. It has a large biodiversity due to its location between Africa and Eurasia.

**Mesoamerica** has many ancient abandoned cities that were once thriving centers. However the overuse of soil resources, possible drought, and overpopulation caused the collapse of many of these cities around 800 -1000 AD.


**ISBN: 978-1-927309-37-7**

The image below is a composite showing the Earth at night. The bright areas are cities and population centers. In general, the brighter the area, the bigger the human population there.

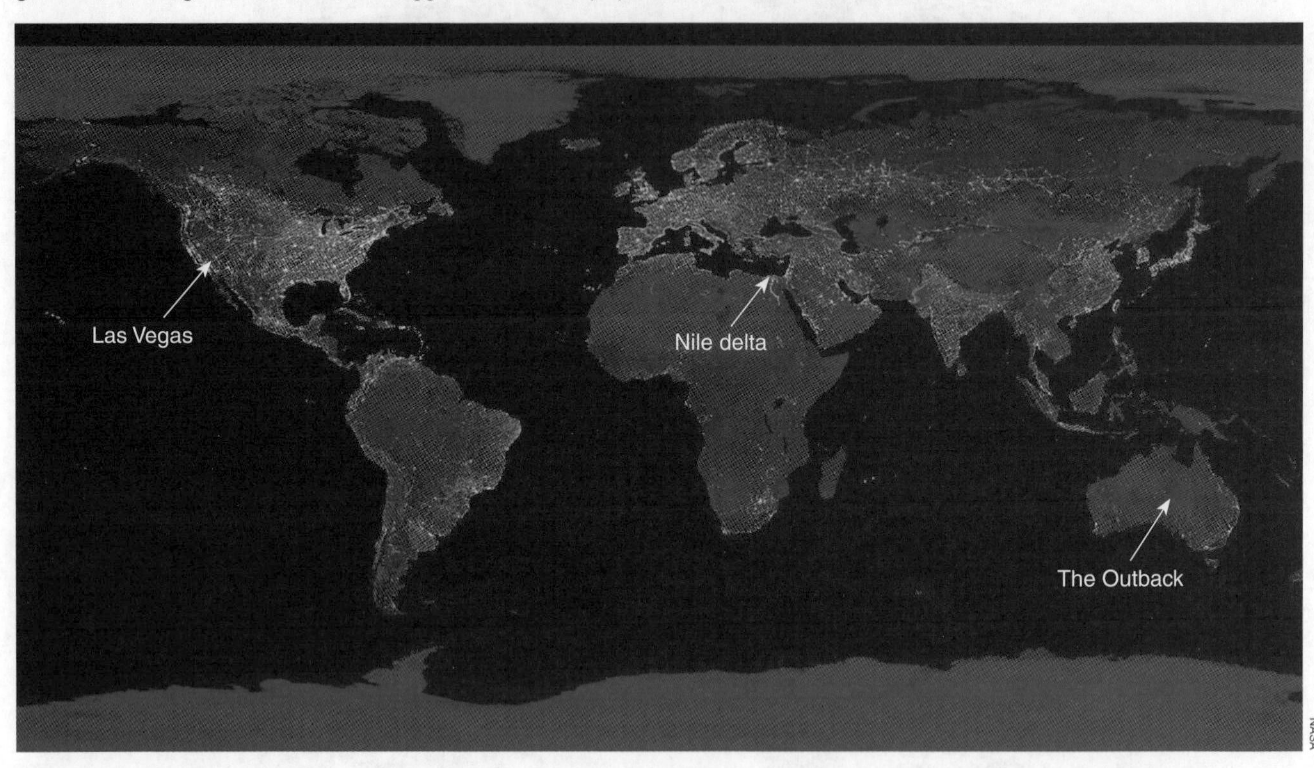


- Notice the location of large populations. For example, the majority of Australia's population live on the east coast. Although Australia's outback is rich in minerals such as bauxite, zinc, uranium, and iron, it has very little water, making it impossible for large populations to live there.
- In Egypt, the bright line of light leading to the Nile delta shows the large populations living around this important water source. Water is the most important natural resource and civilization in Egypt is focused around the Nile delta.
- Places such as Las Vegas in California were only small towns just a few decades ago. The ability of humans to redirect water from rivers and store it in reservoirs has allowed these places to grow to huge population centers.

1. Study the maps on the preceding pages and the image above. Discuss the relationship between resources (soil, minerals, rainfall, fossil fuels) and population centers. Use examples to highlight this relationship:

2. Explain how modern technology now influences where large populations can be become established:


**ISBN: 978-1-927309-37-7**

# 108 Energy, Resources, and Technology

**Key Idea**: The use of resources, transformation of energy, and development of technology are integrally linked with human history.

- There is an argument that the history of human social, technological, and cultural development is linked directly to our ability to harness and transform energy.
- One of the earliest uses of energy was the invention of ways to provide reliable fire. Energy could then be used to cook food, hunt animals, produce weapons, and provide light.
- Domesticating animals and using them to provide mechanical power (horsepower) increased the work that could be done by one person. Using coal and charcoal to provide heat enabled the smelting of iron and metal alloys, expanding the range of tools that could be used.
- The development of the steam and internal combustion engines provided ways for heat to be transformed into reliable and powerful mechanical work. Harnessing other types of mechanical work to drive electrical turbines (e.g. water turbines) produced reliable electrical energy.
- At each of these steps, the ability of one person to do work increased. At each step, the resources that became available to a person increased. But the resources required also increased. Iron production requires coal and charcoal to provide heat and act as reducers. Iron machinery allows the components of engines to be manufactured (again from iron). Engines need fuel - coal and oil. And so there is a continuing increase in resources required and produced.

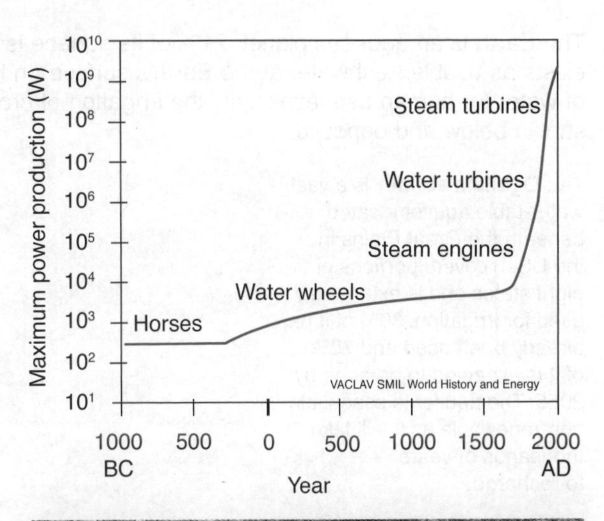

Domesticating horses and cattle allowed one person to plow an entire field, drive a mill, or carry a greater load. This allowed the utilization of more resources and their more efficient use.

More efficient ways of using energy resources increased the power available. A tractor can do the work of many horses. Now many fields can be plowed and planted by one person, allowing more food to be produced.

Transforming energy into electricity makes usable energy available to the masses. 100,000 years ago the average person used about 20,000 kJ of energy a day. Today it is close to a million kJ a day, from using cars, computers, and appliances.

1. In the last 3000 years, which two technological developments signalled sudden increases in the energy available to humans to do work:

2. Discuss how energy and resources have shaped human social and technological development:


**ISBN: 978-1-927309-37-7**

# 109 Global Water Resources

**Key Idea**: Rivers, lakes, and aquifers are key water sources and provide water, transport routes, and energy for human use.

The Earth is an aqueous planet; 71% of its surface is covered by water. Only about 0.0071% of the world's water exists as usable freshwater at the Earth's surface (in lakes, rivers, and wetlands). Rivers and lakes provide sources of water for human use, especially the irrigation of crops. Some of the largest and most important water courses are shown below and opposite.

The **Ogallala aquifer** is a vast water-table aquifer located beneath the Great Plains in the US. It covers portions of eight states and is extensively used for irrigation. 30% of it has already been used and 70% of it is expected to be gone by 2065. The aquifer is essentially non-renewable as it will take thousands of years to recharge.

The North American **Great Lakes** are the largest group of freshwater lakes on Earth, containing 22% of the world's fresh surface water.

The **Colorado River** runs through seven states of Southwestern United States. It has several large dams including the Hoover dam and the Glen Canyon dam. The river is so heavily used for irrigation purposes its flow rarely reaches the sea anymore

The **Mississippi River** drains most of the area between the Rocky Mountains and the Appalachians. A series of locks and dams provide for barge traffic.

The **Amazon River** accounts for 20% of the world's total river flow and drains 40% of South America. The Amazon is the largest rainforest in the world and has the world's highest biodiversity

Rivers have been vital transport routes since ancient times. Cities and towns could transport goods and resources along rivers for trade. Large rivers such as the Mississippi, above, are able to accommodate large ships for much of their length. Adding locks to the river can extend the distance a ship can travel.

Rivers provide vital water for irrigation of crops and act as reservoirs for town supplies. However, overuse of a river for irrigation or building dams can lower its level and reduce the downstream flow. This can cause warming of downstream river waters which can lead to degradation of the riverine habitat.

The flow of water in a river provides a way of generating electricity by driving turbines in dams. The damming of rivers severely affects flow rates downstream. Dams are often used for flood control, but this often means floodplains no longer receive a supply of vital nutrients during floods, lowering the fertility of the soils.

**ISBN: 978-1-927309-37-7**

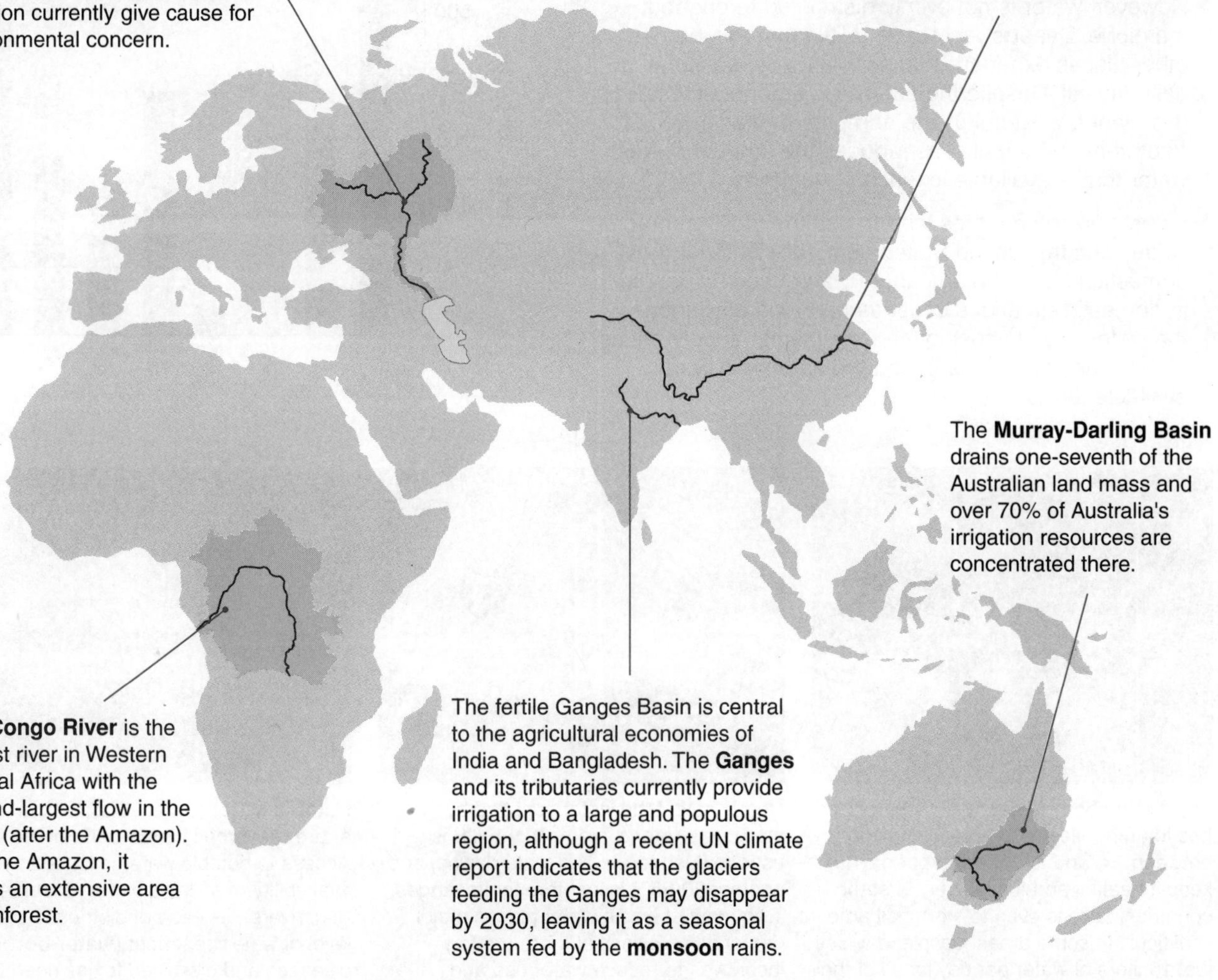

The **Volga River** and its many tributaries form an extensive river system, which drains an area of about 1.35 million $km^2$ in the most heavily populated part of Russia. High levels of chemical pollution currently give cause for environmental concern.

From glacial origins, the **Yangtze River** flows 6300 km eastwards into the East China Sea. The Yangtze is subject to extensive flooding, which is only partly controlled by the massive Three Gorges Dam, and it is heavily polluted.

The **Murray-Darling Basin** drains one-seventh of the Australian land mass and over 70% of Australia's irrigation resources are concentrated there.

The **Congo River** is the largest river in Western Central Africa with the second-largest flow in the world (after the Amazon). Like the Amazon, it drains an extensive area of rainforest.

The fertile Ganges Basin is central to the agricultural economies of India and Bangladesh. The **Ganges** and its tributaries currently provide irrigation to a large and populous region, although a recent UN climate report indicates that the glaciers feeding the Ganges may disappear by 2030, leaving it as a seasonal system fed by the **monsoon** rains.

1. Explain why some deep but extensive aquifers, such as the Ogallala, are considered non-renewable: ______________

______________________________________________

______________________________________________

2. Identify three uses of river by humans:

(a) ______________________________________________

(b) ______________________________________________

(c) ______________________________________________

3. Describe some of the negative aspects of human use of global water resources: ______________

______________________________________________

______________________________________________

______________________________________________


**ISBN: 978-1-927309-37-7**

# 110 Water and People

**Key Idea**: Water is not only important for drinking, but can be used for a range of purposes including generating electricity.

- Water is the most important substance on the planet. Life could not exist without it. There are approximately 1.4 billion trillion liters of water on Earth and Earth is the only planet in the solar system where large volumes of water are found on the surface in liquid form.
- However, water is not evenly distributed throughout the globe. Deserts receive very little rainfall whereas other places experience large volumes of seasonal or daily rainfall. Despite the enormous amount of water on this planet, wasteful usage and poor management of treatment and supply has reduced the amount of fresh water that is available to much of humanity.
- Industrialized countries tend to use the most amount of water, with the United States being one of the largest domestic users. The amount of water used by people in domestic situations depends upon the efficiency of the water use, its cost to the user (higher supply prices usually mean lower use), and the amount of water available for use.

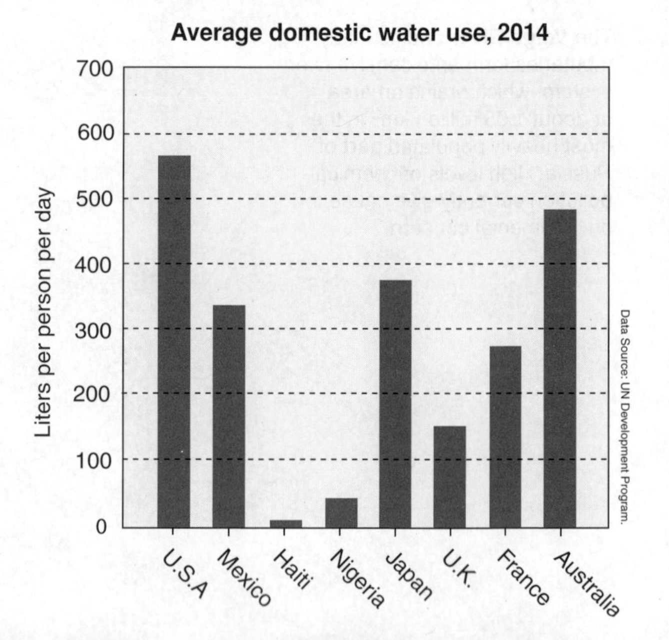

The human body is nearly 70% water and health authorities recommend drinking between 1.8 and 2 L of water per day to keep it healthy and functioning. In some countries, access even to this small amount is difficult. In some areas, people may use just 15 liters of water per day for all of their domestic uses including drinking water (compared to over 570 liters per person per day in the United States).

Nearly half of the water supplied by municipal water systems in the US is used to flush toilets and water lawns, and another 20-35% is lost through leakages. Treatment of waste water places major demands on cities yet there are few incentives to reduce water use and recycle. Providing reliable clean water remains a major public health issue in many poorer regions of the world.

**Potable water** (water suitable to drink) is a rare resource in large parts of the world. Access to potable water is limited by the availability of suitable water resources (e.g. rivers), its ease of distribution, and level of water treatment (**water-borne** diseases and dissolved toxins need to be removed). In many countries, distribution of water is difficult and storage of large amounts is almost impossible.

1. What factors limit water supply to some countries? ______________________________

2. (a) Which type of countries tend to use the most amount of water? ______________________________

   (b) What is the major use of water in these countries? ______________________________

3. Why is potable water such a rare resource in many countries? ______________________________

**ISBN: 978-1-927309-37-7**

## Costs and benefits of using water: hydroelectric dams

- Water has been used to generate usable energy for hundreds of years. The water wheel allowed the gravitational potential energy in water at the top of a waterfall to be converted to the kinetic energy of a spinning wheel. The spinning wheel could then be used to drive mechanical devices such as grain mills. This principle is still used today. Water turbines are used in dams to spin giant electric generators to produce electricity.
- Electricity production is of fundamental importance in our world today. Without electricity, our standard of living would plummet. As we use more high demand devices we require more and more electricity generating schemes (even with more efficient devices).
- Fossil fuel fired power stations produce by far the most amount of electricity in the world today. However they are also responsible for producing huge volumes of $CO_2$, which is linked to global warming.
- Hydroelectric power reduces the need to use coal by using a renewable resource (water flow) and thus reducing $CO_2$ production and the consumption of a non-renewable resource (coal).
- However, there are important environmental costs to be considered when building hydroelectric dams.

| Costs | Benefits |
|---|---|
| Construction costs | Electricity generation |
| Flooded land | Reduction in $CO_2$ production from fossil fuels. |
| Disruption of river habitat | Water storage for irrigation and domestic use. |
| Relocation of families | |

Damming a river...

...helps burn less coal.

Chongkian

- Cost/benefit analyses are carried out when planning dams. To do this, monetary values are often assigned to resources, giving them a quantitative value. For example, a forest that will be flooded by the reservoir behind a dam may be worth $1 million a year in tourism or $10 million in clear-felled timber. The loss of this monetary value can then be compared to the monetary benefit of the dam, e.g. the dam will produce $5 million worth of electricity per year and save $2 million of coal a year. Comparing these cost and benefit values over the lifespan of the dam can help in making decisions about its economic viability. However, cost-benefit analyses are by no means the only consideration and it may not be possible to assign an economic value to some factors.

4. Explain how water can be used to generate electricity: ______________________

5. (a) Identify some costs and benefits of hydroelectric dams: ______________________

   (b) Explain how these costs and benefits can be evaluated and compared to produce a cost/benefit ratio: ______________________

6. What costs and benefits might be difficult to quantify in monetary terms? ______________________


**ISBN: 978-1-927309-37-7**

# 111 Non Renewable Resources

**Key Idea**: Mineral resources are non renewable. Some are in relatively short supply (e.g. coal and oil) while others are essentially limitless (e.g. iron).

- The Earth contains enormous mineral resources, which can be obtained and used with relative ease to produce usable energy. The most commonly used of these are the **fossil fuels**, i.e. **coal**, **oil** and **natural gas.** These can be burned immediately to produce heat energy or they can be refined to provide a variety of energy or material needs. Around 85% of the world's energy needs comes from burning fossil fuels.
- The use of fossil fuels as an energy source is a relatively cheap and efficient way of producing energy. Coal and oil have high energy densities, containing about 30 MJ per kilogram (about double that of wood).

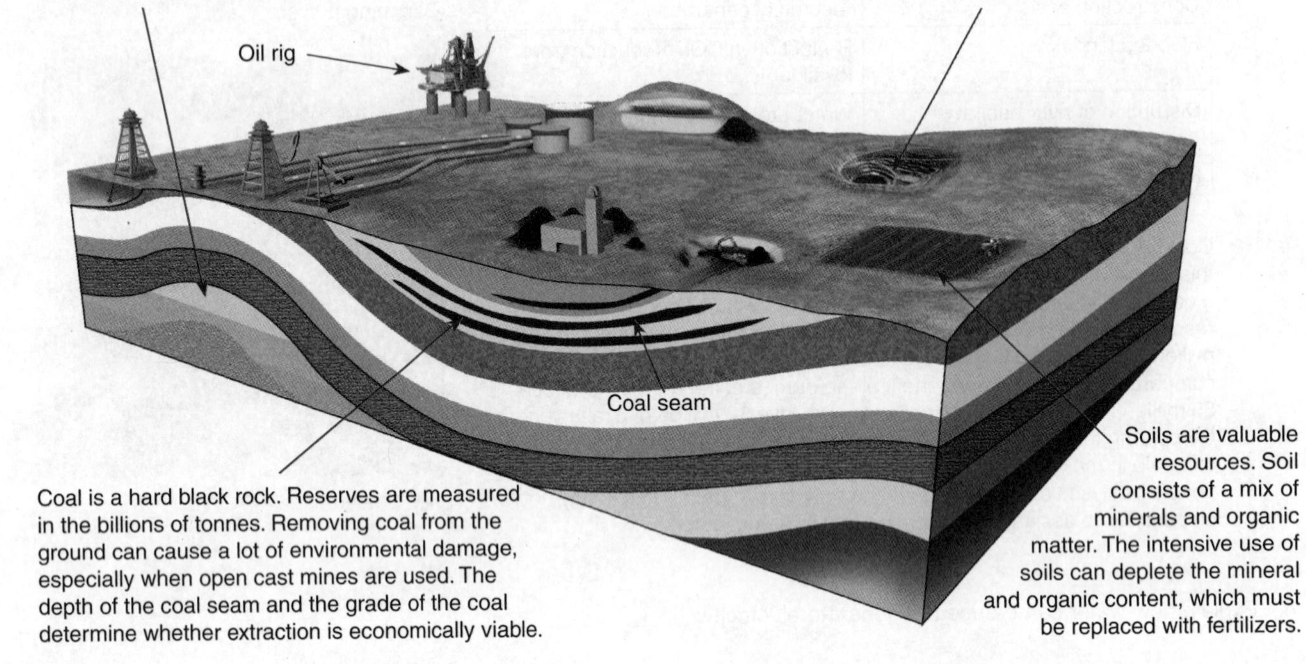


1. Explain why coal and oil are favored as energy resources over many other sources of energy: ___

2. Describe some differences between coal and oil and their extraction from the ground: ___

3. What factors determine whether or not it is worth mining a mineral? ___


**ISBN: 978-1-927309-37-7**

# 112 Coal

**Key Idea**: Coal is a valuable mineral. Its mining produces economic and energy benefits, but also social and environmental costs.

- Coal is formed from the remains of terrestrial plant material buried in vast shallow swamps during the Carboniferous period (359 to 299 mya) and subsequently compacted under sediments to form a hard black rock.
- Burning coal produces about a third of the world's energy needs. Nearly eight billion tonnes of coal are produced globally each year. Burning coal produces vast quantities of greenhouse gases and pollutants, contributing to smog and global warming.

| Using coal | |
|---|---|
| **Benefits** | **Costs** |
| Huge supplies (billions of tonnes) | High $CO_2$ production when burned |
| High net energy yields. | High particle pollution from soot |
| Can be used to produce syngas and converted to other fuels. | Low grade coals produce high pollution and contribute to acid rain |
| Relatively easy to extract when close to the surface | High land disturbance through mining, including subsidence |
| Important in industry as coke (reducer) | Noise and dust pollution during mining |

## Best practice coal production

- When plans are put in place for opening a coal mine, best practice dictates that economic, environmental, social, and even political costs and benefits are considered.
- By considering these aspects, the net benefit (or cost) of the mine can be identified. Some costs and benefits of coal production are indicated above right, but there are many more including economic benefits such as the creation of jobs and building of industry, or social costs, such as the loss of land use and effects on health.
- It is preferable (and appropriate) that a mine's operators allocate some of the mine's revenue to pay for some of the environmental and social costs, including repairing environmental damage and restoring the original environment (if possible) when the mine closes. Large mines can cost tens of millions of dollars to close and rehabilitate.

**Preparing the mine**

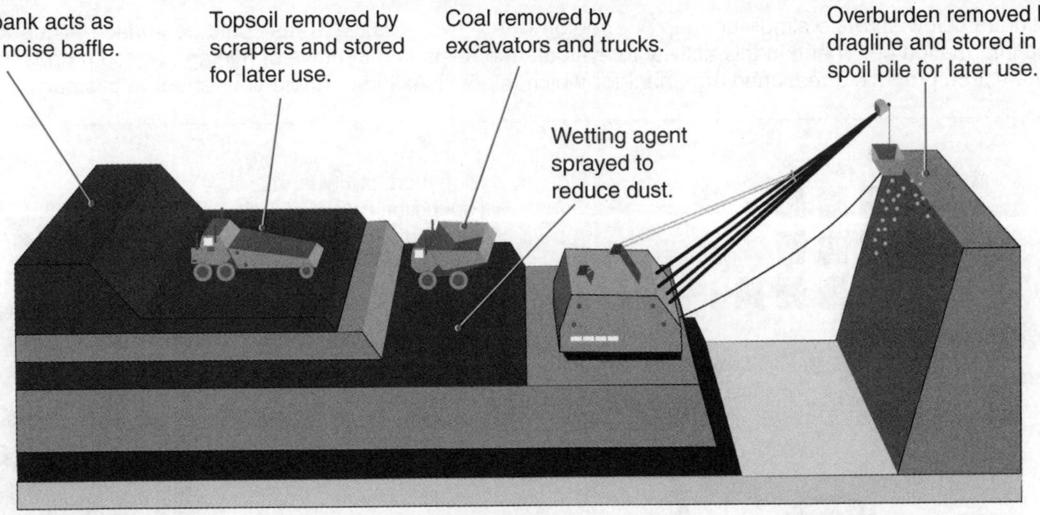

Graded bank acts as dust and noise baffle.
Topsoil removed by scrapers and stored for later use.
Coal removed by excavators and trucks.
Wetting agent sprayed to reduce dust.
Overburden removed by dragline and stored in a spoil pile for later use.

**Remediating the mine**

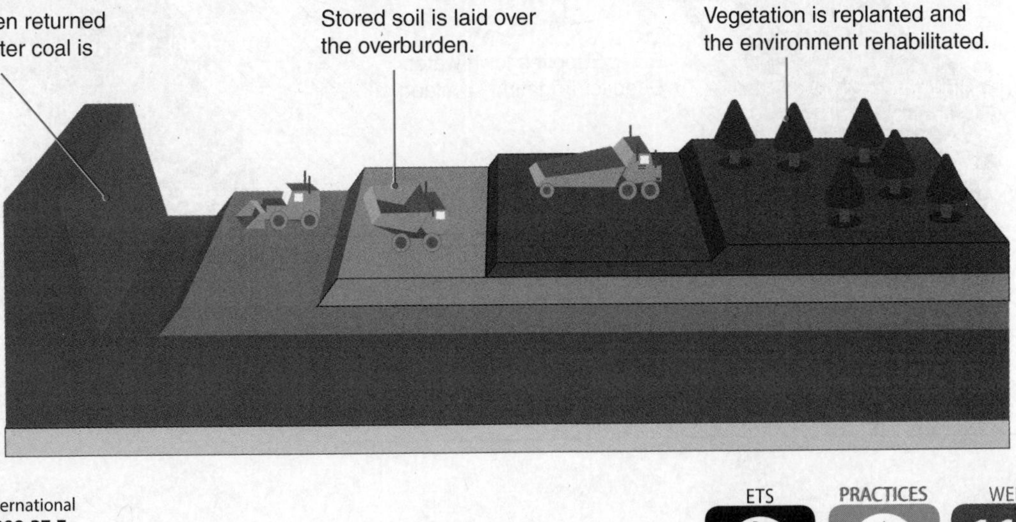

Overburden returned to mine after coal is removed.
Stored soil is laid over the overburden.
Vegetation is replanted and the environment rehabilitated.


**ISBN: 978-1-927309-37-7**

**Major world coal reserves**

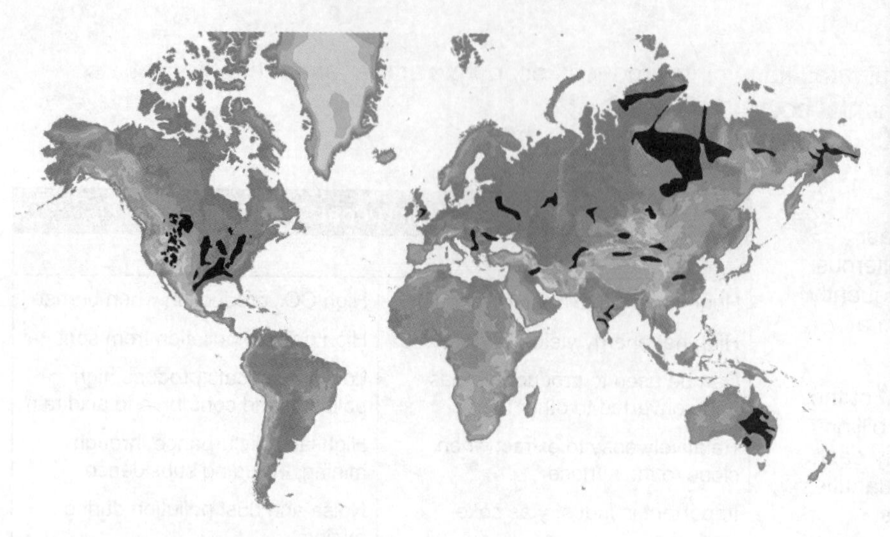

Coal is pulverized and used to fuel thermal power stations. In developing countries, it is often used for home heating and cooking. This can lead to health problems if furnaces or stoves are not properly vented or coal ash is handled improperly, as coal can contain many toxic substances.

1. For each of the following, list two benefits of coal production:

   (a) Economic benefits: ______

   (b) Social benefits: ______

2. For each of the following, list two costs of coal production:

   (a) Environmental costs: ______

   (b) Economic costs: ______

   (c) Social costs: ______

3. The diagram below shows a simplified map of a region where two possible areas could be surface-mined for coal. Use the information below and in this activity to evaluate the costs and benefits of mining at each proposed site. On a separate sheet, produce a reasoned argument for which mine site is best. Staple your sheet to this page.

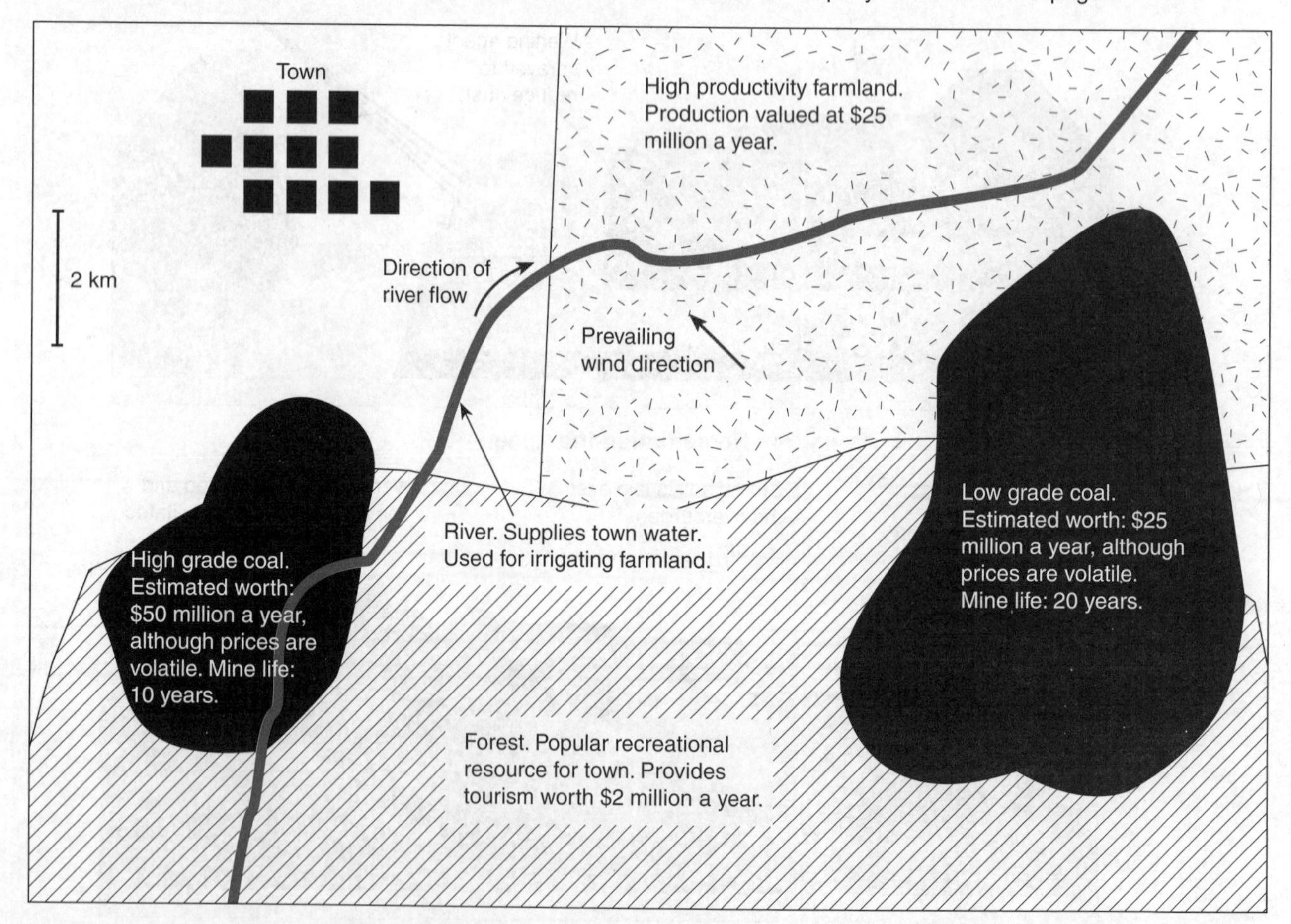



**ISBN: 978-1-927309-37-7**

# 113 Oil and Natural Gas

**Key Idea**: Oil and natural gas are valuable for their energy content. Extraction can be costly and returns on investment dependent on the location and type of oil.

- Oil is formed from the remains of algae and zooplankton which settled to the bottom of shallow seas and lakes about the same time as the coal-forming swamps. These remains were buried and compressed under layers of nonporous sediment. The process, although continuous, occurs so slowly that oil (like coal) is essentially non-renewable.
- Oil and natural gas are both composed of a mixture of hydrocarbons and are generally found in the same underground reservoirs. Natural gas is generally defined as a mixture of hydrocarbons with four or fewer carbon atoms in the chain (as these are gaseous at standard temperatures and pressures). Oil is defined as the mixture of hydrocarbons with five or more carbon atoms in the chain.
- Crude oil can be refined and used for an extensive array of applications including fuel, road tar, plastics, and cosmetics.

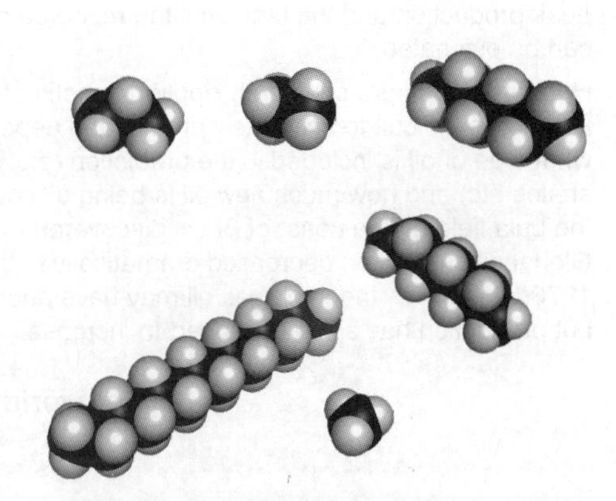

**Natural gas** is often found in the same reservoirs as oil. Drilling rigs require specialized facilities to store the gas. Because of this, much natural gas is either vented, or reinjected to maintain pressure in the reservoir.

Transport of natural gas requires specialized equipment. Liquid natural gas (LNG) tankers are able to cool the gas to -162 °C and transport it as a liquid (saving space). Gas can also be piped to shore if facilities are nearby.

USDE

Oil may be found in materials that make extraction using conventional drilling impossible. These **non conventional oils** (e.g. oil shale) are often mined in the same way as coal and the oil washed from them.

Crude and heavy oils require refining before use. Crude oil is separated into different sized fractions by a distillation tower. Heavy oils may be heated under pressure to break them into smaller more usable molecules.

## Energy return on energy investment

- The energy invested to produce a barrel of oil or the **energy return on energy invested** (ERoEI) can be expressed as a ratio of energy expended to energy gained.
- In the early 1900s, the ratio was around 100:1 (100 barrels of oil were produced from 1 barrel of oil invested). The ratio has reduced over time to between 30:1 and 10:1, as the resources become increasingly difficult to harvest and process.

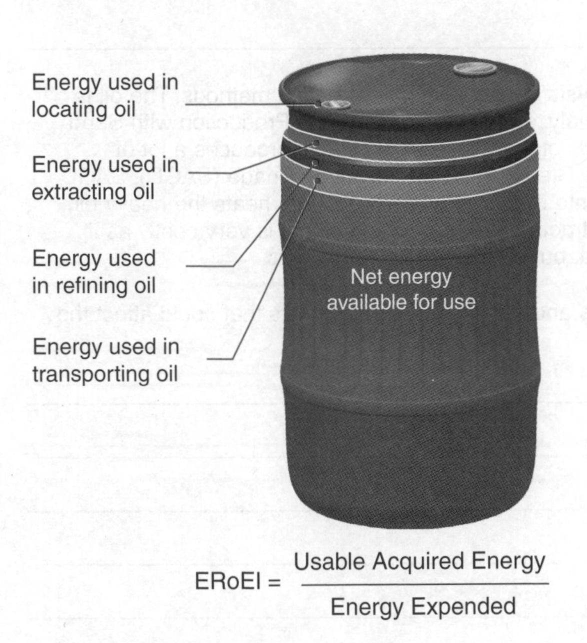


$$\text{ERoEI} = \frac{\text{Usable Acquired Energy}}{\text{Energy Expended}}$$

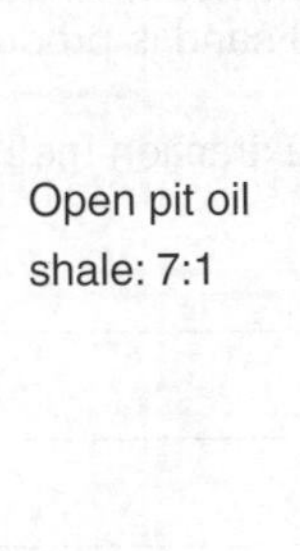

**Ratio: barrels of oil output/input**

Crude oil: 10:1

Open pit oil shale: 7:1

*In situ* oil sands: 5:1


**ISBN: 978-1-927309-37-7**

## Peak oil

The future of the production of any mineral can be estimated from its current and historical production rates. Called a **Hubbert Curve** after its discoverer, it shows that the extraction rate of any non-renewable resource will fit into a bell shape curve (right). Plotting the amount of mineral extracted on a graph gives the shape of part of the curve, from which the rest can be estimated. Using these curves, peak production and the time until the resource runs out can be estimated.

Hubbert curves are constantly applied to estimate peak oil. Peak oil is difficult to accurately predict and depends on what type of oil is included in the prediction (e.g. crude, oil shales etc) and how much new oil is being discovered (e.g. the Lula field off the coast of Brazil discovered in 2006). Global oil production decreased dramatically in the mid 1970s and it was thought peak oil may have been reached, but production has since continued to increase.

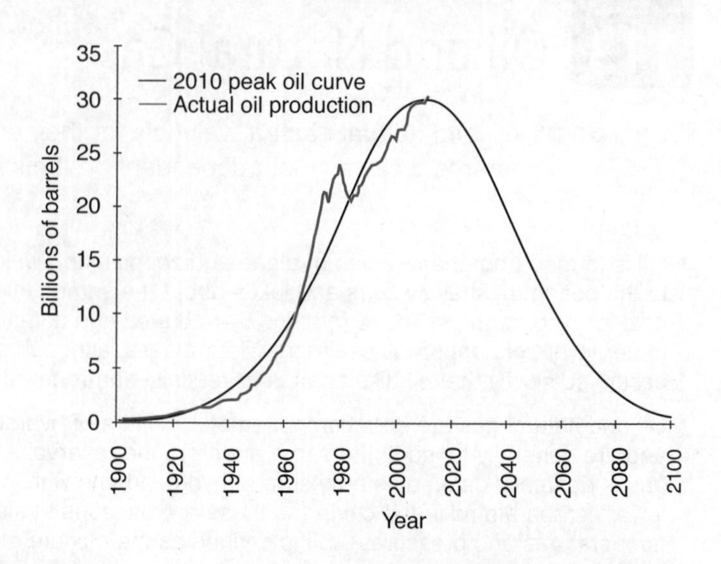


**Major world oil reserves**

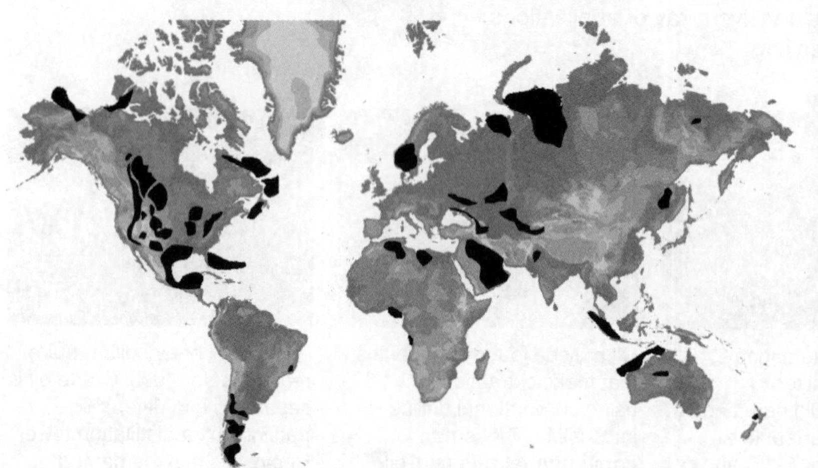

| Oil | |
|---|---|
| **Benefits** | **Costs** |
| Large supply | Many reserves are offshore and difficult to extract |
| High net energy gain | High $CO_2$ production |
| Can be refined to produce many different fuel types | Potential for large environmental damage if spilled |
| Easy to transport | Rate of use will use up reserves in near future |

1. Describe the difference in the composition of natural gas and oil: ______________________

2. (a) Discuss the significance of peak oil: ______________________

(b) What factors affect our estimate of peak oil? ______________________

3. An oil company located a new area of oil sands and began an analysis of two possible extraction methods. The oil is heavy oil, making pumping difficult. The first method of extraction analyzed was Cold Heavy Oil Production with Sand (CHOPS). The system is relatively cheap but extracts only about 10% of the available oil. It also produces a lot of contaminated sand that must be disposed of. The second method is Steam Assisted Gravity Drainage (SAGD). SAGD requires two parallel wells, one above the other. Steam is pumped into the upper well. The steam heats the heavy oil, which drains into the lower well and is pumped out. This method extracts up to 80% of the oil but is very costly as it requires heating water to steam. No contaminated sand is produced, but oil-contaminated water is.

Evaluate the costs and benefits of each of these extraction methods and identify any other factors that could affect the decision of which method to use:

______________________


**ISBN: 978-1-927309-37-7**

# 114 Environmental Issues of Oil Extraction

**Key Idea**: The method used to extract oil depends on where the oil is found. Oil extraction has many environmental costs.

- Oil extraction is associated with a number of environmental issues, even before considering its transportation and use. Drilling for oil on land risks groundwater pollution, oil spills from drilling offshore affect vast areas, and the mining of oil sands and oil shales destroys thousands of hectares of forests.
- However, oil is still the most important fuel for the world's transportation industry and is integral to our daily lives. Some of the issues associated with oil extraction are described below.

### *In situ* extraction

*In situ* extraction is a method of removing oil from oil sands that are too deep to mine. It uses large amounts of energy and water. The water must be stored in tailings ponds for decontamination. Extraction produces up to three times as much $CO_2$ as the same quantity of conventional oil.

### Offshore oil platform

Oil platforms disrupt the seabed when wellheads and pipelines are laid down. There is the potential for oil spills to affect large areas of the ocean. The flaring of the gases contributes to global warming.

### Hydraulic fracturing well

In this method, oil-containing rock is fractured by high pressure fluid. The fracking fluid is mostly water with numerous additives to enhance oil mobility. The water must be stored and storage ponds may contaminate groundwater. Groundwater can be contaminated by additives and oil if the well casings are not sealed correctly or fissures link fractured layers to groundwater.

### Land based oil platform

Land based platforms cause disruption due to the construction of pads for pumps, storage and pipelines. Runoff and leaks from wells can contaminate ground and surface water. Accidental release of air pollutants, such as methane, contributes to global warming. There is the possibility of land subsidence above oil or gas fields.

### Oil sands and oil shale mining

This requires the removal of vast tracts of forest. It produces large volumes of solid and liquid toxic tailings which require huge tracts of land. Leakages of tailings ponds contaminate water.

1. What environmental issues do hydraulic fracturing and the mining of oil sands and oil shales have in common?

___

___

___

2. Explain why an oil spill from offshore oil extraction can potentially affect huge marine areas: ___

___

___

3. In groups, discuss the following scenario and evaluate the costs and benefits associated with it: "*In situ* extraction of oil uses enough clean burning natural gas in one day to heat 3 million homes. It has a ERoEI of 5:1, but produces large amounts greenhouse gases compared to other extraction methods". Include reference to energy inputs and outputs, environmental issues, and pros and cons of energy intensive oil extraction. Attach your group's summary to this page.


**ISBN: 978-1-927309-37-7**

# 115 Intensive Agricultural Practices

**Key Idea**: By maximizing the efficiency of resource use and energy conversion into product, agriculture can produce the maximum yield possible from minimum land use.

- Producing food from a limited amount of land presents several challenges: to maximize yield while minimizing losses to pests and disease, to ensure sustainability of production, and (in the case of animals) to meet certain standards of welfare and safety.
- Industrialized intensive agricultural systems meet these demands by using high inputs of energy to obtain high yields per unit of land farmed. Such systems apply not just to crop plants, but to animals too, which are raised to slaughter weight at high densities in confined areas (a technique called factory farming).

Intensive farming

## Some features of industrialized agriculture

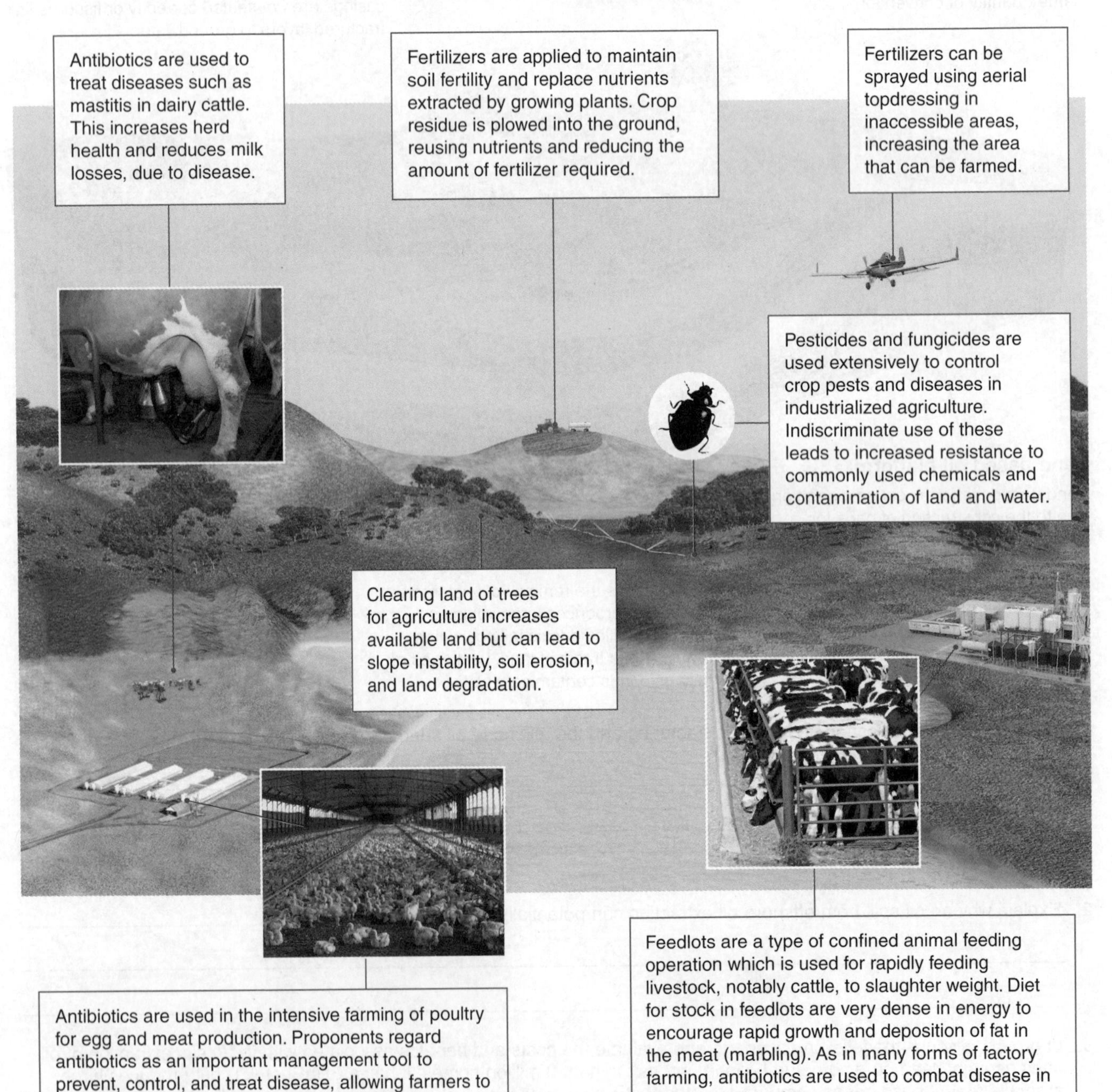





ISBN: 978-1-927309-37-7

## Agriculture and productivity

Increasing net productivity in agriculture (increasing yield) is a matter of manipulating and maximizing the conversion of energy into useful product, e.g. maximizing the grass grown to feed to livestock. On a farm, the simplest way to increase the net productivity is to produce a monoculture. Monocultures reduce competition between the desirable crop and weed species, allowing crops to put more energy into biomass. Other agricultural practices designed to increase productivity in crops include pest control and spraying to reduce disease. Higher productivity in feed-crops also allows greater secondary productivity (production of consumer biomass) in livestock. Here, similar agricultural practices make sure the energy from feed-crops is efficiently assimilated by livestock.

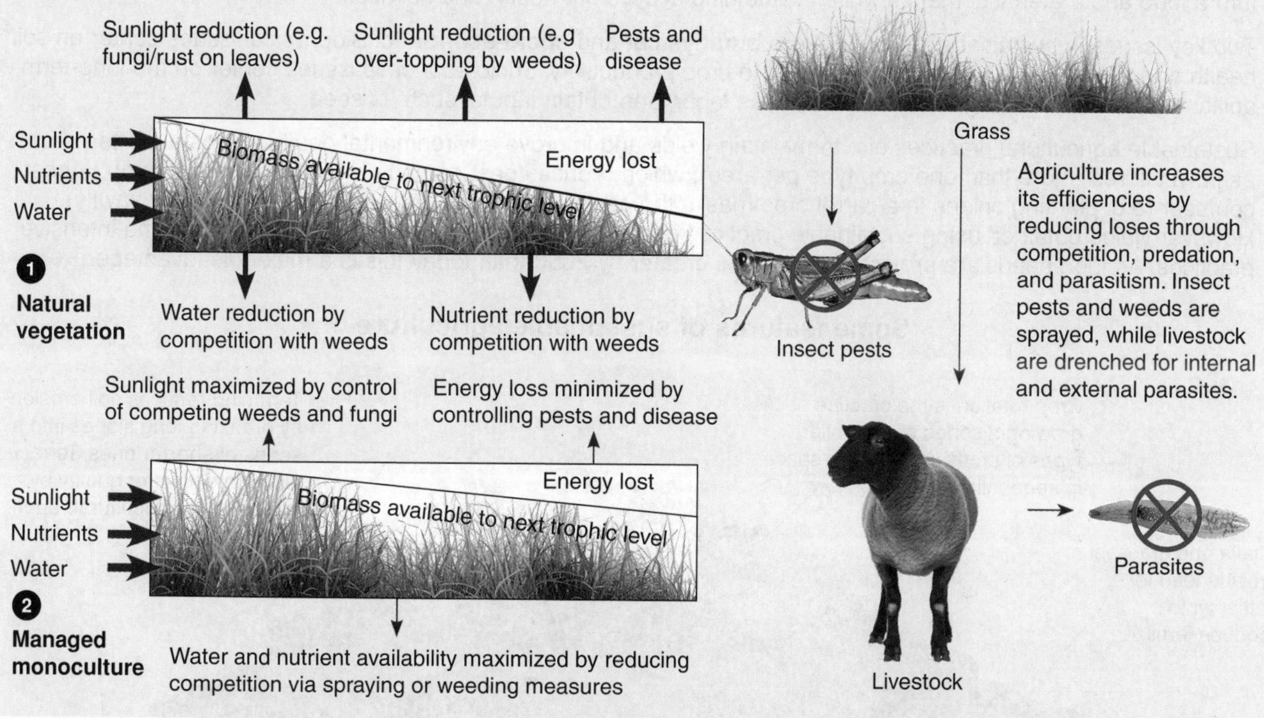


1. Explain the need for each of the following in industrialized intensive agriculture, including how they help to maximize production while minimizing land use.

(a) Pesticides: ______

(b) Fertilizers: ______

(c) Antibiotics: ______

2. Conventional intensive agriculture requires the input of large amounts of energy. Identify places where this energy input occurs and how it might be reduced, considering any effects this might have on production:

______


**ISBN: 978-1-927309-37-7**

# 116 Sustainable Agricultural Practices

**Key Idea**: Sustainable agricultural practices involve an ecosystem approach, and focus on maintaining crop yields while maintaining ecosystem health.

- **Sustainable agriculture** refers to farming practices that maximize the net benefit to society by meeting current and future food and material demands while maintaining ecosystem health and services.
- Two key issues in sustainable agriculture are **biophysical** and **socio-economic**. Biophysical issues center on soil health and the biological processes essential to crop productivity. Socio-economic issues center on the long-term ability of farmers to manage resources, such as labor, and obtain inputs, such as seed.
- Sustainable agricultural practices aim to maintain yields and improve environmental health. Crops are often grown as polycultures (more than one crop type per area), which reduces pest damage by providing a trap crop or pest confuser (e.g. planting onions in a carrot crop masks the carrots' odor and reduces damage by carrot sawfly). However yields obtained using sustainable practices can be up to 25% lower than those obtained using intensive practices. As food needs are projected to be 50% greater by 2050 than today this is a major disadvantage.

## Some features of sustainable agriculture

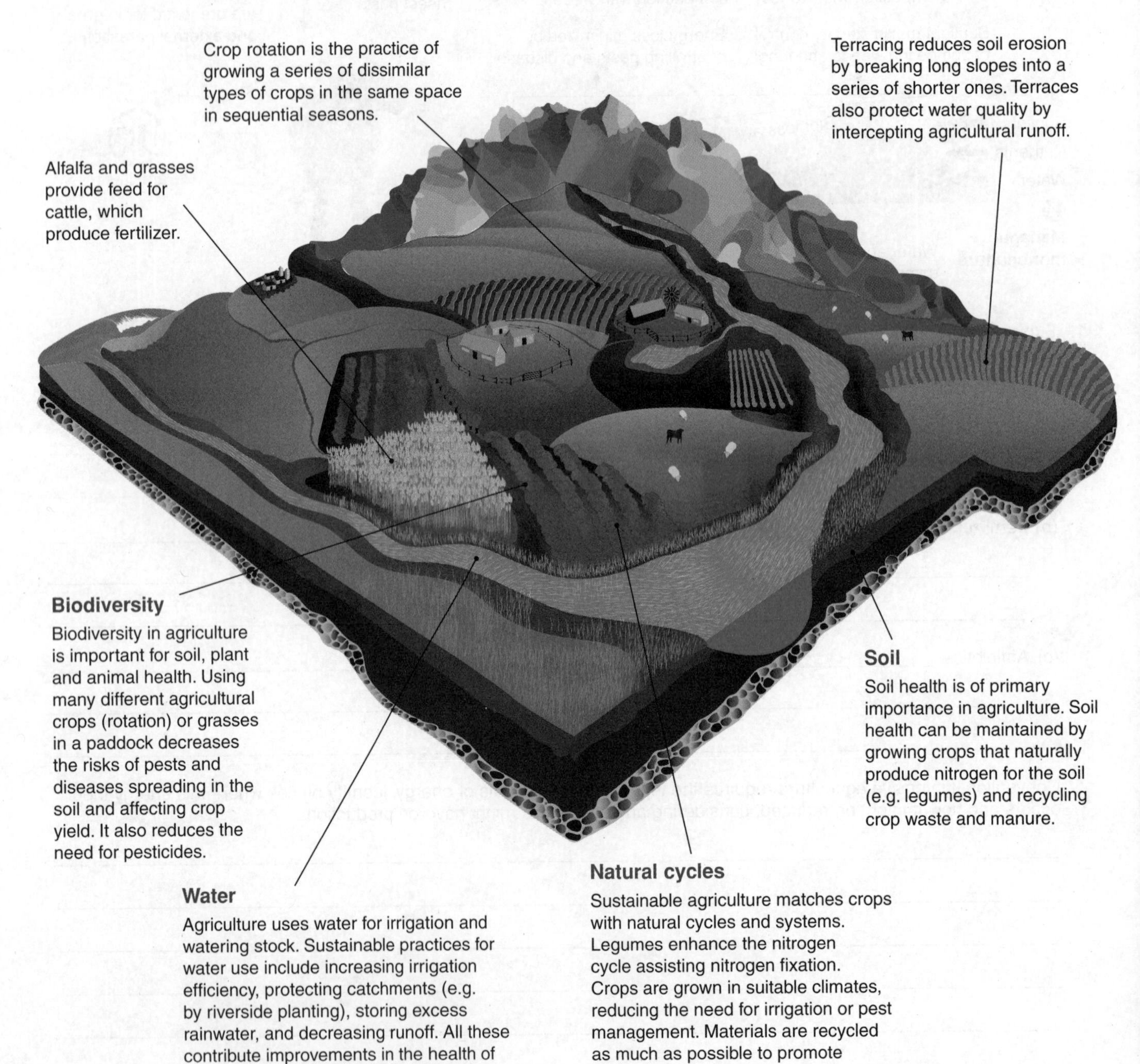

ISBN: 978-1-927309-37-7

Many sustainable practices don't yet include the use of high yielding genetically engineered organisms, such as rice and wheat. GMOs that can maximize yield with minimum resource use or in marginal growing environments will have to be a serious option to feed the world's rapidly growing population.

Phosphate rock mine

Mineral resources, such as rock phosphate and calcium sulfate, supply many of the nutrients needed to maintain agricultural production. Peak phosphorus is expected in 2030. After this, we will need to find new ways to produce phosphorus for fertilizers, such as reclaiming it from plant and animal matter.

Sharon Loxton

Nitrogen fertilizers require ammonia which is made from nitrogen and hydrogen via the Haber process. This requires large amounts of energy and hydrogen, both of which usually come from fossil fuels. To make fertilizers truly sustainable, renewable energy sources and hydrogen supplies are needed.

1. What is sustainable agriculture?

2. Explain how sustainable agriculture manages each of the following resources to meet its goals of long term sustainability:

   (a) Biodiversity:

   (b) Water:

   (c) Soil:

3. Discuss the issues associated with resource use and crop yield in agriculture. Produce a reasoned argument as to the best agricultural practice to feed the growing human population over the next 30 years:


**ISBN: 978-1-927309-37-7**

# 117 Soil Management

**Key Idea**: The type of soil makes a big difference to how it should be used and managed.

- Soil is a valuable but rather fragile resource and can be easily damaged by inappropriate farming practices. For example, attempts to clear rainforest and bring it into agricultural production have not proved sustainable. These soils are thin and nutrient poor. After only a few years farming they may be abandoned due to poor production.
- Overgrazing and deforestation may cause **desertification**. When cultivating soils, farmers must be careful not to compact soil, which would leave it lacking structure. The continual use of heavy machinery can cause soil compaction.
- Healthy soils are 'alive' with a diverse community of organisms, including bacteria, fungi, and invertebrates. These organisms improve soil structure and help to create **humus**.
- Soils known as loams consist of around 20% clay, 40% sand, and 40% silt. Loams are generally considered the most ideal soil, as they retain water and nutrients, but drain freely (right).

**Soil type diagram**

Percentage clay: 100, 90, 80, 70, 60, 50, 40, 30, 20, 10
Percentage silt: 10, 20, 30, 40, 50, 60, 70, 80, 90, 100
Percentage sand: 100, 90, 80, 70, 60, 50, 40, 30, 20, 10
Read clay in this direction
Read silt in this direction
Read sand in this direction
Clay
Sandy clay
Silty clay
Clay loam
Silty clay loam
Sandy clay loam
Loam
Loamy sand
Sand
Sandy loam
Silt loam
Silt

| | Clay | Silt | Sand | Loam |
|---|---|---|---|---|
| Nutrient holding capacity | ++ | + | 0 | + |
| Water infiltration capacity | 0 | + | ++ | + |
| Water holding capacity | ++ | + | 0 | + |
| Aeration | 0 | + | ++ | + |
| Workability | 0 | + | ++ | + |

0 = low + =medium ++= high

Loams are easily worked...

...while silts and clays can be muddy.

The capacity of soil to be worked and produce viable crops depends on its mixture of particles. Silt provides a moderate capacity in all areas due to its intermediate particle size. By itself, it is a poor soil as it too easily turns to mud when wet and is blown away by winds when dry. Loam consists of a variety of particle sizes and maintains a more consistent texture when both wet and dry.

1. Explain the term loam and how it applies to the soil properties: ______________________

______________________

______________________

2. Using the scale on soil samples 1 and 2 (right), calculate the percentage of sand, silt and clay in each sample and then use the soil triangle to identify the type of soil:

Soil sample 1: % sand: __________ % silt: __________ % clay: __________

Soil type: ______________________

Soil sample 2: % sand: __________ % silt: __________ % clay: __________

Soil type: ______________________

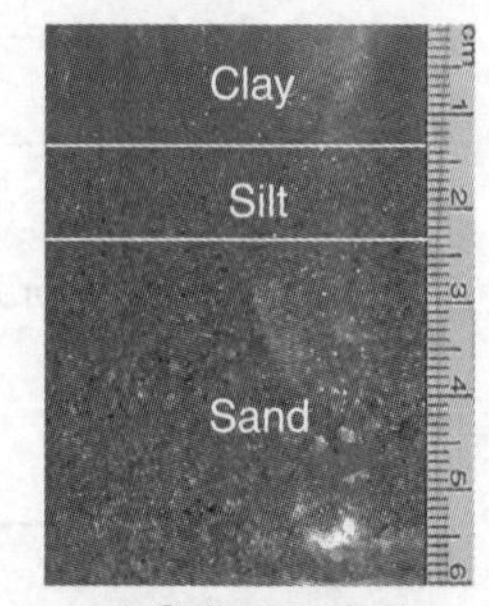

Soil sample 1

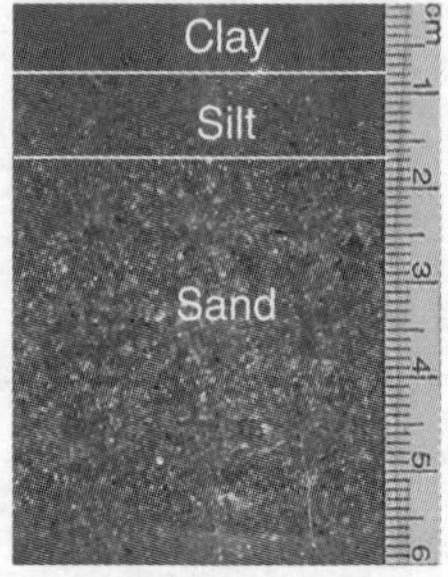

Soil sample 2

3. How does poor soil management lead to soil degradation? ______________________

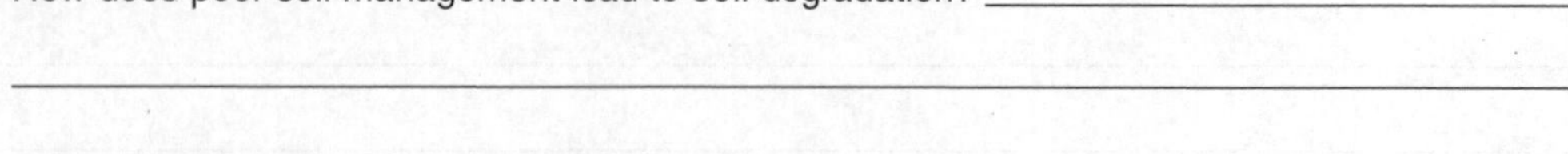

______________________

______________________

ISBN: 978-1-927309-37-7

## No till farming

- Tillage is the mechanical agitation of the ground by plowing and overturning the soil. It has several benefits including the aeration of the soil, mechanical destruction of weeds, and plowing the nutrients in green crops into the ground ready for the next crop.
- However there are several disadvantages to tillage, including the exposure of soil to the wind, increasing erosion and loss of water. One method of reducing soil erosion is the no-tillage system (right and below) in which the residue of the previous crop is left on the surface and seed is planted beneath the ground with special machinery.

No-till farming

USDA

### No-till

Vegetation after harvest is mowed, rolled, or sprayed to begin break down. Seeds are planted directly into the ground with seed drills. Residue from the previous harvest helps to reduce water and soil loss from exposed or vulnerable soils, and prevents weeds becoming established.

### Conservation tillage

Up to 50% of the crop residue is plowed and buried. Cultivators are used to break up the soil clods. After planting, cultivators and herbicides are used to remove weeds.

### Intensive tillage

Up to 90% of crop residue is plowed and buried. Fields are rotary hoed, disked, and harrowed to break up soil and produce a smooth surface for planting. This disturbance to the soil can help if soils are wet or compacted. After planting, cultivators and herbicides are used to prevent weeds growing.

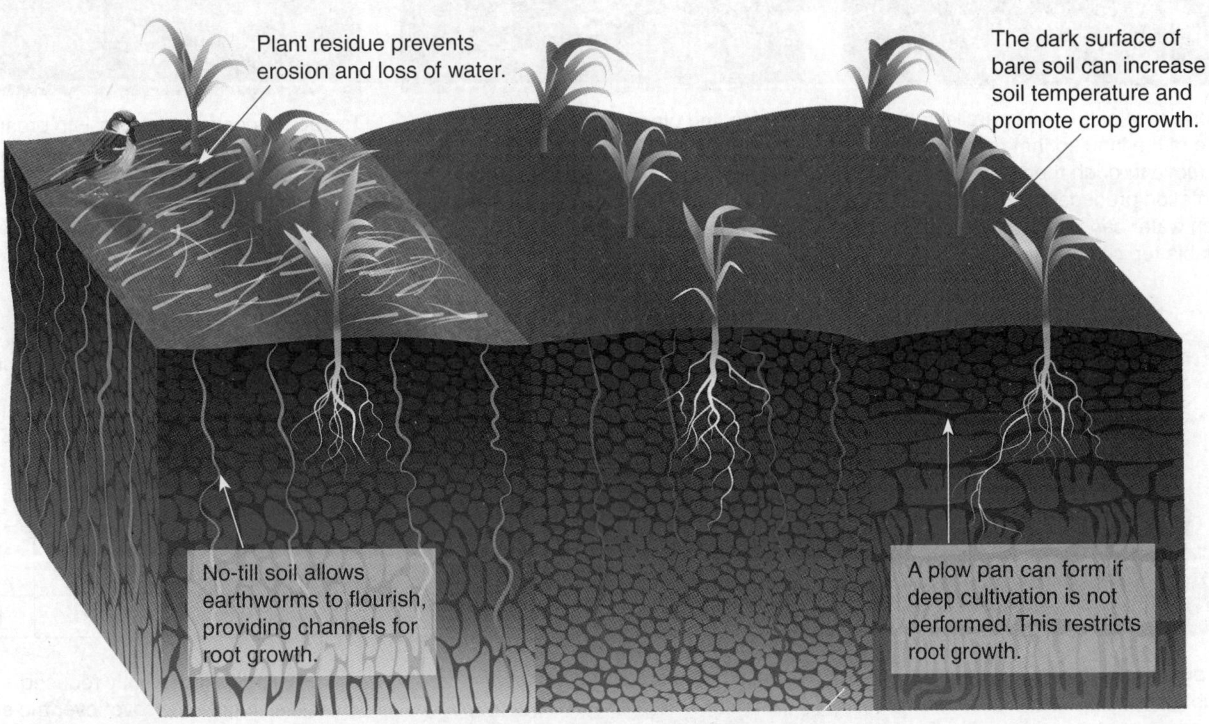


4. Explain why loamy soils are more easily worked and produce higher crop yields than other soil types:

5. (a) Describe the main difference between no-till and tillage farming methods:

   (b) Contrast the effort required for no-till and intensive tillage methods and suggest when each of these methods might be advantageous.


**ISBN: 978-1-927309-37-7**

# 118 Reducing Soil Erosion

**Key Idea**: Huge volumes of topsoil are lost every year. Soil conservation practices are required to make sure this valuable resource is not lost.

- Good soil is vital for productive agriculture. Most modern cropping techniques use heavy machinery to turn over the remnants of harvested crops, break up the soil, and smooth it flat to form a planting surface. This leaves the soil exposed, and large volumes of topsoil can be lost through wind or rain before there is sufficient crop cover to protect it.
- Even when a crop is fully established, there may still be exposed ground from which soil can be lost. Alternative planting techniques such as **no-till farming**, **terracing**, **contour plowing**, **windbreaks**, and **intercropping** reduce the exposure of soil to the elements.

Crops are often planted parallel to the slope of the land so that machinery can move through them easily. This orientation produces channels down which water can easily flow, taking valuable top soil with it.

USDA

Plowing and planting across, rather than down, slopes produces contours that slow water runoff and reduce soil loss by up to 50%. Water has more time to settle into the soil, reducing the amount of irrigation required.

**Terracing** converts a slope into broad strips, slowing or preventing water and soil runoff and reducing erosion. This technique is commonly used in paddy fields. Terraces also help to control flooding downstream.

USDA

**Windbreaks** reduce soil erosion by reducing wind speed close to the ground. They also reduce water loss, and so lower irrigation requirements. Windbreaks placed near drainage ditches help to reduce erosion because the tree roots stabilize soil at the edge of the ditch.

Wawny

**Agroforestry** is a combination of agriculture and forestry. Crops or stock are raised on the same land as a stand of woody perennials. Biodiversity levels are often higher than in conventional agricultural systems and soil loss is much reduced.

Cunningham- Saigo

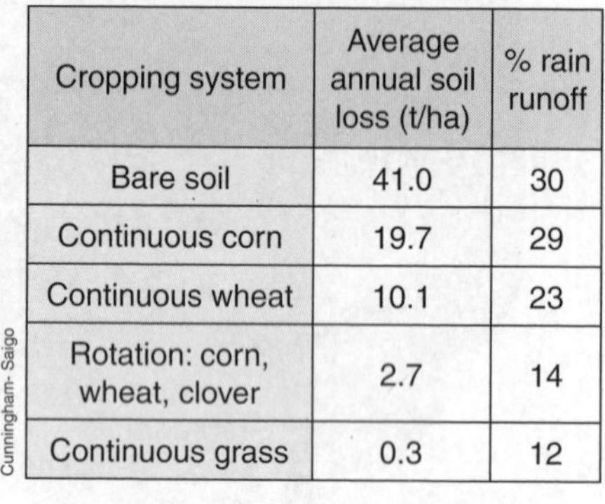

| Cropping system | Average annual soil loss (t/ha) | % rain runoff |
|---|---|---|
| Bare soil | 41.0 | 30 |
| Continuous corn | 19.7 | 29 |
| Continuous wheat | 10.1 | 23 |
| Rotation: corn, wheat, clover | 2.7 | 14 |
| Continuous grass | 0.3 | 12 |

Soil erosion is significantly reduced when the vegetative cover over the soil is maintained (above). Continuous cover can be achieved by using machinery to plant crops directly into the soil beneath the existing ground cover, often along with fertilizers and pesticides.

1. Explain how terracing and contour plowing reduce soil loss compared to plowing parallel to the slope: ______________________________

2. Explain how maintaining vegetative cover reduces soil erosion: ______________________________

ISBN: 978-1-927309-37-7

# 119 Managing Rangelands

**Key Idea**: Rangelands for grazing livestock cover half of the Earth's land. However they are fragile ecosystem and must be carefully managed.

- **Rangelands** are large, relatively undeveloped areas populated by grasses, grass-like plants, and scrub. They are usually semi-arid to arid areas and include grasslands, tundra, scrublands, coastal scrub, alpine areas, and savannah.
- Globally, rangelands cover around 50% of the Earth's land surface. The USA has about 3.1 million $km^2$ of rangeland, of which 1.6 million $km^2$ is privately owned. Rangelands cover 80% of Australia, mostly as the outback, but only 3% of Australia's population live in rangeland areas.
- Rangelands are often used to graze livestock (e.g. sheep and cattle) but they regenerate slowly because of low rainfall. Careful management is required to prevent damage and soil loss as a result of overgrazing.

- Grasses grow continuously from a plant meristem (dividing cells) close to the ground, so the leaf can be cropped without causing growth to stop. This allows a field to be grazed in a near-continuous fashion. Grazing by animals stimulates grass to grow and removes dead material. Grasslands cropped at their optimum capacity can be much more productive than if left uncropped (right).
- Overgrazing occurs when too many animals are grazed for too long on a section of pasture and the grass does not have enough time between cropping to regrow. Overgrazing may destroy the meristem, in which case plant regeneration stops. Exposed soils may become colonized by invasive species or eroded by wind and rain.

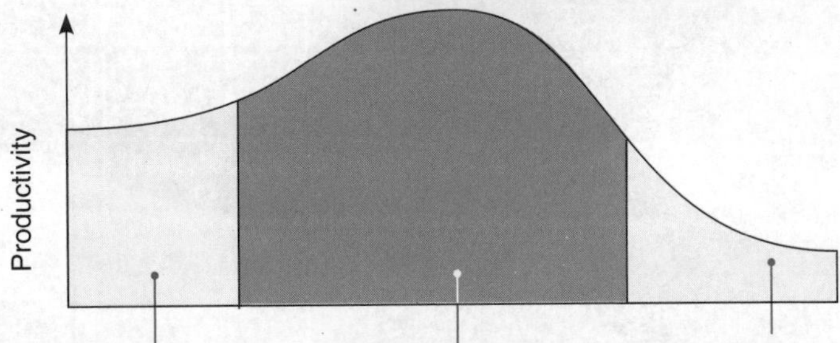


**Undergrazing**
Net productivity is reduced because standing dead material leaves little room for new growth to come through.

**Optimum grazing**
Old material is removed so new growth can come through, but enough growing material is left to allow recovery.

**Overgrazing**
Too much material is removed and new growth can not become established. Plants die and soil is lost through erosion.

**Total net primary production and efficiency of grazed and ungrazed grasslands**

| | | Net production ($kcal/m^2$) | Efficiency (%) |
|---|---|---|---|
| Grazed | Desert | 1081 | 0.13 |
| | Shortgrass plains | 3761 | 0.80 |
| | Mixed grasslands | 2254 | 0.51 |
| | Prairie | 3749 | 0.77 |
| Ungrazed | Desert | 1177 | 0.16 |
| | Shortgrass plains | 2721 | 0.57 |
| | Mixed grasslands | 2052 | 0.47 |
| | Prairie | 2220 | 0.44 |

From Ecology and Field Biology, R. Smith

1. Using evidence from this page, explain how carefully managed grazing on a rangeland can increase its productivity:

2. Why can rangelands easily become overgrazed?


ISBN: 978-1-927309-37-7

## Rotating livestock

Sustainable management of natural resources such as soil and grasslands is necessary if those resources are to be available in the future. Calculating stocking rates for an area to be grazed is important to ensure that the land is not overgrazed and stock rotations are sustainable. Livestock are selective grazers, so care must be taken when grazing only one livestock species that particular grasses or shrubs are not eaten to local extinction.

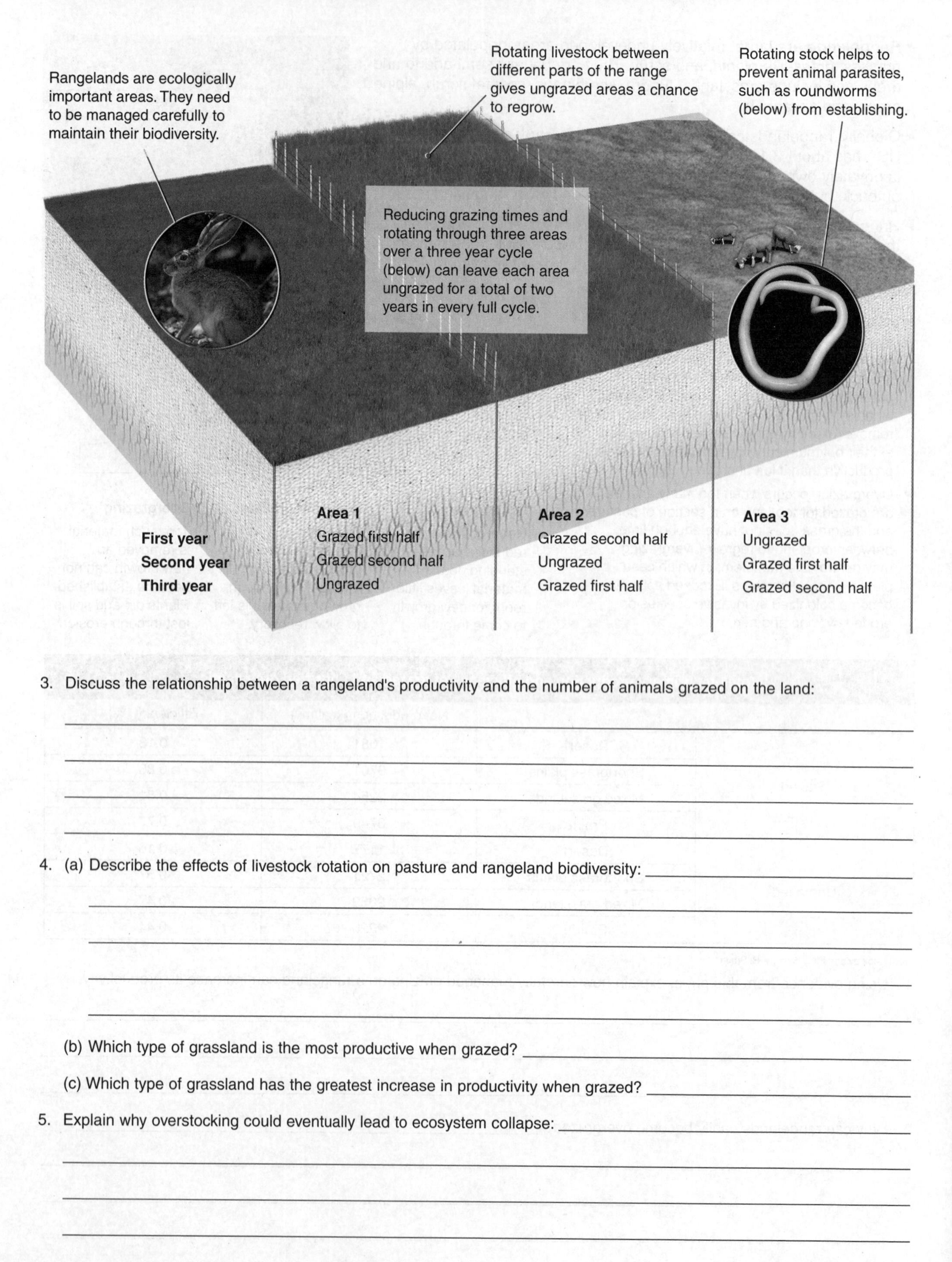


3. Discuss the relationship between a rangeland's productivity and the number of animals grazed on the land:

4. (a) Describe the effects of livestock rotation on pasture and rangeland biodiversity:

(b) Which type of grassland is the most productive when grazed?

(c) Which type of grassland has the greatest increase in productivity when grazed?

5. Explain why overstocking could eventually lead to ecosystem collapse:


**ISBN: 978-1-927309-37-7**

# 120 Living With Limited Resources

**Key Idea**: Reprocessing reduces the energy required to produce resources, reduces costs, and effectively increases the stock of otherwise limited resources.

## Reusing waste

- Much of the waste produced by industrialized countries contains valuable resources, which could be used again if properly processed. As resources become scarcer and competition for them grows, both individuals and companies are beginning to explore ways of using resources and reusing waste material as efficiently as possible.
- Although most materials can theoretically be reused or recycled, there are always some situations where it is impractical. In those cases, waste may be burned to extract energy or taken directly to a landfill (rubbish dump).
- The schematic below shows an integrated resource recycling scheme that reduces waste as much as possible.

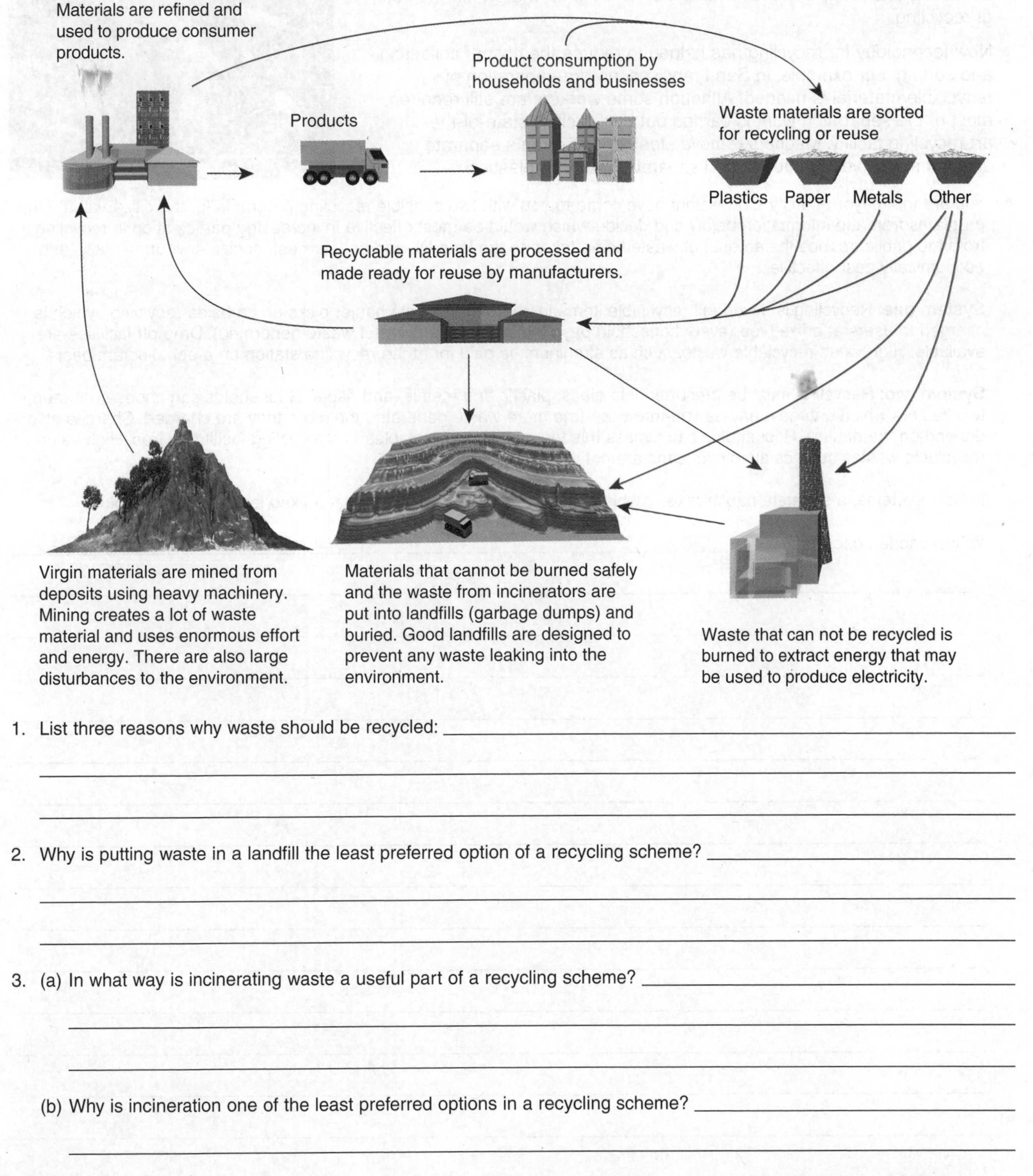


1. List three reasons why waste should be recycled: ______________________________

______________________________

______________________________

2. Why is putting waste in a landfill the least preferred option of a recycling scheme? ______________________________

______________________________

______________________________

3. (a) In what way is incinerating waste a useful part of a recycling scheme? ______________________________

______________________________

______________________________

(b) Why is incineration one of the least preferred options in a recycling scheme? ______________________________

______________________________


ISBN: 978-1-927309-37-7

- Recycling, like any resource extraction method, costs energy. This may be in the form of collecting and sorting material, transport to manufacturers, and the energy used in melting or remelting a material. In order for recycling to be advantageous, it must cost less energy than producing a new item or material from the original resource.
- Some materials returned from recycling do not return the same quality of product as new material (e.g. wood and aluminum) and are often made into inferior product (a process called downcycling). In these cases, new material needs to be added, so recycling is not a closed loop or a total solution. However, it does significantly reduce the need for new materials, extending the life of the resource and reducing the energy spent on producing new material.

- Aluminum requires 96% less energy to make from recycled cans than it does to process from bauxite (ore). Recycled glass uses about 21% less energy, recycled plastic bottles use 76% less energy, and newsprint about 45% less. In all cases, what matters the most is the cost of extracting and refining the raw materials, which is far more than the cost of recycling.
- New technology for recycling has helped to reduce the cost of collecting and sorting. For example, in San Francisco, no pre-separation of recyclable material is needed. Although some workers are still required, most of the separation work is carried out by the city's state-of-the-art recycling facility. Magnets remove steel, eddy currents separate aluminum, and vacuum tubes and separators remove plastics.

4. You are the mayor of a city. Counselors have come to you with two possible recycling programs for the city. Evaluate the programs from the information below and decide which would be most effective in increasing participation in recycling from the public, reduce the amount of waste being taken to the land fill, and be the most, social, environmentally, and economically cost effective.

   **System one**: Recycling is mixed (all recyclable items in one bin). A single hauler picks up curbside recycling, which is charged to users at a fixed fee (every household pays the same regardless of waste generated). Drop off facilities are available. High value recyclable waste, such as aluminum, is paid for at the recycling station on a per kilogram basis.

   **System two**: Recycling must be presorted into glass, plastic and metals, and paper. Households can choose between two haulers which charge a pay-as-you-throw fee (the more waste generated the more they are charged. Charges also depend on the hauler). Households can bypass this fee by taking recyclables to a recycling facility for free. High value recyclable wastes such as aluminum cans are not paid for.

   In both systems, a separate hauler takes rubbish to the landfill. This is charged at a fixed fee for all households.

   Your reasoned decision:


**ISBN: 978-1-927309-37-7**

# 121 Chapter Review

Summarize what you know about this topic so far under the headings provided. You can draw diagrams or mind maps, or write short notes to organize your thoughts. Use the introduction and the images and hints included to help you:

Resources, people, and water

HINT: Relationships between resource locations and major population centers, both historic and present. How has water shaped human civilization?

Fossil fuels

HINT: Environmental, social, and economic costs and benefits of oil and coal extraction.

Soil resources

HINT: Managing soil to maintain fertility.

ISBN: 978-1-927309-37-7

# 122 KEY TERMS AND IDEAS: Did You Get It?

1. Test your vocabulary by matching each term to its definition, as identified by its preceding letter code.

|  | |  |
|---|---|---|
| coal | A | A naturally occurring solid, inorganic substance, with a crystal structure representable by a specific chemical formula. |
| ERoEI | B | Black sedimentary rock, consisting primarily of carbon. Formed from the buried and compressed remains of ancient swamps and now used as a high energy fuel. |
| fossil fuel | C | Fuel produced millions of years ago by the burying and compression of organic matter. |
| mineral | D | A resource that cannot be replaced unless over geologic time scales. |
| non-renewable resource | E | Liquid made from hydrocarbons that was formed from the buried remains of marine or lake living planktonic organisms. |
| oil | F | A resource that can be replaced or regenerated within a short time span. |
| renewable resource | G | The amount of energy returned by a product compared to the amount of energy invested in the extraction or production of that product. |

2. The table below shows some properties of three metals, iron (as steel), aluminum, and titanium, which are used today for various purposes:

| | Metal | | |
|---|---|---|---|
| **Property** | **Iron (as carbon steel)** | **Aluminum (alloy 6061)** | **Titanium (grade 5)** |
| **Strength (MPa)** | 841 | 300 | 950 |
| **Density (g/cm$^3$)** | 7.58 | 2.7 | 4.5 |
| **Resistance to corrosion** | Medium-low | High | Very high |
| **Price per tonne** | $500 | $2000 | $30,000 |
| **Abundance in crust (ppm)** | 63,000 | 82,000 | 6,600 |
| **Ease of refinement from ore** | Easy | Difficult | Difficult |

Using the table above explain why the metals are used in the following ways:

(a) Iron is commonly used in the construction industry to build large scale buildings (e.g. factory sheds or sky scrapers):

(b) Aluminum and titanium are commonly used in the aerospace industry (e.g. building parts of planes):

3. When used in the production of electricity in power stations, coal has an efficiency of 37%, oil 37%, and gas 45%. Considering the cost and effort of producing these resources, provide an opinion on which is the best one to use in electricity production:


**ISBN: 978-1-927309-37-7**

# 123 Summative Assessment

Resource Island is located 35° South, 1540 km from the nearest land mass. It is 21 km wide at its widest point and 843 m above sea level at its highest point, the table-topped Mount Kiilua. The prevailing winds come from the west. Rainfall is around 1000 mm per year, 65% of which falls during March to September. Currently undeveloped, Resource Island is to be colonized. Scientists think that if the island is managed carefully, it may one day be able to sustain up to 15,000 people.

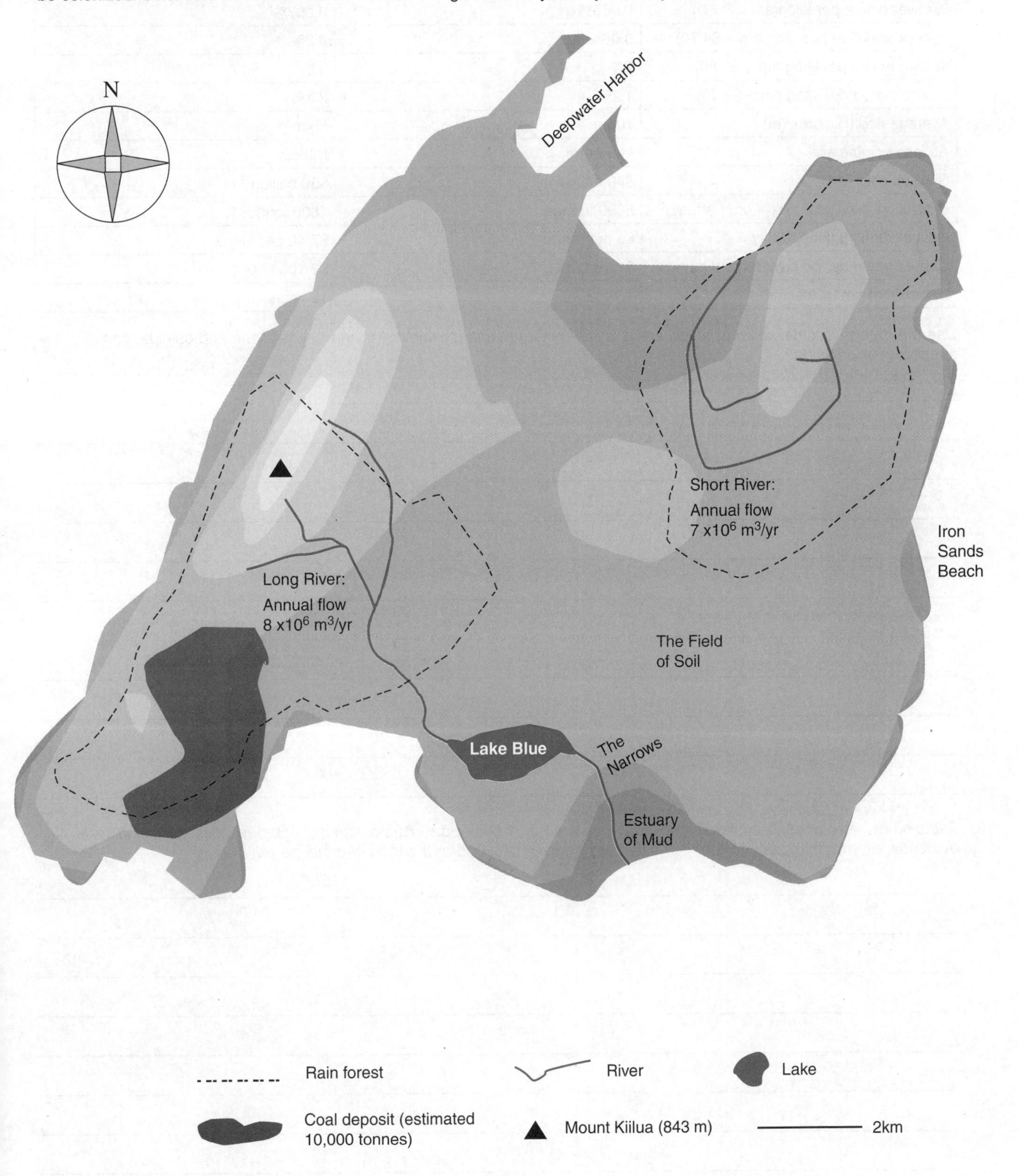


1. Resource Island will eventually host a population of 15,000 people. Mark out where you would place the main city and fields for crops and livestock. Suggest how energy and building resources will be sustained in the long term. Attach a separate sheet explaining how you would produce the enough power for 15,000 people, deal with both solid and liquid waste, and ensure adequate water supplies for drinking and irrigation. Your decisions must take into account and justify the impact on the environment.


**ISBN: 978-1-927309-37-7**

2. A mining company explores two potential surface mining areas using a drill to provide core samples. The costs and mineral content of each site is shown below:

| | Area 1 (surface area approximately 20 km$^2$) | Area 2 (surface area approximately 14 km$^2$) |
|---|---|---|
| % gold (value per kilogram = $43,000) | 0.00008% | 0.00005% |
| % silver (value per kilogram = $650) | 0.005% | 0.007% |
| % copper (value per kilogram = $4.70) | 0.08% | 0.2% |
| % lead (value per kilogram = $1.80) | 2% | 3% |
| % zinc (value per kilogram = $2.15) | 1.2% | 2.1% |
| Average depth of ores (m) | 100 m | 50 m |
| Access to mine site | Moderate | Difficult |
| Start up cost | $40 million | $50 million |
| Extraction rate (total rock + ore per day) | 5000 tonnes | 4800 tonnes |
| Cost of running mine facility | $9 per tonne | $7.40 per tonne |
| Cost of restoring the environment per km$^2$ | $1,200,000 | $2,100,000 |
| Approximate mine lifetime | 15 years | 12 years |

Use the data to decide which of these areas is the most suitable for mining, giving any reasons and calculations to support your decision:

3. Explain how resources (fuel, water, soil) influence human social and technological development and activity. Provide evidence for your explanation. You may use extra paper if required and attach it to this page.


**ISBN: 978-1-927309-37-7**

ESS3.B

# Natural Hazards

**Key terms**

climate change
displacement
drought
earthquake
El Niño
eruption
fire
flood
global warming
hurricane
migration
mitigation
natural hazard
tsunami

## Disciplinary core ideas

Activity number

***Show understanding of these core ideas***

**Natural hazards have shaped human history**

- ☐ 1 Natural hazards are environmental events that could potentially cause damage or threaten human lives and property. Natural hazards may arise as a result of the Earth's internal processes (e.g. earthquakes, volcanic eruption), surface processes (e.g. tsunamis, landslides), or severe weather (hurricanes, floods). — 124 125
- ☐ 2 The occurrence of natural hazards and other geological events has shaped the course of human history, influencing the size and location of populations and acting as drivers for human migration. — 124 125 126
- ☐ 3 Climate change data supports the link between global warming and an increased frequency of extreme weather events, such as flooding, droughts, hurricanes, and storm surges. — 125 130
- ☐ 4 Extreme weather events are associated with the displacement or long term migration of human populations. — 126
- ☐ 5 Sea level rise and changes in predicted patterns of temperature and rainfall will determine where crops can be grown and where livestock can be raised. Climate change may alter vegetation distribution and make some regions less suitable for particular types of agriculture and others more suitable. — 125 127
- ☐ 6 Climate change mitigation will involve new technologies and renewable energies, as well as design solutions that will alleviate the impact of climate change. — 128

## Crosscutting concepts

Activity number

***Understand how these fundamental concepts link different topics***

- ☐ 1 **CE** Empirical evidence enables us to explain how the occurrence of natural hazards has influenced patterns of human settlement and migration. — 126 127
- ☐ 2 **CE** Empirical evidence enables us to make a claim about the link between global warming and the increased frequency of extreme weather events. — 125 126 130
- ☐ 3 **CE** Empirical evidence enable us to make a link between realized and predicted rates of human displacement and climate change. — 126
- ☐ 4 **SPQ** The significance of a natural hazard depends on the scale at which it occurs. — 125 127

## Science and engineering practices

Activity number

***Demonstrate competence in these science and engineering practices***

- ☐ 1 Construct an explantation based on evidence for how the occurrence of natural hazards has influenced human activity. — 125-127
- ☐ 2 Construct an explantation based on evidence for the past and predicted effects of climate change on the location and mass movements of human populations. — 126
- ☐ 3 Evaluate solutions for mitigating the risks and impact of natural hazards. — 128
- ☐ 4 Evaluate the claims or evidence behind explanations for natural disasters. — 131

# 124 What Are Natural Hazards?

**Key Idea**: Natural hazards are environmental events that could potentially cause damage or threaten human lives and property.

- The Earth can be a hazardous place. The processes that shape the surface of the Earth can also produce surface disturbances on vast scales, e.g. tropical cyclones. **Natural hazards** are any natural event that may cause damage on a small or large scale. They can be grouped into biological (e.g. disease) and geophysical hazards. Geophysical hazards include geological events (ground occurring, e.g. earthquakes) or meteorological events (atmospheric, e.g. hurricanes).
- It is important to distinguish between a natural hazard, e.g. a volcanic eruption, and a **natural disaster**, e.g. the wide scale loss of life and property caused by a volcanic eruption.

**Storms** can produce large scale effects. Hazards include lightning strikes, wind and water damage, flooding, and landslides due to water -logged soil.

**Rivers** and **lakes** present possible flood hazards. People living on the floodplains of large rivers are often inundated when rivers burst their banks. Many rivers near towns have levees or stopbanks to contain the water and prevent flooding.

1. Identify the natural hazards associated with each of the following phenomena:

   (a) Storm: ______________________

   (b) Volcano: ______________________

   (c) High mountains: ______________________

2. Identify ways to reduce the risk of damage from each of the following hazards:

   (a) Drought: ______________________

   ______________________

   (b) River: ______________________

   ______________________

   (c) Seaside: ______________________

   ______________________




**ISBN: 978-1-927309-37-7**

Hazards in high mountains include rock falls, blizzards, and avalanches. Avalanches occur when the snowpack loses adhesion and slides down the mountain face. More than 100 people a year are killed in avalanches despite the large effort that is put into reducing the damage they cause.

Tropical cyclones, typhoons, and hurricanes present hazards around the tropical regions of the world. Damage may occur due to 200 km/h winds and flooding from rainfall and storm surges. Large cyclones can cause damage up to 40 km from the coast. Climate change may increase the frequency of cyclones.

Many areas of land used for grazing or cropping are prone to drought. Drought causes enormous economic damage as it seriously reduces farm productivity. Efforts to reduce drought damage include building dams to store water and planting drought tolerant crops. Climate change may increase the occurrences of drought.

**Landslides** are common on mountains and steep terrain. Every year, many thousands of homes are destroyed by landslides.

**Tsunamis** result from movements of the sea floor caused by earthquakes or subsea landslides. Waves may reach more than 10 m high when they reach the shore. Many high risk areas (such as Japan) have high sea walls to reduce tsunami damage.

**Earthquakes** result from the sudden movement of the ground along fault lines. They present a hazard with the potential to cause enormous damage. Billions of dollars a year is put into recovery from earthquakes or planning and building to minimize damage risk.

3. (a) Identify one significant natural hazard in your local area: ______________________

______________________________________________________________________

(b) How does your local area prepare for or reduce the risks posed by this natural hazard? ______________________

______________________________________________________________________

______________________________________________________________________

______________________________________________________________________

(c) Why do you think people might have settled in the area despite the presence of this hazard? ______________________

______________________________________________________________________

______________________________________________________________________


**ISBN: 978-1-927309-37-7**

# 125 The Effects of Natural Hazards

**Key Idea**: The impact a natural hazard has depends on the features of the natural hazard itself, the natural features of the land affected, and a region's economic development.

- Natural hazards are not recent phenomena. They have been (and will continue to be) caused by the same processes that have always occurred (e.g. movement along a fault line produces earthquakes). However, some natural hazards have been made more common through human activity (e.g. forest fires, or flood events caused by changing the course of a river).

## Many factors influence the impact of a natural hazard

- **Natural features**
  Natural features of the Earth and the environment can have a significant influence on how much damage a natural hazard can cause. For example, the angle of the coastline (steep or gentle incline) will influence how much damage will be caused by a storm surge (right).
- **Magnitude**
  The size (or magnitude) of the event has a significant impact on the damage that the natural hazard causes. For example a 2.5 magnitude earthquake is not usually felt, but a 7.0 magnitude earthquake in the same location can cause serious damage.
- **Frequency**
  The frequency of a hazard (how often it occurs) will directly affect how a particular area responds and recovers from the most recent event. For example, areas prone to frequent flood events may be severely damaged with each event and also do not have time to recover between floods.

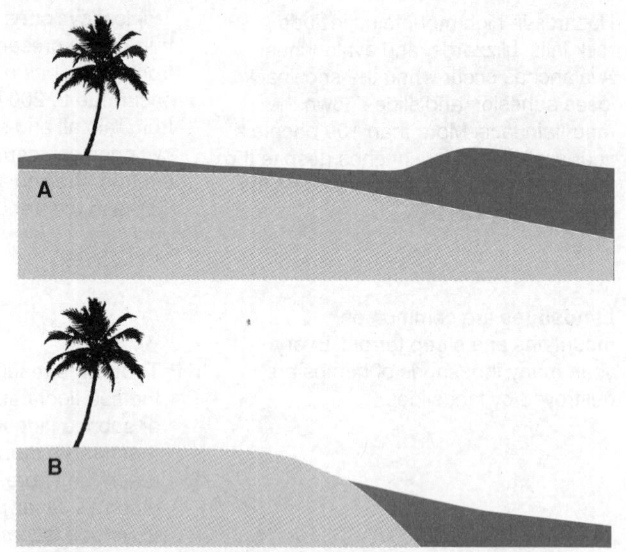


The effect of storm surge on a gently sloping coastline (A) and a steep coastline (B) is shown above. The storm surge travels much further inland when the coastline has a gentle slope, potentially damaging more land and property. A steeper slope at the coastline prevents or limits inundation, and acts as a natural protective barrier.

## Many factors influence the impact of a natural disaster

When a natural hazard event affects human life or property, the event is called a **natural disaster**. The extent of the natural disaster depends not only on the features of the natural hazard, but social factors too.

- **Level of development**
  The level of development (e.g. infrastructure and money) of an area will contribute to how well that region can respond to a natural disaster. Better economically developed regions will have the resources to respond more quickly and more efficiently than lesser economically developed regions.
- **Preparedness**
  How ready the population is to respond to a natural disaster will influence how well they cope immediately after the event. Factors include constructing buildings to withstand earthquakes in vulnerable areas (e.g. San Francisco) or having resources ready to be distributed after the event. Early warning systems, such as the tsunami warning system, are designed to give coastal residents time to evacuate to higher ground.
- **Accessibility**
  Remote or severely damaged areas can be difficult for disaster teams to access. If support cannot be provided to the affected population, death rates may rise. For example, if the water supply has been contaminated, disease may spread through the population. Without access to clean water and medical supplies, people may die from causes secondary to the disaster event itself.

**Haiti earthquake (2010): Magnitude 7.0**

United nations Logan Abassi cc2.0

**Kobe earthquake (1995): Magnitude 7.2**

Kobe (1995) and Haiti (2010) were both struck by similar sized earthquakes. The damage was less severe in Kobe because Japan's strict building codes reduced the number of collapsed buildings. Regular earthquake drills meant the population were well rehearsed for an earthquake and emergency services were able to deliver supplies quickly. The death toll was much lower in Kobe and recovery was much faster because Japan has a higher level of economic development and preparedness than Haiti.

**ISBN: 978-1-927309-37-7**

## Natural hazards can cause natural disasters

The Atlantic hurricane season refers to a period (June to November) when hurricanes usually form in the Atlantic Ocean. During this period, the East Coast of the US can expect a number of hurricanes ranging in intensity from category 1-5. Anything above category 3 is a major hurricane. An average of 6 hurricanes develop during the Atlantic season, with 2-3 of these developing into category 3 or greater hurricanes.

Hurricanes are a regular natural hazard, but some have greater impact than others. The devastation caused in New Orleans by Hurricane Katrina in 2005 was due to a combination of the intensity of the hurricane and the physical land features of New Orleans. Hurricane Katrina varied in intensity reaching category 5 before reducing in intensity to category 1-2 when it reached New Orleans. The prolonged heavy rain and storm surge (up to 9 meters) meant that many of the levees and floodwalls failed, resulting in flooding of up to 80% of the city (right). Many of the 1464 deaths associated with Hurricane Katrina were a result of levee failure.

US Coastguard Public domain

### A history of flooding in New Orleans

New Orleans was originally built on natural levees along the Mississippi River. New Orleans is completely surrounded by water, the Mississippi River on one side and Lake Pontchartrain on the other. As the city grew and demand for land increased, people settled the lower lying land, which was more prone to flooding. Houses were built elevated above ground to cope with the frequent flood events. Over time, a series of drains, levees and floodwalls were developed to help protect New Orleans from flooding. Today, much of the city lies below sea level (right), relying on the levees and floodwalls to protect it. During Hurricane Katrina, the strong storm surges and prolonged rain caused most of the levees to fail, resulting in widespread flooding and destruction. (NGVD is the National Geodetic Vertical Datum, a reference point for elevation).

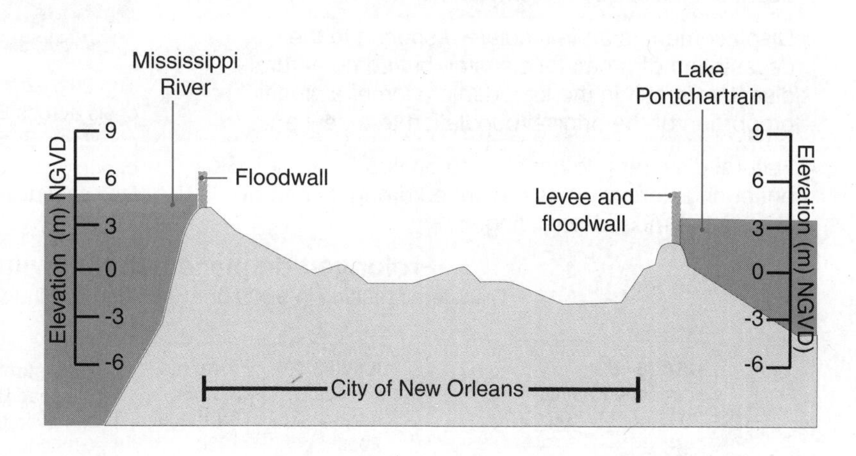


1. Describe some features of natural hazards that can increase their impact: ______________________

2. Using an example, explain how a country's level of economic development can affect the impact of a natural disaster?

3. What factors contributed to Hurricane Katrina significantly damaging New Orleans? ______________________


**ISBN: 978-1-927309-37-7**

# 126 Natural Hazards and Human Migration

**Key Idea**: Migration is a survival response of people facing the prospect, impact, or aftermath of natural disasters. Some displaced people are never able to return.

## Natural hazards can force migration

Throughout history, natural hazards or the natural disasters that may arise as a result of them have significantly altered the size of human populations or forced movements away from the affected area. Sometimes, these movements are temporary and areas affected by the hazard are eventually repopulated. However, in some cases, people never return, because they settle permanently elsewhere or because the damaged area is no longer capable of sustaining them.

- The impact of a natural disaster depends on a variety of factors, including the time of day that the hazard strikes, the vulnerability of the population, and the social and economic factors in the environment.
- Displacement is an immediate response to the devastation of a natural disaster, but some natural disasters result in the forced migration of a significant proportion of the original population to a new area.
- Natural disasters do not have to be sudden onset. Slow onset disasters such as prolonged droughts can be among the most devastating.

Oxfam delivers water in southern Ethiopia

The 2011-2012 sub-Saharan drought caused a severe food crisis across East Africa and affected 9.5 million people in four countries. Such slow onset disasters force mass movements of people in search of food and water. Many die and survivors face uncertain futures in makeshift refugee camps.

### Prolonged displacement following disasters

Disasters resulting in 9000 or more people still displaced in 2015 noted

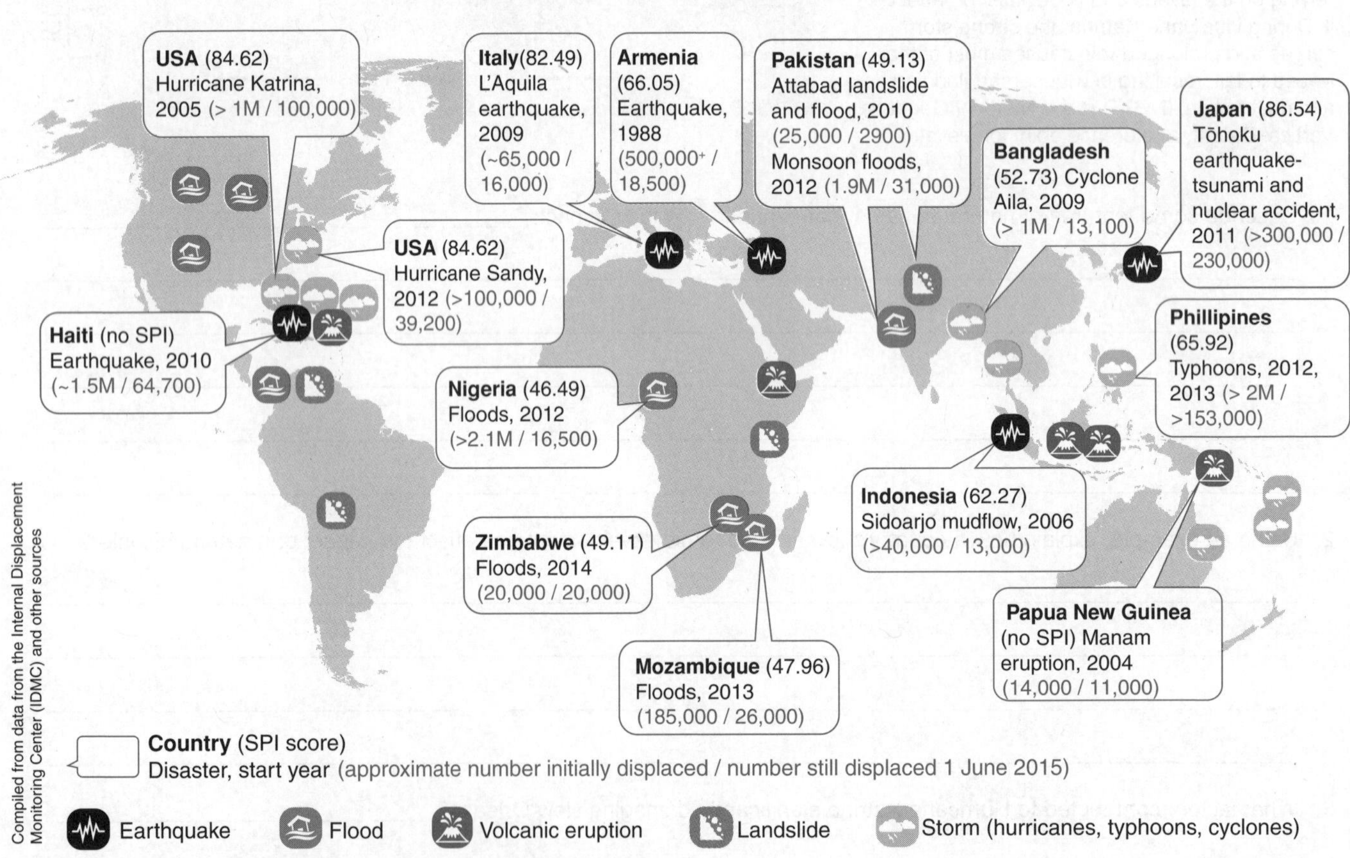


## People displaced after disaster may never return

- The map above indicates the location of some natural disasters in the last decade or so, with some specific events identified. The blue numbers indicate the number of people still displaced as a result of the disaster (as of 2015).
- The Social Progress Index (SPI) score is given for the countries indicated. This index is based on 54 indicators in the areas of basic human needs (e.g. sanitation, water, safety, freedom). It therefore indicates the extent to which countries provide for the social and environmental needs of their citizens.

ISBN: 978-1-927309-37-7

## Climate change and the fate of island nations

Even under the most conservative projections of climate change, rising sea levels will place many coastal and low lying regions of the world at risk of inundation. Many of these at-risk island nations are located in the Pacific and Indian Oceans and, for many populations, permanent relocation is the only viable option for the future.

Mean sea level rose by about 15 cm during the 20th century and a further rise of up to 58 cm is projected before 2090. A rise in global mean sea level of 1 m would inundate many island groups and coastal communities.

Kiribati's capital and most populated region on Tarawa atoll

The island nation of Kiribati is made up of 33 atolls and reef islands and one raised coral island. More than 33% of its 100,000+ inhabitants live in an area of 16 km$^2$. Although atolls and reef islands can respond to sea level by increasing in surface area (through greater coral growth), there is no increase in height, so they are still vulnerable to inundation and salt water intrusion.

Storm dunes, Funafuti atoll, Tuvalu, the highest point on the atoll.

Some 2800 km south of Kiribati, the tiny island nation of Tuvalu (maximum elevation 4.6 m) is also under threat from climate change, being vulnerable to tropical cyclones, storm surges, and king tide events. A sea level rise of 20-40 cm will make Tuvalu unhabitable for its population of around 11,000 and already its leaders are making plans for evacuation, probably to nearby Fiji.

1. Study the map opposite. Suggest three reasons why those initially displaced by natural disasters may not have returned:

   (a) ______________________________

   (b) ______________________________

   (c) ______________________________

2. How might the Social Progress Index indicate vulnerability of the population to the effects of natural hazards?

   ______________________________

   ______________________________

3. Slow onset disasters, such as drought and associated famine, are wide-reaching humanitarian crises affecting millions. Explain why those displaced are less likely to return home than those affected by sudden-onset disasters:

   ______________________________

   ______________________________

4. What social, economic, and cultural problems are likely to result in the forced migration of people from their homes because of natural hazards?

   ______________________________

   ______________________________

   ______________________________

5. (a) Explain why the people of low lying island nations are at high risk of forced migration: ______________________________

   ______________________________

   (b) Coral atolls can be relatively resilient to sea level rise by increasing in surface area. Why is this unlikely to help the people of island nations threatened by sea level rise and increased air and sea surface temperatures?

   ______________________________

   ______________________________


**ISBN: 978-1-927309-37-7**

# 127 Case Study: The East African Drought

**Key Idea**: The frequency and severity of droughts in East Africa is increasing. Drought has been a major factor forcing people from their homes.

## The 2011 East Africa drought

Between July 2011 and mid-2012, a severe drought affected the entire East African region resulting in a humanitarian crisis affecting around 11 million people. The crisis was worst in the countries of Somalia, Djibouti, Ethiopia, and Kenya (right) because it caused severe food shortages, but neighboring countries were also affected.

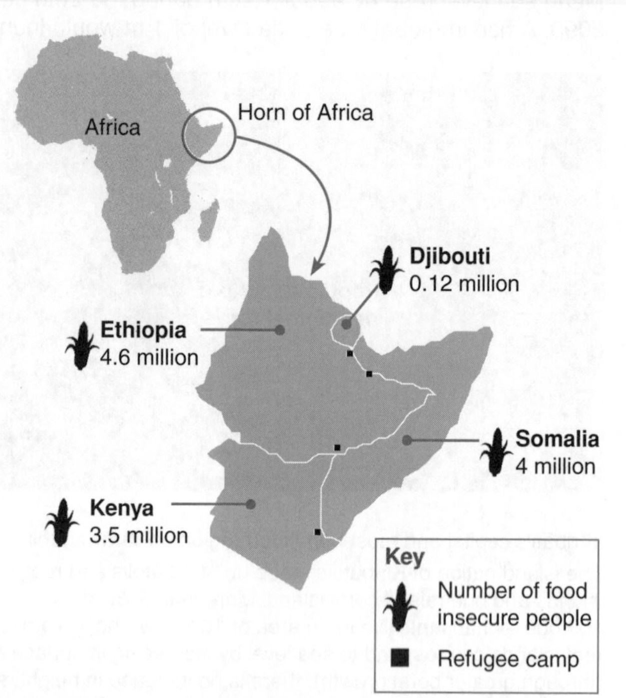


### What caused the 2011 drought?

East Africa is prone to droughts, but the drought of 2011 was the worst in 60 years. A strong La Niña weather pattern caused seasonal rains to fail for two consecutive years in East Africa. Rainfall was around 30% of the average of the previous 15 years.

### What effect will climate change have on East Africa?

East African countries have always been prone to droughts. However, the number and severity of droughts have increased since 1900. Simulations suggest that future climate change will increase the frequency or severity of drought events. In a region still struggling to recover, further droughts will worsen food insecurity and poverty and will likely result in high numbers of deaths.

## The 2011 East Africa drought triggered mass migration

The drought caused crops to fail and in some regions up to 60% of livestock died due to a lack of water or pasture. The overall shortage of food pushed up food prices significantly and many families could not afford to buy what little food was available.

Famine was declared in some regions of south-central Somalia. Hundreds of thousands of people fled to refugee camps in Kenya and Ethiopia in search of food and shelter. It often took many weeks of walking to reach the camps and many did not survive the trip.

Drought continues to affect regions of East Africa today. Hundreds of thousands of people to continue to live in refugee camps like the one above. Continued drought, food insecurity, and political instability mean these people may remain displaced indefinitely.

1. What was the main cause of the 2011 East Africa drought? ______

2. What affect did the drought have on food supplies in the affected countries? ______

3. How does a natural disaster, such as drought, result in mass migration? ______



WEB 127

PRACTICES


ISBN: 978-1-927309-37-7

# 128 Reducing the Impact of Drought

**Key Idea**: Forecasting and preparing for drought can reduce the impact of its effect.

## Droughts affect nations

In Africa, up to 80% of the population is employed in agriculture, meaning millions of people are dependent on agriculture for their income. Drought brings food insecurity, a loss of income, malnourishment, and susceptibility to disease, which can place the lives of these people in jeopardy. The effects of drought in developing nations can be widespread and affect a large proportion of a country.

In more developed countries, such as the US, the effects of drought are mostly limited to farmers, as these nations have other industries to support them. However, the current prolonged Californian drought is affecting not only the state's agricultural production, but also the urban water supply, and it is driving up the cost of food.

Dry river bed in California (2009).

## Preparation and preparedness can reduce the impact of a drought

Drought is sometimes called a creeping natural hazard because it develops over a period of time. It is therefore possible to plan ahead to reduce its impact. Often being well prepared for a drought is more effective and costs less money than an emergency response (e.g. supplying aid). The African Climate Policy Center has been established to provide information and strategies to help African countries adapt to shifting rainfall patterns and drought. Some of these strategies are described below.

SuSanA cc2.0

Rainwater harvesting from rooftops (above) during the rainy season allows water to be stored and used in times of shortage. The water can be used for drinking and cooking, and also to supplement the watering of crops or livestock.

In areas of Kenya, the use of drought resistant strains of sorghum and millet has seen harvest yields double. Yields can also be increased by using fast maturing crops, or through planting less traditional crops which are quite resilient to a range of conditions (e.g. potato).

Infiltration pits are trenches to capture and store water when it rains. The simplest are holes dug around a plant, others are larger, lined, and filled with rocks. Rainwater soaks down through the pit instead of running off across the soil. These pits also reduce erosion.

1. Why does drought affect a larger proportion of the population in less developed nations than in more developed nations?

___

2. (a) Why is it possible to prepare for drought? ___

(b) Describe the ways in which people can prepare for drought and reduce its impact: ___


ISBN: 978-1-927309-37-7

# 129 Chapter Review

Summarize what you know about this topic so far under the headings provided. You can draw diagrams or mind maps, or write short notes to organize your thoughts. Use the introduction and the images and hints included to help you:

**Natural hazards and natural disasters**

HINT: Describe examples of natural hazards. What is the difference between a natural hazard and a natural disaster?

**The effects of natural hazards**

HINT: What features can increase the impact of a natural hazard? How can the effects of natural hazards be reduced?

**Natural disasters and migration**

HINT: Give examples of how natural disasters can cause mass migration.




ISBN: 978-1-927309-37-7

# 130 KEY TERMS AND IDEAS: Did You Get It?

The number of hurricanes forming during the North Atlantic hurricane season are monitored. The Atlantic Database Reanalysis Project (NOAA), is a project that aims to correct and add new information about past North Atlantic hurricanes. Going back to 1851 and revisiting storms in more recent years, information on tropical cyclones is revised using an enhanced collection of historical meteorological data in the context of today's scientific understanding of hurricanes and analysis techniques. The plots below represent data collected during the project.

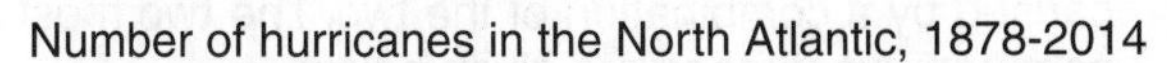


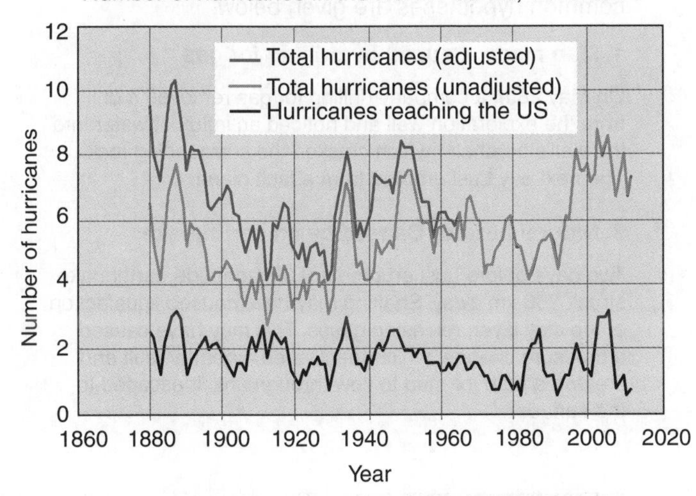


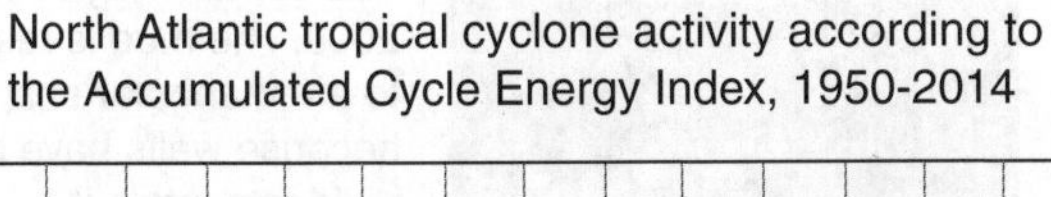


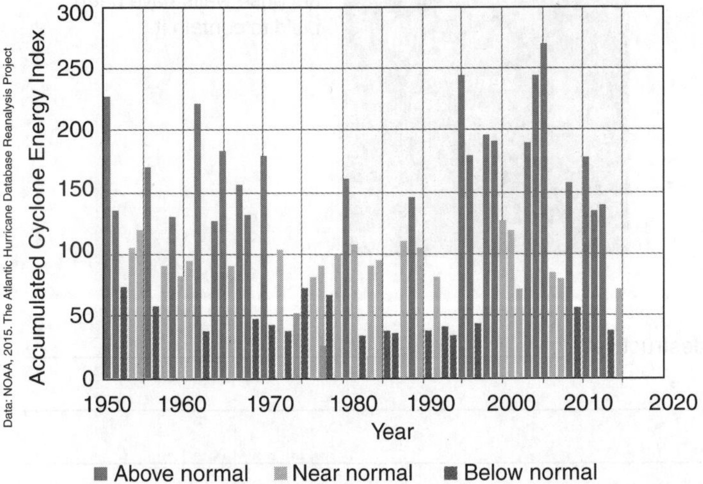


Data: NOAA, 2015. The Atlantic Hurricane Database Reanalysis Project

1. Study the graph (top left) of the number of hurricanes forming in the North Atlantic. Why do you think the number of adjusted hurricanes differs from the unadjusted number?

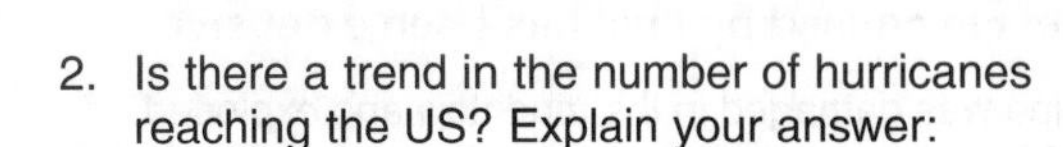

2. Is there a trend in the number of hurricanes reaching the US? Explain your answer:

3. Study the plot of cyclone energy (bottom left).

   (a) How many cyclones were above normal for the period 1950-1983:

   (b) How many cyclones were above normal for the period 1984-2015:

   (c) Has there been a change in the cyclone energy between the two periods?

4. Malé (right) is the capital of the Republic of Maldives. The city is built on the North Malé Atoll, and is elevated 2.4 m above sea level. The city is one of the most densely populated in the world, with 133,412 people living within 5.8 km$^2$.

Shahee Ilyas CC3.0

   (a) Identify a natural hazard that could affect Malé: ____________

   (b) Identify a human induced hazard that could affect Malé:

   (c) What is the likely physical effect on Malé of the hazard identified in 4(b): ____________

   (d) Predict what would happen to the population of Malé as a result of the hazard identified in 4(b): ____________


**ISBN: 978-1-927309-37-7**

# 131 Summative Assessment

## Case study: The Sidoarjo mud flow

Mud volcanoes are vents or fissures in the ground that discharge hot mud. They are fairly common and vary in size from one meter high and two meters wide, to 700 meters high and 10 kilometers wide.

On May, 2006, mud began erupting from the ground in the Sidoarjo district in Jakarta, Indonesia. This newly formed mud volcano was named Lusi, and to date is the most destructive mud volcano recorded.

Usually mud volcanoes are only active for a short period, but Lusi continues to erupt today. Mud flow has slowed from the peak discharges of 180,000 $m^3$ of mud per day to discharges of 10,000 $m^3$ per day in 2011. Discharges are predicted to continue for at least another 25 years.

### The damage caused by Lusi has been extensive

A gas pipe was damaged in the mud flow and exploded, killing 12 people. Around 7 $km^2$ of land is covered in mud, up to 40 m thick in some places. Over 39,000 people have been displaced. Agriculture, business, roads, and buildings have been destroyed. The economic cost is currently $2.7 billion. Attempts to stop the flow by pumping concrete into the vent have been unsuccessful, but the mud has been contained using large concrete walls and trenches.

A school destroyed by the mud flow

## What caused the Sidoarjo mud flow?

There is controversy about what caused the Sidoarjo mud flow. Some people suggest it was a natural event whereas others think the disaster was caused by human activity. Still others believe it was caused by a combination of the two. The two most common hypotheses are given below.

### 1. Man made: Caused by drilling for gas

On May 28th, a company drilling for gas removed a drill from the exploration well and noticed an influx of water into the well's borehole, which cracked the surrounding rock. The next day Lusi erupted from a fault plane.

### 2. Natural hazard: Caused by an earthquake

Two days before Lusi erupted a 6.3 magnitude earthquake struck 250 km away. Shaking may have caused liquefaction of the clay layer, releasing gases. This may have caused a pressure change, which reactivated a nearby fault and created a path for mud to flow through until it escaped to the surface.

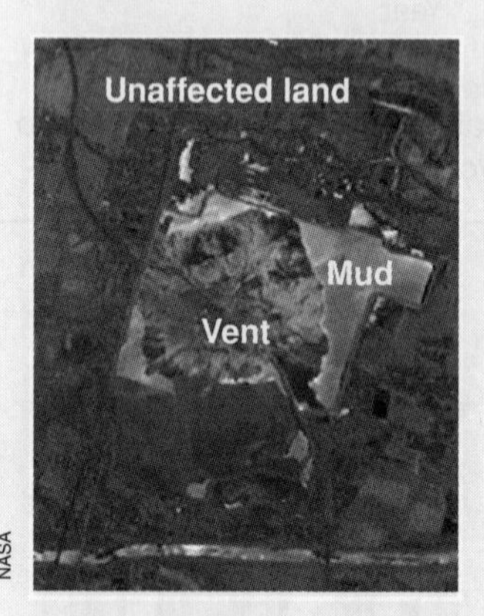


This image of Lusi (left) was captured by NASA in 2009. The mud is restricted to a rectangular shape because walls have been build to contain it.

1. What features of the Sidoarjo mud flow have made it so destructive? ______________________

2. Identify design solutions that have been utilized to stop or restrict the flow of mud. Comment on their success: ______________________

3. Thousands of people have been displaced by the mudflow. What impact is this likely to have on the development and economic viability of the region?

4. Two hypotheses are given as to what may have caused the Sidoarjo mud flow. Research both and summarize your findings on a separate piece of paper. Decide which hypothesis you think is most likely and justify your reasons.

PRACTICES

TEST


**ISBN: 978-1-927309-37-7**

# ESS3.C ETS1.B Human Impacts on Earth's Systems

**Key terms**

biodiversity

deforestation

ecosystem services

pollution

resource

remediation

sustainability

sustainable development

## Disciplinary core ideas

***Show understanding of these core ideas***

| | Activity number |
|---|---|
| **Natural resources must be managed responsibly** | |
| ☐ 1 Humans depend on the services provided by ecosystems. Ecosystems services include provisioning services such as food and water, and essential regulating and supporting services such as climate regulation and nutrient cycling. | 132 |
| ☐ 2 The biodiversity of ecosystems influences their ability to provide the essential services on which human depend. Maintaining biodiversity is therefore essential to the sustainability of human societies. | 132 |
| ☐ 3 Sustainability refers to the longevity or endurance of systems and processes. The sustainability of human societies demands that humans live within the carrying capacity of supporting ecosystems. This depends on responsible management of natural resources so that needs are met into the future. | 132 133 135 |
| ☐ 4 Pollution, deforestation, and poor land management and urban planning compromise the sustainability of human populations. | 134 136-139 |
| **New technologies can contribute to sustainability** | |
| ☐ 5 Sustainability will depend on the contributions of scientists and engineers in developing new technologies to reduce human impact on natural systems while still providing for the needs of the human population. | 135-137 139-142 |
| ☐ 6 Science and engineering contribute to sustainability by developing technologies for sustainable agriculture, renewable energy, reduced pollution and waste, and systems that allow for use of resources without environmental degradation. | 135-137 139-142 |
| ☐ 7 Sustainability will also depend on changes to laws, urban planning and transport, and changes to individual lifestyles. | 132 140 |

## Crosscutting concepts

***Understand how these fundamental concepts link different topics***

| | Activity number |
|---|---|
| ☐ 1 SC The relationship between the management of resources, biodiversity, and the sustainability of human populations can be modeled. | 135 145 |
| ☐ 2 SC Changes or rates or change in resource use, the sustainability of human populations, and biodiversity use can be quantified and modeled. | 135 142 145 |
| ☐ 3 SC Human activities can act to stabilize or destabilize ecosystems. | 134-136 138-142 |

## Science and engineering practices

***Demonstrate competence in these science and engineering practices***

| | Activity number |
|---|---|
| ☐ 1 Create a computational model or simulation to illustrate the relationship between the management of natural resources, the sustainability of human populations, and biodiversity. | 145 |
| ☐ 2 Evaluate or refine a technological solution that reduces the impact of human activities on natural systems. | 135-137 140-142 |

# 132 Humans Depend on Biodiversity

**Key Idea:** Humans rely on ecosystems and the services they provide for health, well being, and livelihood. The biodiversity of an ecosystem affects its ability to provide these services.

## Ecosystems provide services

- Humans depend on Earth's ecosystems for the services they provide. These ecosystem services include resources such as food and fuel, as well as processes such as purification of the air and water. These directly affect human health.
- Biologically diverse and resilient ecosystems that are managed in a sustainable way are better able to provide the ecosystem services on which we depend.
- The UN has identified four categories of ecosystem services: supporting, provisioning, regulating, and cultural.
- Regulating and provisioning services are important for human health and security (security of resources and security against natural disasters).
- Cultural services are particularly important to the social fabric of human societies and contribute to well being. These are often things we cannot value in monetary terms.

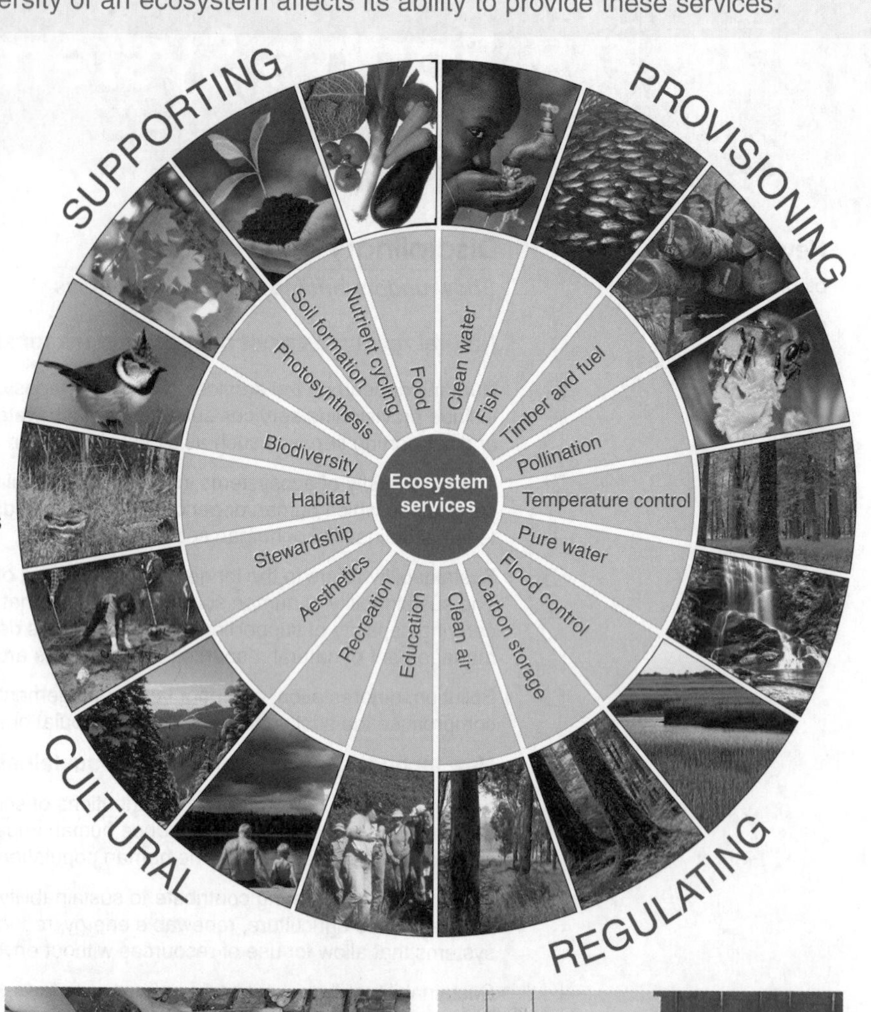

Biodiversity is important in crop development, e.g. promoting disease resistance. Many medical breakthroughs have come from understanding the biology of microbes and wild plants and animals.

High biodiversity creates buffers between humans and infectious diseases (e.g. Ebola) and increases the efficiency of processes such as water purification and nutrient cycling.

Biodiversity and ecosystem health are essential for reducing the impact of human activities (e.g. pollution) and the effects of environmental disasters (e.g. eruptions and landslides).

1. What are ecosystem services and why are they important to humans? ______________________

______________________________________________________________________

______________________________________________________________________

______________________________________________________________________

2. What is the relationship between biodiversity and the ability of an ecosystem to provide essential ecosystem services?

______________________________________________________________________

______________________________________________________________________

______________________________________________________________________




ISBN: 978-1-927309-37-7

## Ecosystem services: a case study

Doug Kerr

- It is difficult fully quantify the value of ecosystem services. One way of doing so is to estimate what people would pay for the services the ecosystem provides as the following case study illustrates.
- The Peconic Estuary in Long Island has many wetlands and mudflats. Development of the area has caused these to degrade. Management programs needed to estimate the benefits of rehabilitating the estuary. A study was carried out to estimate the contributions of the estuary to the production of wildlife (e.g. shellfish, fish, and birds).
- It was estimated that a hectare of eelgrass is worth \$2631 per year, a hectare of saltmarsh is worth \$835 per year, and a hectare of intertidal mudflat is worth \$168 per year in terms of commercial values for the fish, viewing values for birds, and hunting values for waterfowl.

## Human sustainability, resources, and biodiversity

- The sustainability (longevity) of human populations depends on the responsible management of the resources provided by the natural environment. Healthy, biodiverse ecosystems are essential to sustainability as these provide the essential services on which humans depend, e.g. clean air, fresh water, and carbon storage.
- If the quantity and quality of essential resources, such as water, are diminished, human sustainability is detrimentally affected. This also applies to biodiversity. If this decreases, the essential ecosystem services on which humans rely are negatively affected.
- We can express this relationship by saying that human sustainability (HS) is approximately equal to the resources available (RA) and the biodiversity (B) of the area. This could be written as a simple equation: $HS \sim RA + B$.
- Using this simple equation, we can see that any decrease in RA or B will cause a decrease in HS. However, human technology can compensate to a certain degree for reduction in resources or biodiversity. Technology (T) can help efficiency and reduce resource use, or it can help conservation programs to improve biodiversity. These factors can be put into our first equation so that now: $HS \sim (RA + T_1) + (B + T_2)$.
- Here the equation now shows HS can remain stable even if RA or B decrease, provided that $T_1$ or $T_2$ increase appropriately.
- It is important to remember that technological solutions to reduced resources or biodiversity require both effort and money. They come at a cost (C), so that $C \sim T_1 + T_2$. Benefits (Bt) from the system described are effectively equal to human sustainability (i.e. the benefit is that humans survive) so that $HS \sim Bt$.
- These simple equations by no means show the complexity of the human relationship with the environment but they do provide a way to visualize and compute the effects of simple changes to a system.

3. (a) Describe a way of putting a value on ecosystem services: ______________________

(b) What is the purpose of putting a value on ecosystem services? ______________________

4. Use the equations above to describe what would need to happen in the following scenarios:

(a) Keeping HS stable while allowing B to decrease: ______________________

(b) Keeping HS stable while reducing the need for resources (RA): ______________________

(c) The effect on cost (C) of allowing either (a) or (b) to occur: ______________________

(d) The effect of reducing both B and RA on HS and C: ______________________


**ISBN: 978-1-927309-37-7**

# 133 Human Sustainability

**Key Idea**: The key to humans living sustainably on this planet is finding a way to prolong the life of resources in the face of a continually growing population.

## The human population

- In the last 60 years, the human population has increased from fewer than 3 billion people to over 7.4 billion. Since the 1950s, improvements in medicine and access to more food have allowed the world's population to grow at rate of almost 2% each year.
- Many scientists believe growth of this magnitude is not sustainable and that the human population has already surpassed the planet's carrying capacity. They predict the inevitable collapse of food supplies and populations in the near future. Current predictions suggest the human population will reach at least 9.7 billion by 2050.
- As the human population grows it uses more resources. Even with careful resource management and more efficient use the rate of resource use will continue to increase. This means that eventually either the resource will run out or it has to be replaced with another resource before that happens.
- In many countries, initiatives have been taken to lower birth rates in an attempt to relieve pressure on resources.

| Year | Population (billions) | Year | Population (billions) |
|---|---|---|---|
| 1850 | 1.26 | 1960 | 3.01 |
| 1900 | 1.65 | 1970 | 3.68 |
| 1910 | 1.75 | 1980 | 4.44 |
| 1920 | 1.86 | 1990 | 5.31 |
| 1930 | 2.07 | 2000 | 6.12 |
| 1940 | 2.30 | 2010 | 6.93 |
| 1950 | 2.52 | 2016 | 7.4 |

## Population growth

- Births, deaths, immigrations (movements into the population) and emigrations (movements out of the population) are events that determine the number of individuals in a population.
- Population growth depends on the number of individuals being added to the population from births and immigration and the number being lost through deaths and emigration (right).

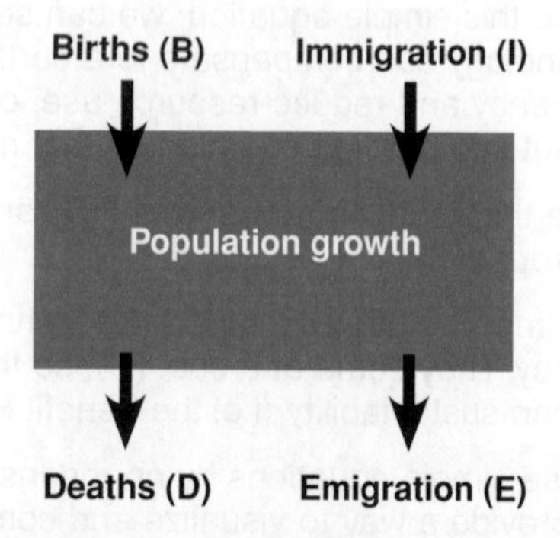


## Resource use

- An example of the human population affecting resources is the consumption of coal of the last 200 years (right). The majority of the energy consumed by humanity originates from coal, either from using it directly for heat or using it to fuel power stations that produce electricity.
- As the human population has increased, so has the consumption of coal (and every other energy resource for that matter). Some of this comes from use of technology with high energy demands, but that technology has itself helped increase food supplies, living standards, life expectancy, and the human population itself.

| Year | ~ Coal consumption (x $10^{18}$ J) | Year | ~ Coal consumption (x $10^{18}$ J) |
|---|---|---|---|
| 1840 | 1 | 1940 | 41 |
| 1860 | 5 | 1960 | 50 |
| 1880 | 12 | 1980 | 70 |
| 1900 | 25 | 2000 | 100 |
| 1920 | 39 | 2015 | 150 |

1. What is the rate of the human population growth since 1950? ______

2. What will the human population be in 2050? ______

3. Produce an equation that could be used to calculate the population growth of a certain population (e.g a country's population growth).

______

4. (a) How would the equation for the entire global human population differ? ______

______

(b) Write the equation for the entire global human population: ______

______

WEB 133 PRACTICES 

DATA


**ISBN: 978-1-927309-37-7**

5. Use the data on the opposite page to produce a graph to show the growth of the human population and the consumption of coal since 1840. You will need a left and right Y axis.

NEED HELP?
See Activities 3 and 7

6. What type of growth curve do both human population and coal consumption show? ______________________

7. Use your equation from question 3 to complete the following:

(a) A town has a population of 70,230 in the year 2010. Over the next five years, 6556 people move into the town for work, but 4096 move to other parts of the country. A baby boom sees the birth of 5225 babies but there are also 4978 deaths. What is the population of the town in 2015?

(b) What is the percentage growth over the 5 years? ______________________

(c) The town uses water from a reservoir that holds 200 million L and is replenished at a rate of 60 million L a day. In 2010, the town used 46 million L per day. Calculate the average water use per person per day.

(d) How much water would be used per day in 2015 if each person used the same amount of water as in 2010?

(e) Assuming the town underwent the same rate of growth every 5 years into the future, when would the reservoir begin to fill more slowly than it was being used (i.e. the tipping point at which the water supply become unsustainable?)

(f) At the end of 2025, the town introduces measures to extend the life of the water supply by asking people to cut their water use by 10%. Under these new measures, when will the water supply become unsustainable?


**ISBN: 978-1-927309-37-7**

# 134 Human Impact on the Ocean

**Key Idea**: Humans have had a significant impact on the oceans, including the stocks of fish and the quality of the environment.

Less than two centuries ago the oceans must have seemed to people to be an inexhaustible resource. However, after years of exploiting this resource we are finally realizing that even the ocean has its limits and that we, as a species, have begun to exceed them.

The decline of whales is an example of humans overexploiting a resource. During the 1700s-1900s the slaughter of whales became so efficient that ships were needing to sail further from land and for longer voyages to find any whales at all. By 1930, 50,000 whales were being killed a year. However whales breed very slowly and by 1980 many species were (and still are) on the brink of extinction. It is estimated that 90% of blue whales were killed by 1980 and over 2.9 million whales killed since 1900.

Fishing has provided food for people for thousands of years. However fish stocks have plummeted as the human population has increased and fishing has become a major global industry. Many fish species have been fished so intensively that they are no longer economic. Others are on the brink of collapse. Fishing techniques have become so sophisticated and efforts are on such a large scale that hundreds of tonnes of fish can be caught by one vessel on one fishing cruise. Fishing vessels can reach over 100 m long.

Pollution is a major problem in parts of the oceans. Activities causing pollution range from deliberate dumping of rubbish from ships, to runoff from the land and contaminated discharge into rivers that lead to the sea. An estimated 8 million tonnes of plastic finds its way into the sea each year. Plastic can have severe detrimental effects on marine life, especially those that mistake plastic bags for jellyfish or other prey species. Some areas are so polluted it is dangerous to eat fish caught there.

Cargo ships carry ballast water in their hulls to keep them stable when empty. As the cargo is loaded, the ballast water is pumped out to maintain stability and buoyancy. This has led to many marine organisms being unintentionally transported about the globe. This can be disastrous for the local marine environment when the organism is an invasive species. Newer ships are able to discharge ballast and replace it out at sea (where potentially invasive organisms would die in the open ocean).

Runoff from the land into the ocean has caused problems with eutrophication (nutrient enrichment). Combined with increasing surface water temperatures this has led to large algal blooms along coastlines. In some cases, these blooms make it dangerous for swimming. The algae often produce toxins that are concentrated by filter feeding organisms (e.g. clams). This makes them dangerous to eat and the public is often warned against collecting shellfish, taking away an important food resource.

Sea bed mining and drilling has had an impact on bottom dwelling marine organisms. Oil spills can affect the seabed and coastline for hundreds of kilometers. Although technology has made oil spills while drilling less likely, it is still a significant risk, and accidents can still occur, with catastrophic consequences, as the 2010 blowout of the oil well drilled by the Deepwater Horizon illustrated. Sea bed mining often uses dredges to vacuum up material from the sea floor, disturbing bottom dwelling organisms.

1. Describe three impacts of human activity on the ocean:

(a) ______________________________

______________________________

(b) ______________________________

______________________________

(c) ______________________________

______________________________




**ISBN: 978-1-927309-37-7**

# 135 Fishing and Sustainability

**Key Idea:** Fisheries globally have a history of unsustainable stock management. The depletion of fish stocks has made it necessary to implement careful management strategies.

- Fishing is an ancient human tradition. It provides food, and is economically, socially, and culturally important. Today, it is a worldwide resource extraction industry. Decades of overfishing in all of the world's oceans has pushed commercially important species (such as cod, right) into steep decline.
- According to the United Nation's Food and Agriculture Organization (FAO) almost half the ocean's commercially targeted marine fish stocks are either heavily or over-exploited. Without drastic changes to the world's fishing operations, many fish stocks will soon be effectively lost.

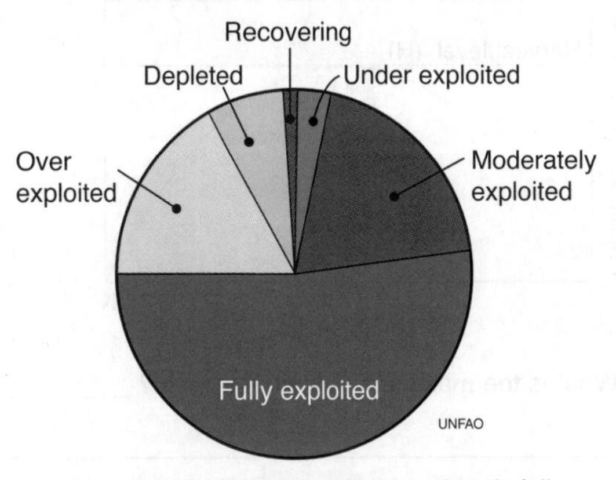


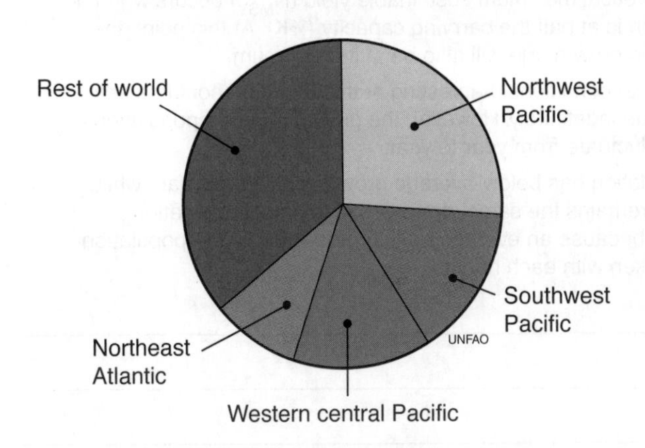


The single largest fishery is the Northwest Pacific, taking 26% of the total global catch.

**Percentage exploitation of fisheries**

Recovering

Depleted

Under exploited

Over exploited

Moderately exploited

Fully exploited

UNFAO

52% of the world's fished species are already fully exploited. Any increase in catch from these species would result in over-exploitation. 7% of the fish species are already depleted and 17% are over-exploited.

Part of the problem with fishing is by-catch, i.e. the fish that are not wanted. Even if thrown back, these fish or other marine organisms often don't survive. Techniques to reduce bycatch include changing hook design and attaching devices called pingers to the line or net that frighten away non-target organisms.

Different fish species are fished with different kinds of lines or nets. In particular, the mesh size of the net can be changed. This can be set so that small fish can swim through it and larger fish are caught. This can help to ensure that young fish survive to breeding age or that the wrong species of fish are not caught in the net.

Precision Seafood Harvesting

New types of net designs are constantly being tested. One of the newest net designs is called Precision Seafood Harvesting. It consists of a PVC liner towed by a trawler and forms a tunnel of water that reduces stress and damage to target fish, increasing catch efficiency. Holes allow unwanted fish to escape.

1. What percentage of fish stocks are depleted, overexploited, or fully exploited? ______________________

2. (a) What is bycatch? ______________________

(b) Name two ways bycatch can be reduced: ______________________

______________________


**ISBN: 978-1-927309-37-7**

## Calculating sustainable yields

- The sustainable harvesting of any food source requires that its rate of harvest is no more than its replacement rate. If the harvest rate is higher than the replacement rate then it follows that the food source will continually reduce at ever increasing percentages (assuming a constant harvest rate) and thus eventually be lost.
- **Sustainable yield** (SY) refers to the number or weight of fish that can be removed by fishing without reducing the stock biomass from year to year. It assumes that the environmental conditions remain the same and do not contribute to fluctuations in biomass levels.
- The **maximum sustainable yield** (MSY) is the maximum amount of fish that can be taken without affecting the stock biomass and replacement rate. Calculating an MSY relies on obtaining precise data about a population's age structure, size, and growth rate. If the MSY is incorrectly established, unsustainable quotas may be set, and the fish stock may become depleted. The equation for a sustainable yield is shown below:

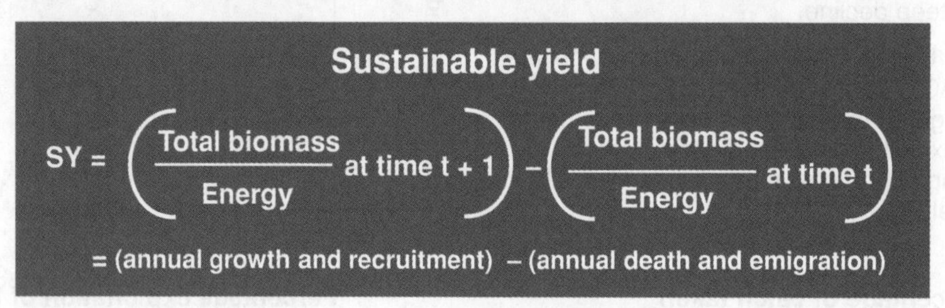


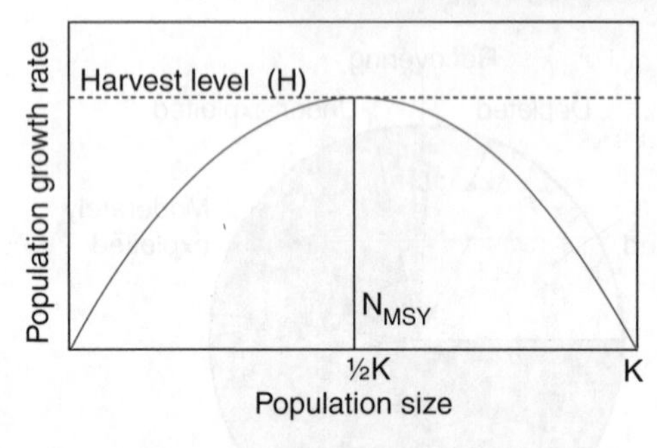


- The theoretical maximum sustainable yield ($N_{MSY}$) occurs when a population is at half the carrying capacity (½K). At this point, the population growth rate will also be at its maximum.
- Under ideal conditions, harvesting at this rate (H) should be able to continue indefinitely. However, the growth rate of a population is likely to fluctuate from year to year.
- If a population has below-average growth for several years while the take remains the same, there is a high risk of population collapse because an ever-increasing proportion of the population will be taken with each harvest.

3. What is the maximum sustainable yield? ______________________________

______________________________

______________________________

4. A fish population consists of about 3.5 million individuals. A study shows that about 1.8 million are of breeding age.

(a) Researchers want to know the maximum sustainable yield for the population so that it can be fished sustainably. What factors will they need to know to accurately determine the MSY?

______________________________

______________________________

______________________________

(b) Calculate the number of non-breeding individuals: ______________________________

(c) Should these smaller non-breeding individuals be included in the catch? Explain your reasoning: ______________________________

______________________________

______________________________

______________________________

(d) It is found that the larger a breeding individual is, the more fertile it is. What implications might this have on the harvesting method for these fish and the viability of the fishery?

______________________________

______________________________

______________________________

______________________________


**ISBN: 978-1-927309-37-7**

# 136 Plastics in the Ocean

**Key Idea**: Much of the plastic waste produced by humans ends up in the seas by deliberate or accidental dumping. Research into plastics is addressing this problem.

## Plastic is a problem

- The problem with plastic is its stability. In nature, organic material is broken down by enzymes and microbes that have evolved over billions of years to deal with the chemical bonds found in nature. Very few organisms can degrade plastic because the chemical bonds in most plastic are not similar to the chemical bonds found in nature.
- As a result, plastic can remain in the environment for hundreds of years, and the vast quantities of plastic products thrown away over the last half a century are now causing large environmental and waste management issues. Even the most remote parts of the world are affected by plastic wastes.

The albatross chick (above) died on Midway Island after eating numerous plastic items, including a felt pen and a number of bottle caps.

## Concentrating the problem

The surface water of the oceans circulate in giant whirlpools called gyres. In the same way that you can concentrate debris in a small pool by swirling the water around, these gyres concentrate floating debris. When this happens with floating plastic, giant areas of the ocean become plastic "garbage patches".

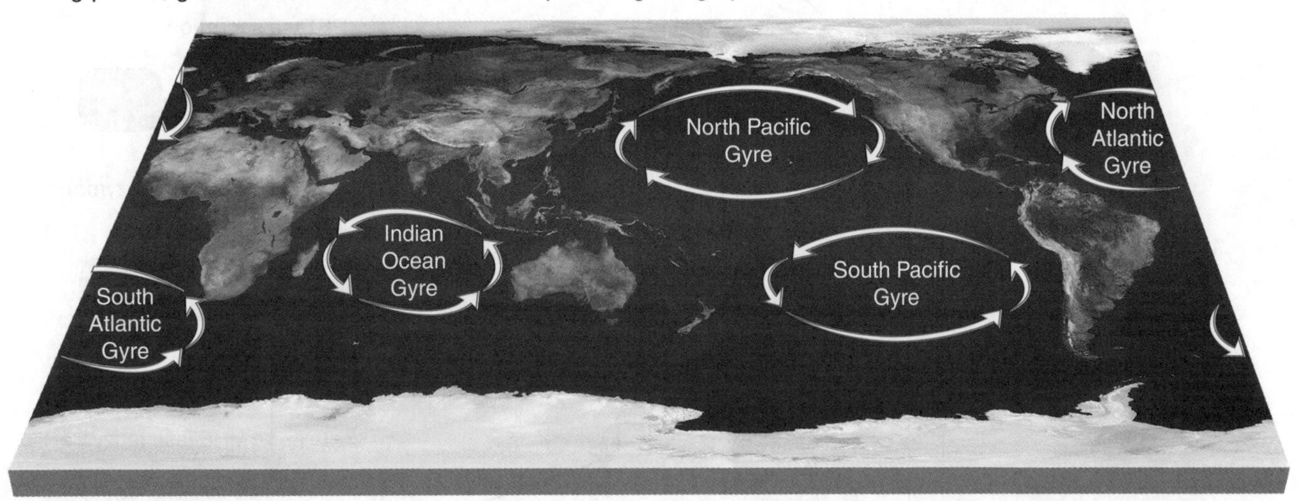


The **Great Pacific Garbage Patch** is an area within the North Pacific Gyre. Although given the name "garbage patch", most of the debris is not easily visible. The patches contain concentrations of waste and debris (mostly plastic) above that normally found in the ocean. The area covers around 1.2 million km$^2$, although definitions of the extent vary and the concentration of debris changes with the seasons. This makes an accurate estimate difficult.

1. Why do plastics persist in the environment? ______________________

______________________

2. (a) How does plastic waste become concentrated in certain areas of water? ______________________

______________________

______________________

(b) Midway Island is in the middle of the North Pacific Ocean and very isolated, yet it has a massive problem with plastic washing up on its beaches. Explain how this happens:

______________________

______________________

3. Describe an advantage and a disadvantage of biodegradable plastic: ______________________

______________________

______________________


**ISBN: 978-1-927309-37-7**

## Solving the problem

- Increasingly, degradable plastics are being produced. These may be photodegradable (break down in sunlight) or made with blends of sugars and other chemicals, which facilitate break down within 45 days. However, although many degradable plastics break down into smaller pieces, those pieces still persist in the environment.
- Most plastic products are stamped with a number (1-7) to indicate what the material is made of and help recycling. Plastic types 1 and 2 are the most commonly recycled plastics. Type 4 is less commonly recycled, while types 3, 5, 6, and 7 are unlikely to be recycled.

### Plastic eating bacteria

- PET (Polyethylene terephthalate) plastic is widely used to make bottles. A few fungal species are know to digest PET, but until recently no bacteria were known to do so.
- However, in 2016 a Japanese research group collected 250 samples of sludge from a PET bottle recycling plant. They incubated these samples with very thin PET film. After 15 days they found that the PET film had vanished in some of the samples, indicating something was breaking it down.
- Further analysis found that the bacterium *Ideonella sakaiensis* was responsible, the first bacterium to be shown to digest PET plastic. It does so by secreting the enzyme PETase, which breaks the PET molecule down into its single monomers. The enzyme MHETase then breaks the monomers down into compounds the bacteria can use.
- The technology is still in its early stages. If used commercially, it is predicted the bacteria could degrade more than 50 million tonnes of PET plastic annually. This is close to the global production of PET plastic. Currently only 2.2 tonnes is recycled annually.

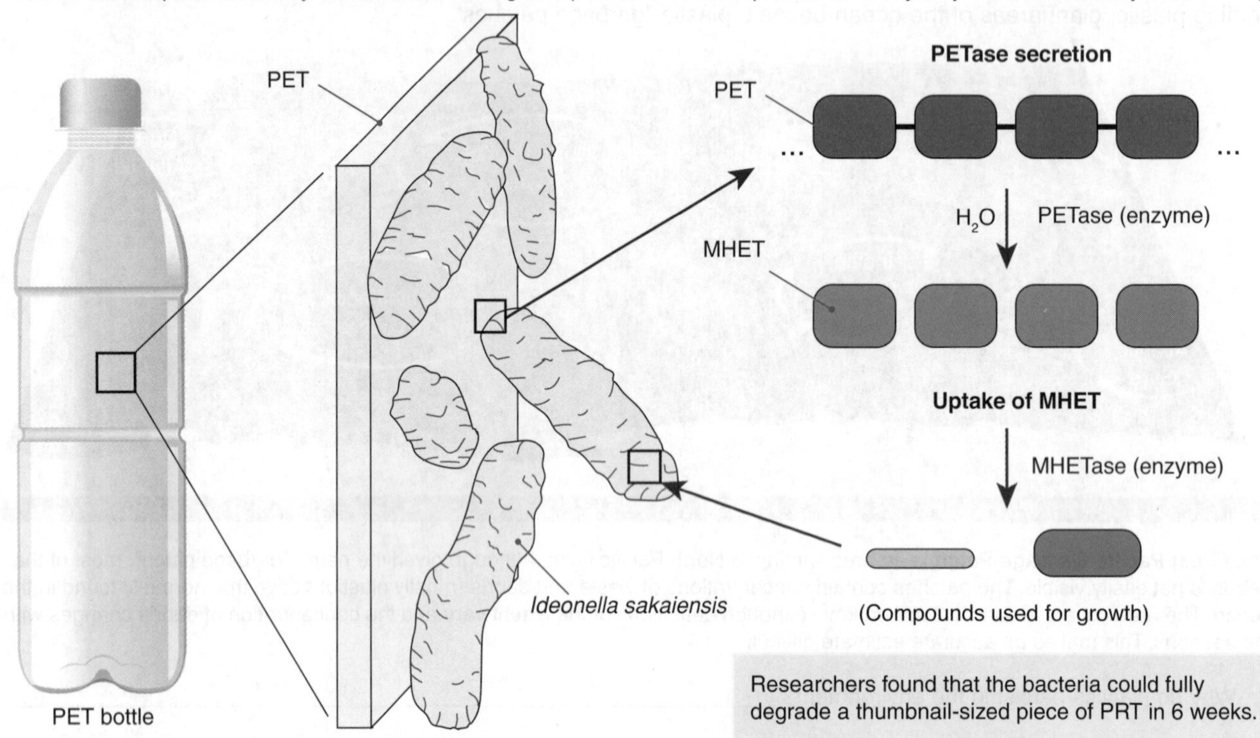


Researchers found that the bacteria could fully degrade a thumbnail-sized piece of PRT in 6 weeks.

4. (a) Briefly describe how the bacteria *Ideonella sakaiensis* was found and isolated: ______

(b) Evaluate the potential of this technology in reducing PET plastic waste: ______


ISBN: 978-1-927309-37-7

# 137 The Problem With Oil

**Key Idea**: Oil is an important resource, but oil spills can cause enormous environmental damage. New technologies are helping to quickly clean up oil spills and reduce their harm.

## Oil and oil spills

- Oil is arguably one of the most important chemicals in human economics. It provides energy for transport and electricity and the raw materials for many consumer products, including plastics. Billions of dollars a year are spent on removing it from the ground and billions more in revenue made from its sale.
- However, crude oil is a very toxic substance and removing it from reservoirs is fraught with difficulty and danger. Some of the greatest man-made environmental disasters have occurred because of the search for and transport of oil.
- For example, the Deepwater Horizon disaster in 2010 produced an oil slick that, at its peak, covered 6500 $km^2$ of ocean.

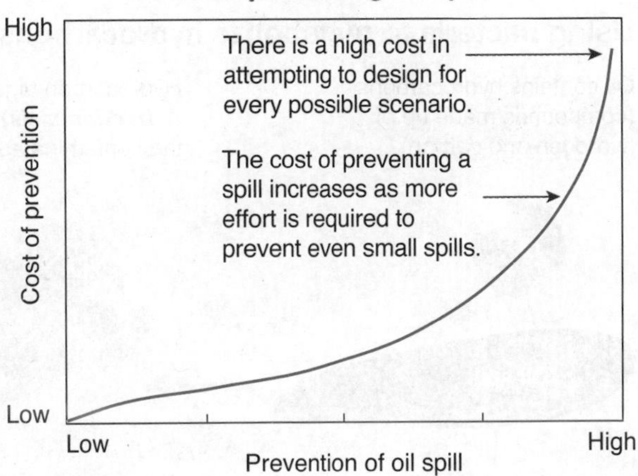


## Cost of preventing oil spills

- Preventing oil spills is important as oil is extremely damaging to the environment. However the cost of prevention is high. It is more costly to build a double hulled tanker than a single hulled tanker. These extra costs are eventually passed on to the consumer.
- At some point, the extra cost of prevention measures become too high for the oil company and the consumer. At this point the cost of prevention measures outweighs the benefit of preventing spills (the reduction of damage to the environment).
- A cost-benefit analysis done soon after the Exxon Valdez grounding in 1989 found the benefit to the environment of requiring the oil industry to use double hulled oil tankers was less than half the cost to the oil industry of upgrading oil tankers, even in the most favorable scenarios. In other words, it was less costly to clean up the mess than prevent the spill. Of course this doesn't take into account that any oil spill is detrimental to the environment.

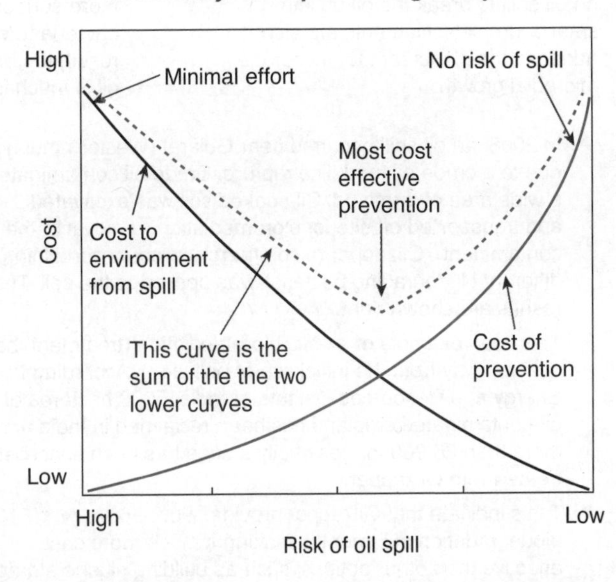


1. Why is oil an important resource? ______________________________

2. Why does the cost of preventing an oil spill increase rapidly close to 100% prevention? ______________________________

3. How does a cost-benefit analysis help us decide what level of risk is acceptable when dealing with oil and oil spills? ______________________________


ISBN: 978-1-927309-37-7

## Cleaning up oil spills

- Oil is difficult to remove once in the environment. In water, oil slicks can be contained using floating booms and chemical dispersants can help break up the slick. Oil eating bacteria can also be used. The oceans contain microbes that use the hydrocarbons found in oil as their energy source.
- The bacterium *Alcanivorax borkumensis* is one such organism. Its numbers quickly increase after an oil spill and it breaks down the oil into harmless compounds ($H_2O$ and $CO_2$). Bioremediation was used in the Deepwater Horizon oil spill in the Gulf of Mexico in 2010.

## Using bacteria to metabolize hydrocarbons

Oil contains hydrocarbons (compounds made up of hydrogen and carbon).

Hydrocarbon digesting bacteria (e.g. ***A. borkumensis***) are introduced into the contaminated area.

The microbes metabolize the hydrocarbons. Hydrogen and carbon from the oil are added to oxygen to form water and carbon dioxide.

$CH_4$

Oil spill

$H_2O$

$CO_2$

**Chemical dispersants** are added to an oil spill to break the oil up into smaller droplets. Nutrients are also added in the dispersant to encourage microbial growth.

The smaller oil drops provide more surface area for the bacteria to work on. As a result, the breakdown of the oil is much faster.

Not all the oil can be broken down by the microbes but, because there is less, it is more easily dispersed by ocean currents and the wind.

- In 2008, an oil spill occurred near Gujarat (Western India) due to a crude oil trunk line rupture. Crude oil contaminated a wide area of farmland. Oil-soaked soil was excavated and transported off site for bioremediation (removal of the contaminant). Oilzapper (a commercial product containing five different oil degrading bacteria) was applied to the soil. The results are shown right.
- This is an example of *ex-situ* bioremediation (treatment that occurs away from the initial site of pollution). According to the Energy and Resources Institute of India, 5000 hectares of oil-contaminated cropland has been reclaimed in India and more than 26,000 tonnes of oily sludge has been successfully treated with Oilzapper.
- Trials indicate that Oilzapper provides 90.2-96% biodegradation in 5 months making it much more cost effective than other options, such as building sludge storage pits and waiting for *in-situ* degradation in the soil.

Total petroleum hydrocarbon (TPH) in contaminated soil after treatment with Oilzapper

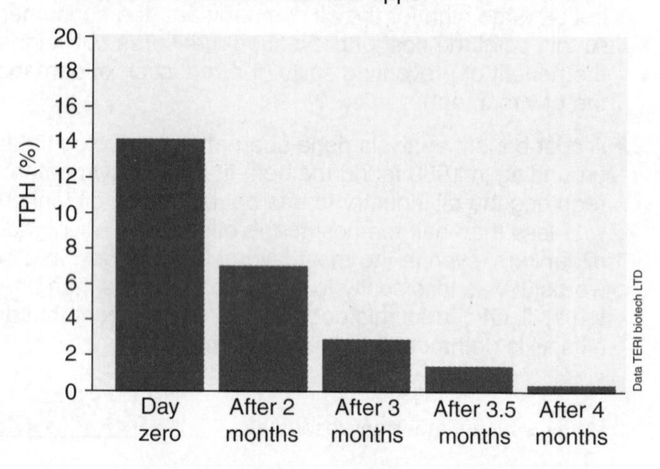


4. List some technological solutions to oil spills: ______

5. Outline how hydrocarbon-metabolizing bacteria are used to clean up an oil spill: ______

6. (a) How long did it take for the bacteria in Oilzapper to remove the oil in the Gujarat soil? ______

   (b) Evaluate the effectiveness of oilzapper as a bioremediation tool: ______


**ISBN: 978-1-927309-37-7**

# 138 Human Impact on the Land

**Key Idea**: Humans have had significant impacts on land. These include removing forests for agriculture, polluting and degrading soils, and disturbing large tracts of land.

Before ten thousand years ago, humans were a hunter-gatherer species. The change to a sedentary agricultural lifestyle meant that land needed to be cleared and protected from wildlife for the exclusive use of humans and their livestock. This was the start of humans dominating the planet. One third of the world's land area is now dedicated exclusively to producing food for humans. Other parts are mined for resources and forests are cleared for timber.

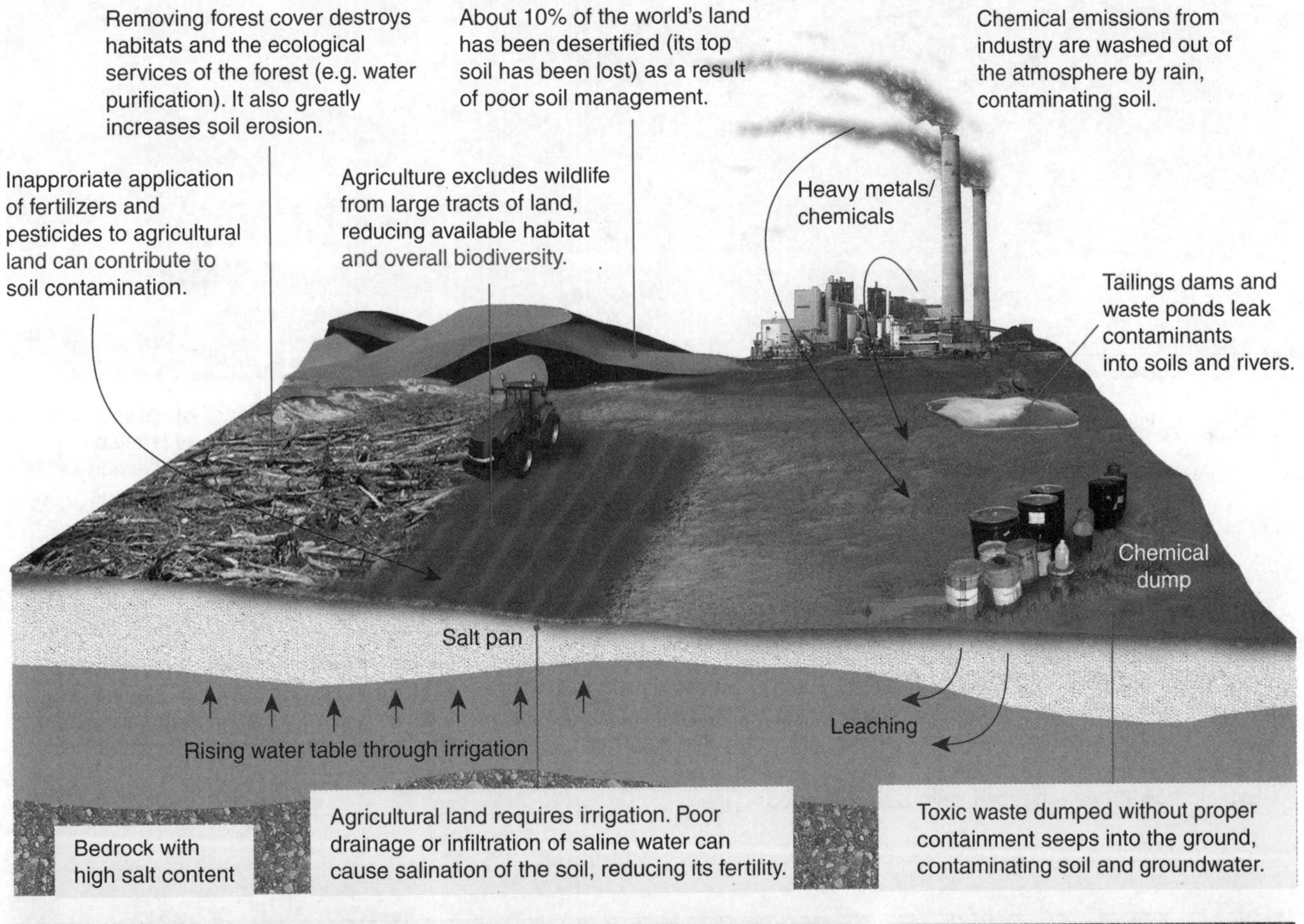

Deforestation is a major threat to wildlife. Deliberate destruction, such as logging, removes thousands of hectares of habitat a year, but accidental destruction by forest fires also destroys huge tracts of forest each year.

Urban sprawl is the development of large areas of land for housing. As the human population grows, especially around cities and in wealthier countries, more houses are needed. These often use undeveloped or previously agricultural land.

Humans produce a huge amount of waste. Not all is or can be recycled and so it is dumped in landfills. If landfills are not carefully managed, they can leak toxic chemicals into soils and groundwater. Even well managed sites occupy land.

1. Why was the change from humans being a hunter/gatherer society to an agricultural one an important milestone in human-environment interactions?


**ISBN: 978-1-927309-37-7**

## Salinization and soil contamination

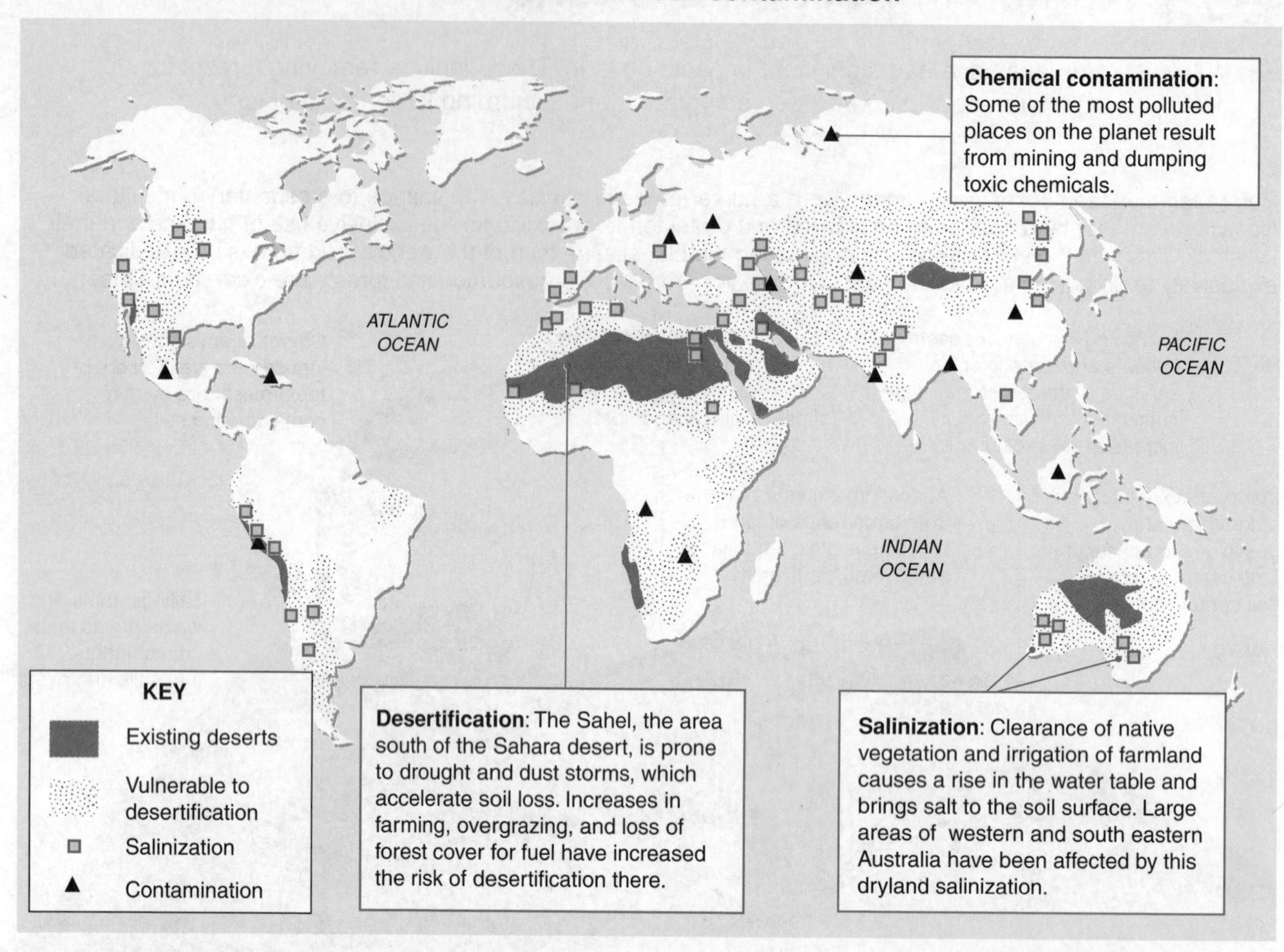


The rise of technology has had some unfortunate environmental results. Most metals are important for modern technology, some more than others. Some are relatively simple to mine and extract (e.g. iron) but others are rare and difficult to extract, and the extraction process often uses highly toxic chemicals. Some of the most contaminated places on Earth are areas around mines and smelters, especially in places where there was or still is very little regulation and toxic waste is dumped into rivers or buried.

2. Explain how human-induced salinization develops: ______

3. (a) What is desertification? ______

   (b) Outline the main causes of desertification? ______

   (c) What risks are associated with desertification? ______

4. Describe a cause of chemical contamination: ______

5. Urban sprawl, the development of land into housing areas, has become a major issue in land use. Describe the impacts of urban sprawl on the environment, including its effect on habitats and the ability of communities to produce food: ______


**ISBN: 978-1-927309-37-7**

# 139 Deforestation

**Key Idea**: Deforestation is the permanent removal of forest from an area. New technology is helping to track and reduce deforestation.

## Deforestation

- At the end of the last glacial period, about 10,000 years ago, forests covered an estimated 6 billion hectares, about 45% of the Earth's land surface. Forests currently cover about 4 billion hectares of land (31% of Earth's surface). They include the cooler temperate forests of North and South America, Europe, China and Australasia, and the tropical forests of equatorial regions. Over the last 5000 years, the loss of forest cover is estimated at 1.8 billion hectares. 5.2 million hectares has been lost in the last 10 years alone. Temperate regions where human civilizations have historically existed the longest (e.g. Europe) have suffered the most but now the vast majority of deforestation is occurring in the tropics. Intensive clearance of forests during settlement of the most recently discovered lands has extensively altered their landscapes (e.g. in New Zealand, 75% of the original forest was lost in a few hundred years).

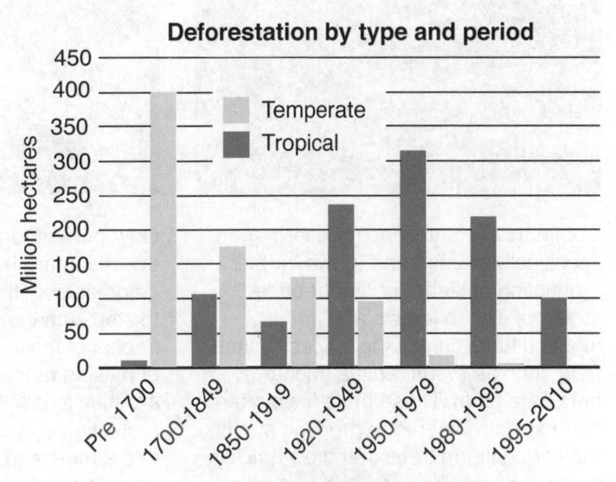


## Causes of deforestation

- **Deforestation** is the end result of many interrelated causes, which often center around socioeconomic drivers. In many tropical regions, the vast majority of deforestation is the result of subsistence farming. Poverty and a lack of secure land can be partially solved by clearing small areas of forest and producing family plots. However huge areas of forests have been cleared for agriculture, including ranching and production of palm oil plantations. These produce revenue for governments through taxes and permits, producing an incentive to clear more forest. Just 14% of deforestation is attributable to commercial logging (although combined with illegal logging it may be higher).

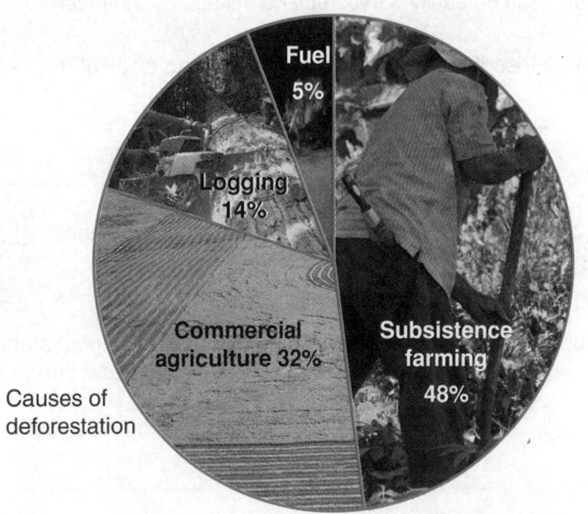


Causes of deforestation

## Tropical deforestation globally

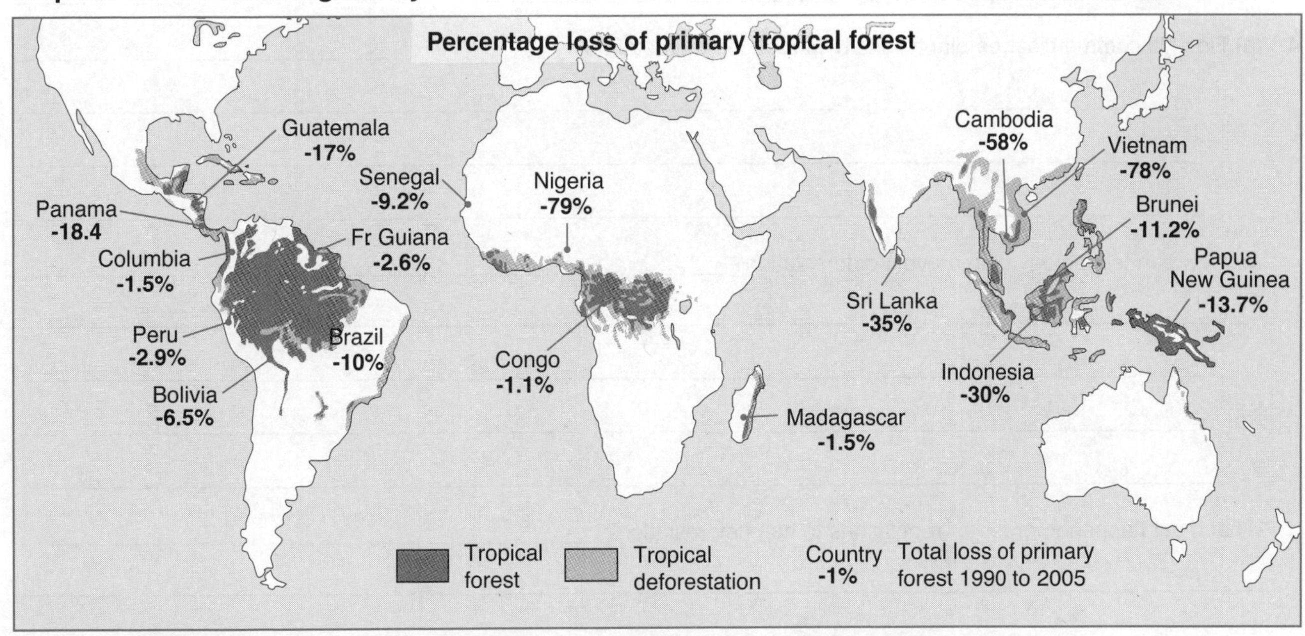


- It is important to distinguish between deforestation involving primary (old growth) forest and deforestation in plantation forests. Plantations are regularly cut down and replaced and can artificially inflate a country's apparent forest cover or rate of deforestation. The loss of primary forests is far more important as these are refuges of high biodiversity, including for rare species, many of which are endemic to relatively small geographical regions (i.e. they are found nowhere else).
- Although temperate deforestation is still a concern, it is in equatorial regions that the pace of deforestation is accelerating (above). This is of global concern as species diversity is highest in the tropics and habitat loss puts a great number of species at risk.


**ISBN: 978-1-927309-37-7**

# Reducing deforestation

Commercial plantations (tree farms) are specifically planted and grown for the production of timber or timber-based products. These forests are virtual monocultures containing a specific timber tree, such as *Pinus radiata* (Monterey or radiata pine). These trees have often been selectively bred to produce straight-trunked, uniform trees that grow quickly and can be easily harvested and milled.

Government regulations and public education are key parts to solving deforestation. It is easy to tell a country to stop cutting down trees, but if the people of that country have no other way of making money or obtaining wood what are they to do? Similarly, government regulation only works if people stick to the rules and the rules are strictly enforced.

Technology can help reduce deforestation by helping to monitor the reduction or growth of forests. Global Forest Watch 2.0 is a real-time interactive forest monitoring system produced by Google, in partnership with the University of Maryland and the UN Environment Program. It uses satellite technology and data sharing to track legal and illegal logging, helping law enforcement officials.

1. Describe the trend in temperate and tropical deforestation over the last 300 years: ____________

2. What are some of the causes of deforestation? ____________

3. Deforestation in temperate regions has largely stabilized and there has been substantial forest regrowth. However, these second growth forests differ in structure and composition to the forests that were lost. Why might this be of concern?

4. (a) How do commercial tree plantations help reduce deforestation? ____________

   (b) How can technology help prevent deforestation? ____________

5. What must happen in order for regulations to halt deforestation? ____________


**ISBN: 978-1-927309-37-7**

# 140 The Availability of Land

**Key Idea**: There is a finite amount of land available on Earth. Solutions must be found so that all humans, their activities, and wildlife have enough room to coexist.

The Earth has a land area of about 148.5 million km$^2$. 57 million km$^2$ is used for agriculture (of which 15.4 million km$^2$ is cropland (as opposed to pasture). About 54% of the world's population live in urban areas, which occupy 2.5 million km$^2$. 46% of the land area is defined as wilderness and has just 2.4% of the world's population of 7.4 billion.

As the population grows, more land will need to be developed for resource extraction, landfill, urban development, or cropping. Alternatively, instead of developing land, cities could become taller and denser, housing more people in high rise towers. There are also plans to develop high rise agricultural towers which would reduce the need to expand farmlands. A tower such the Willis Tower (Sears Towers) has 416,000 m$^2$ of floor space (41.6 hectares) that could be used to grow crops using hydroponics.

1. (a) Calculate the density of humans on the Earth's land surface (in humans per square kilometer):

   (b) 57 million km$^2$ of land is used for feeding the world's population. Taking this into account (people can't live on that land) what is the density of humans per square kilometer?

   (c) Now consider that 46% of the land's surface contains just 2.4% of the human population. What is the density of humans per square kilometer in the land that is left?

   (d) Calculate the density of humans in urban areas:

2. Use the data below to complete the table and graph the available arable land (land used for crop production) per person:

| Year | Arable land (million km$^2$) | Population (billions) | Arable land per person (km$^2$ per person) |
|---|---|---|---|
| 1960 | 14.4 | 3.01 | |
| 1970 | 15.1 | 3.68 | |
| 1980 | 15.2 | 4.44 | |
| 1990 | 16.1 | 5.31 | |
| 2000 | 16.0 | 6.12 | |
| 2010 | 15.8 | 6.93 | |

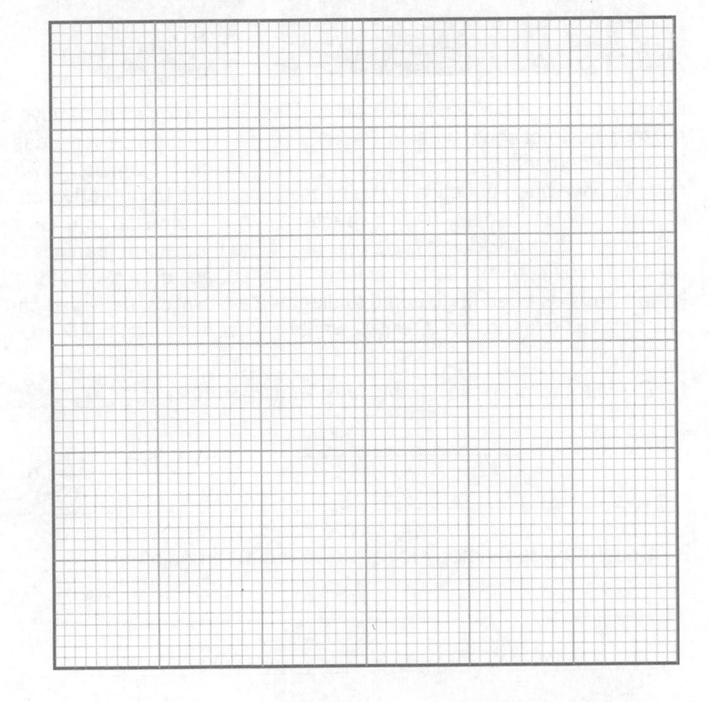


**ISBN: 978-1-927309-37-7**


ETS PRACTICES CCC

SC **KNOW**

# 141 Technology For Remediation

**Key Idea**: New technologies are helping to remediate contaminated sites.

- Land that has been used for industry such as mining must be remediated when it has fulfilled its purpose (e.g. the resource runs out). Remediation is the removal of contaminants in order to make the area safe for human use.
- The method of remediation used depends on the extent and type of contamination (below). For example, polluted top soil can be removed and treated off-site, or plants and bacteria may be placed *in situ* to absorb and break down the contaminants. A treated area is monitored over many years to ensure that no further leaching of contaminants occurs. The remediated land can then be used for other purposes.

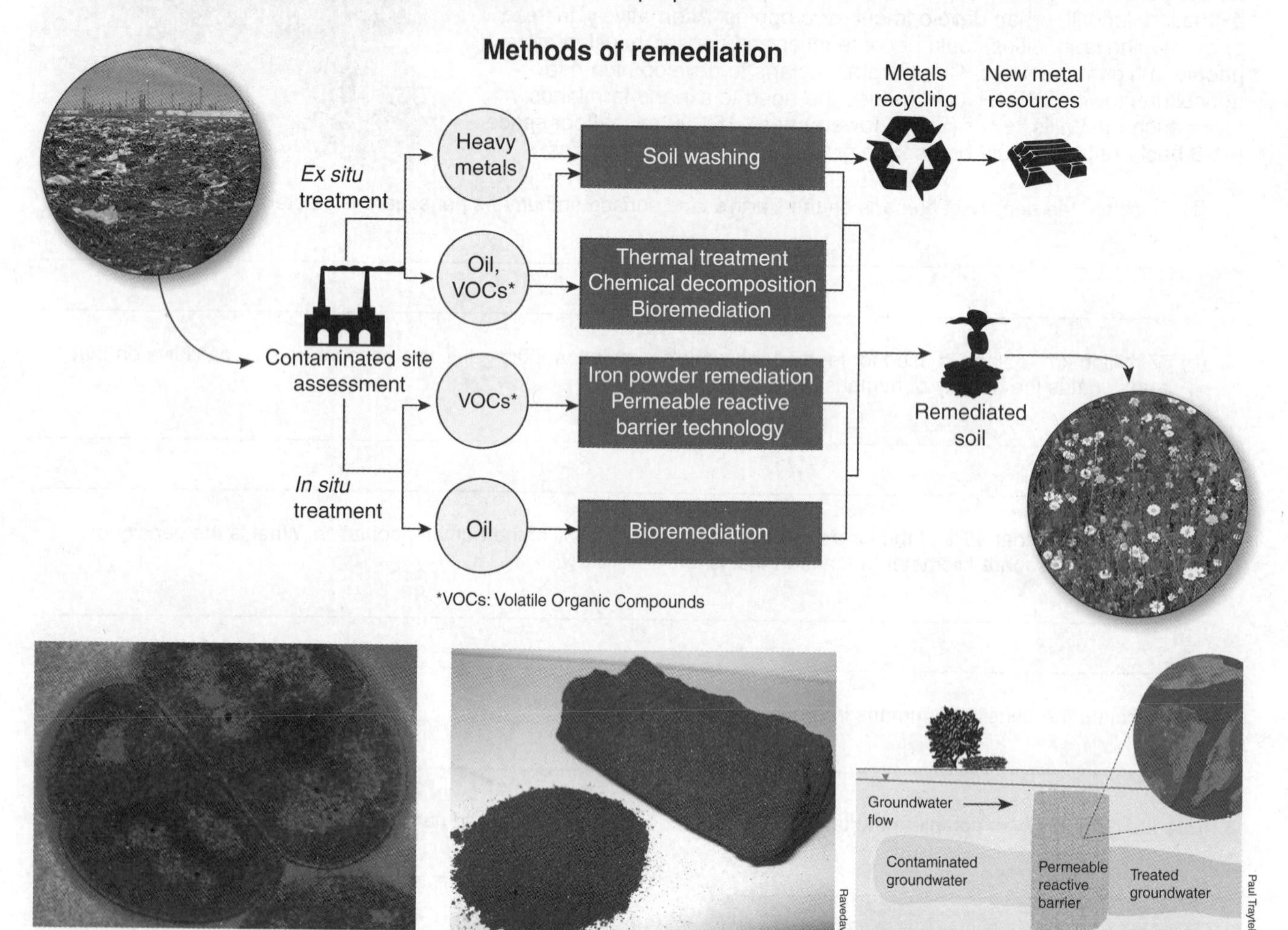


Bioremediation is the use of biological means to extract contaminants. Bacteria have great potential to do this and a number have been genetically engineered to digest contaminants. One such bacteria is *Deinococcus radiodurans*. It is one of the most radiation resistant organisms known and has been engineered to digest mercury and toluene in radioactive waste.

Ravedave

Technologies to remove contaminants can be quite simple. In areas with petroleum-based contaminants, water can be purified using **activated carbon** (highly granulated carbon). Contaminants adhere to the carbon granules and its very high surface area allows for a high rate of adsorption. Activated carbon is commonly used in household water purifiers.

Groundwater flow
Contaminated groundwater
Permeable reactive barrier
Treated groundwater
Paul Traytek

Permeable reactive barriers are new technologies that are a cost effective way of treating contaminated water *in situ*. The barrier is placed between the contaminated site and the groundwater. Water can move through the barrier from the site to the groundwater, but contaminants are either blocked or neutralized by the barrier.

1. Explain the purpose of environmental remediation: ______________________

______________________

______________________

2. Describe a technology for environmental remediation: ______________________

______________________

______________________



**ISBN: 978-1-927309-37-7**

# 142 Land Reclamation

**Key Idea**: Previously used or unusable land can be reclaimed and used again. In some cases, reclamation can help mitigate sea level rise.

## Designing environmentally sound landfills

As populations grow, the waste produced also grows. What cannot be recycled is dumped in landfills. Landfills today are designed to be repurposed once they are full. Before a landfill is established, ground engineers test the ground for stability and the presence of groundwater. A layer of high density polyethylene is placed on the prepared ground before rubbish is dumped. Once the first layer is full it is compacted and a layer of earth placed over the top. This continues until the landfill is full. A layer of rock and earth is laid over the final HDPE membrane and then covered with soil. Vents for methane are built into the structure to help stability.

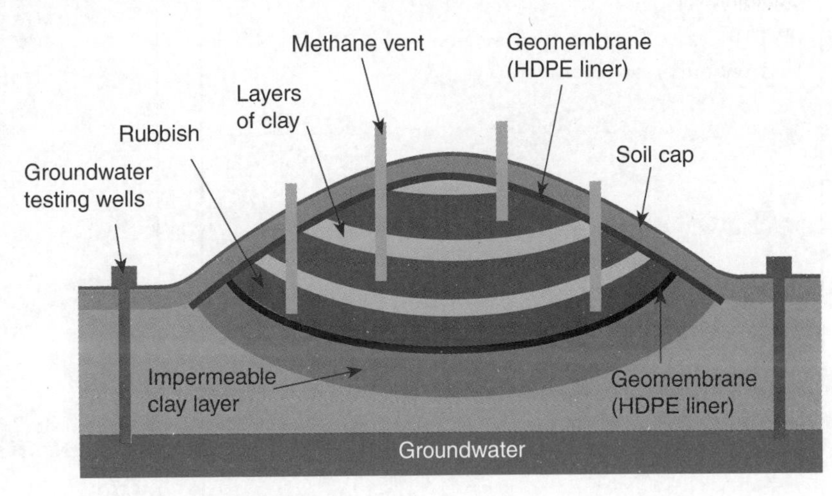


## Reclaiming land in Tuvalu

Funafuti is the main island of Tuvalu. It is an atoll with a land area of just 2.4 $km^2$. During World War II, American forces dug large "borrow pits" for material needed to construct an airfield. The pits remained when the American forces withdrew. The pits were then used as dumping sites, but also filled with water, which in turn affected the groundwater. With the help of the New Zealand government, sand from the lagoon was pumped onto barges and used to fill the pits, which also resulted in an 8% increase in land area. The land was also raised to help reduce the effects of sea level change.

1. What is the purpose of the HDPE liner in a landfill? ______

2. What is the purpose of the wells in the landfill? ______

3. Why does the ground need to be tested for stability before a landfill is developed? ______

4. Name two benefits of filling the borrow pits on Funafuti: ______


**ISBN: 978-1-927309-37-7**

# 143 Chapter Review

Summarize what you know about this topic so far under the headings provided. You can draw diagrams or mind maps, or write short notes to organize your thoughts. Use the introduction and the images and hints included to help you:

**Sustainability**

HINT: Describe factors that influence sustainability and biodiversity.

**Technology and sustainability**

HINT: Describe technologies used to reduce the impact of humans on the environment while helping to maintain resources.

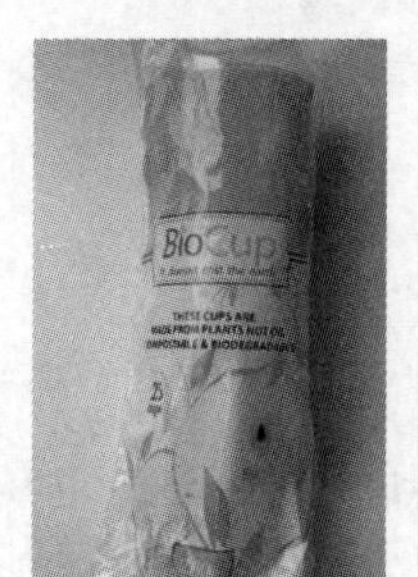





ISBN: 978-1-927309-37-7

# 144 KEY TERMS AND IDEAS: Did You Get It?

1. Test your vocabulary by matching each term to its definition, as identified by its preceding letter code.

|  | |  |
|---|---|---|
| biodiversity | A | The discharge of chemicals into an environment were they are not normally found. |
| deforestation | B | The removal of pollutants and contaminants from an environment so that it can be safely used for another purpose. |
| ecosystem services | C | The economic benefits provided by the ecosystem. |
| pollution | D | The quality of being able to be used into the future without being used up. |
| resource | E | A term describing the variation of life at all levels of biological organization. |
| remediation | F | A substance or article used to satisfy a human need. For example, the use of wood as a building material and for fuel. |
| sustainable | G | The permanent destruction and removal of forest in order to make the land available for another use. |

2. Study the satellite photographs below of deforestation in the Amazon

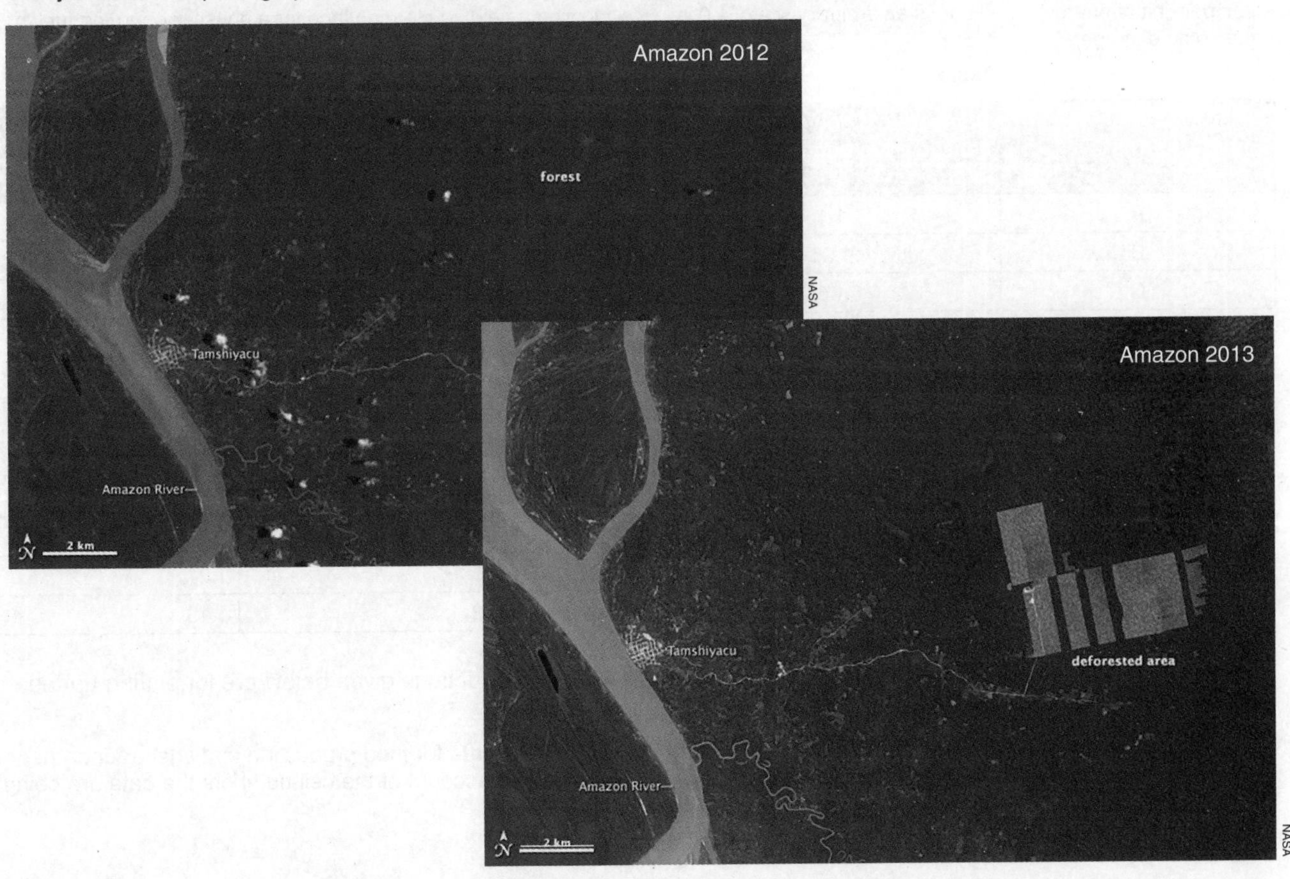


Use the scale to calculate the area of deforestation shown in the 2013 photograph:

3. (a) What is salinization?

(b) Why is it a problem?


**ISBN: 978-1-927309-37-7**

# 145 Summative Assessment

**Computational simulation of the relationship between natural resource use, human sustainability, and biodiversity.**

- The effect of human activities on resources, biodiversity, and sustainability is complex. By looking at specific areas and simplifying their interactions it is possible to produce simple mathematical models that help to show how a sustainable system might work.
- Below are the parameters of a hypothetical system. Your task is to set up a spreadsheet as shown and use it to simulate the effect of using different types of energy resources in the environment. You will do this by entering different combinations of parameters into the spreadsheet you have created and recording the outputs.
- As part of this task you will need to find the model that produces the most amount of food and shelter while remaining sustainable. At the end of the exercise, attach all your notes, answers, and spreadsheet printouts here.
- The spreadsheet is also available on the Teacher's Digital Edition.

## Energy resource

The system has three energy sources. The removal of the energy sources damages the biodiversity of the environment (e.g. by mining) in a different amount (i.e energy source A is easier to mine than energy source C and so mining does less damage the environment and biodiversity). The amount is of damage is the same per unit of energy mined and is an arbitrary scale. 1.0 is equal to no damage.

Table 1

| Energy source | Damage to biodiversity caused by extraction (per energy unit) |
|---|---|
| A | 1.1 |
| B | 1.25 |
| C | 1.6 |

An average person uses **10 units of food a day** and **6 units of shelter/heat**.

The environment and biodiversity can replenish themselves by **15%** each day.

The environment and biodiversity are given a starting health value of 1000 and are deemed to be healthy provided they remain at or above **60% of the original health** value.

## Food and shelter

The two most basic human needs are food and shelter/warmth. The energy sources can be used to produce food and shelter. However the energy sources produce food and shelter with different efficiencies. Also the production of food and shelter damages the biodiversity of the environment in some way (e.g. plowing a field, or clearing land for a house, or burning fuel for heat, as shown below.

Table 2

| Energy source used | Units of energy source needed to produce 1 unit of food | Damage done to biodiversity during production of food |
|---|---|---|
| A | 8 | 1.6 |
| B | 6 | 1.2 |
| C | 2 | 1.1 |

Table 3

| Energy source used | Units of energy source needed to maintain 1 unit of shelter/heat | Damage done to biodiversity during maintenance of shelter/heat |
|---|---|---|
| A | 9 | 1.2 |
| B | 5 | 1.3 |
| C | 3 | 1.15 |

- To carry out the simulation you need to set up a spreadsheet. The instructions given below are for setting up the spreadsheet using Microsoft Excel.

1. Open the spreadsheet and enter the energy source data shown in table 1 above, starting in the top corner (cell A1) of the spreadsheet:

| | A | B |
|---|---|---|
| 1 | **Energy source** | **Extraction effect on biodiversity per unit extracted** |
| 2 | A | 1.1 |
| 3 | B | 1.25 |
| 4 | C | 1.6 |
| 5 | | |

2. Now enter the data for food production and shelter underneath. Be sure to take account of the cell numbers the data are being entered in to:

| | A | B | C |
|---|---|---|---|
| 7 | **Food production (energy source used)** | **Energy units needed** | **Effect on biodiversity per food unit produced** |
| 8 | A | 8 | 1.6 |
| 9 | B | 6 | 1.2 |
| 10 | C | 2 | 1.1 |
| 11 | | | |
| 13 | **Shelter/heat (energy source used)** | **Energy units needed** | **Effect on biodiversity per shelter/heat unit used** |
| 14 | A | 9 | 1.2 |
| 15 | B | 5 | 1.3 |
| 16 | C | 3 | 1.15 |

TEST CCC SC PRACTICES

**ISBN: 978-1-927309-37-7**

3. Enter the number of food units and shelter/heat units used by a person per day (see 4(a) below). Again, these are arbitrary. The number of units can be changed during the simulation to work out the maximum number that is sustainable.

4. (a) To run the simulation you must enter formulae to calculate the effect on the environment of using each energy source into the appropriate cells. Be careful to enter the formulae correctly or the simulation may not run as expected! The formulae are shown below. Make sure that when you have entered the formula that the cell is formatted to show the numerical result.

| | F | G | H | I |
|---|---|---|---|---|
| 1 | | **Units of food used by average person per day** | | **Units of heat/shelter used by average person per day** |
| 2 | | 10 | | 6 |
| 3 | | | | |
| 4 | | | | |
| 5 | | **Food energy source** | | **Shelter/heat energy source** |
| 6 | | C | | C |
| 7 | Extraction effect on environment | =IF(G6=A2,B2, IF(G6=A3,B3, IF(G6=A4,B4))) | | =IF(I6=A2,B2, IF(I6=A3,B3, IF(I6=A4,B4))) |
| 8 | Energy units needed | =IF(G6=$A$8,B8, IF(G6=$A$9,B9, IF(G6=$A$10,B10))) | | =IF(I6=A14,B14, IF(I6=A15,B15, IF(I6=A16,B16))) |
| 9 | Effect on biodiversity | =IF(G6=$A$8,C8, IF(G6=$A$9,C9, IF(G6=$A$10,C10))) | | =IF(I6=A14,C14, IF(I6=A15,C15, IF(I6=A16,C16))) |
| 10 | | | | |

(b) Create a cell where the original health/starting value of the environment can be entered. The rate at which the environment can replenish itself (as a percentage) must also be entered.

| | G | H |
|---|---|---|
| 12 | **Original environment/ biodiversity health** | 1000 |
| 13 | | |
| 14 | **Rate of biodiversity replenishment (%)** | 15 |

(c) Finally the output cells must have their formulae entered. Again make sure that each formula is entered correctly. If you change the layout of the spreadsheet then you must make sure the reference cells in the formula are correct.

| | F | G | H | I |
|---|---|---|---|---|
| 16 | | | | |
| 17 | Effect on environment / biodiversity per day | =(G2*G7*G8)+(G2*G9) +(I2*I7*I8)+(I2*I9) | | |
| 18 | | | | |
| 19 | **Effect on the environment over time** | | | |
| 20 | Day 1 | Day 2 | Day3 | Day 4 |
| 21 | =H12-G17 | =F21*(1+H14/100)-G17 | =G21*(1+H14/100)-G17 | =H21*(1+H14/100)-G17 |
| 22 | | | | |

| | J | K | L | M |
|---|---|---|---|---|
| 16 | | | | |
| 17 | | | | |
| 18 | | | | |
| 19 | | | | |
| 20 | Day 5 | Day 6 | Day 7 | |
| 21 | =I21*(1+H14/100)-G17 | =J21*(1+H14/100)-G17 | =K21*(1+H14/100)-G17 | =IF((H21-G21)>0,"Sustainable","Unsustainable") |
| 22 | | | | |

5. If you have created the spreadsheet correctly, you can now change the energy source for food production and shelter/heat. The image below shows the output for the first 4 days for using energy source A for both food production and shelter/heat, if you have entered everything in correctly.

| **Effect on the environment over time** | | | |
|---|---|---|---|
| Day 1 | Day 2 | Day3 | Day 4 |
| 829.40 | 783.21 | 730.09 | 669.01 |

The sustainability value is tracking down. It will soon be unsustainable (<60% of 1000)

6. (a) Run the simulation by changing the food energy source (G6) and shelter/heat energy source (I6). Find the sustainability of an A and A, B and B, and C and C energy source scenario.

(b) Try combinations of energy sources (e.g. A and C or C and A) to see how sustainability and biodiversity are affected.

(c) Using these combinations, what is the maximum food units and shelter/heat units a person can use and remain sustainable? There may be many possible answers.

(d) Now try changing the rate at which the biodiversity replenishes or the value for the environment's original health. How do these affect the simulation?


**ISBN: 978-1-927309-37-7**

# ESS3.D Global Climate Change

**Key terms**

atmosphere
biosphere
climate change
climate model
cryosphere
geosphere
global warming
greenhouse effect
greenhouse gas
hydrosphere

## Disciplinary core ideas

*Show understanding of these core ideas* — **Activity number**

### Humans have the ability to manage their impact on the Earth

- ☐ 1 Human activities have significant impacts on all of the Earth's natural systems, not just the biosphere and the Earth's biodiversity, but the atmosphere, geosphere, and hydrosphere, and the climate systems associated with them. **149 150**
- ☐ 2 The evidence indicates that the current rise in global temperatures is largely a consequence of human activities and this increase is driving climate change. Greenhouse gases emitted today will drive the Earth's atmospheric gas concentrations, energy balance, and climate well into the future. **148**
- ☐ 3 Although the magnitudes of human impacts are greater than they ever have been, our ability to model, predict, and manage present and future impacts is also increasing. New records and observations around the globe are consistent with predictions from climate change models. The observed changes in the Earth's systems over such a short time period (~100 years) have prompted the development of programs regionally, nationally, and internationally to assess and manage the impacts of climate change. **146 147 148**

### Studies and simulations provide information about Earth's systems

- ☐ 4 Through computer simulations and other information, e.g. recorded extremes and ocean and atmospheric research, important discoveries are being made about how the ocean, atmosphere, and biosphere interact and are modified in response to human activities. **148**
- ☐ 5 Climate modeling is not an exact science. Models must run forward in time faster than the Earth's systems. They do this by making assumptions based on the data available. Different models predict different aspects of the climate. **147 148**

Bob Embleton

Bruno de Giusti

## Crosscutting concepts

*Understand how these fundamental concepts link different topics* — **Activity number**

- ☐ 1 **SC** Rates of global or regional climate change and their associated effects can be quantified and modeled over short or long periods of time. **147-149**
- ☐ 2 **SSM** The effect of human activities on the relationships among Earth's systems can be modeled using mathematical representations. **147-149**
- ☐ 3 **SSM** When using models to describe systems, the boundaries and initial conditions need to be defined, and the inputs and outputs analyzed. **146-148**

## Science and engineering practices

*Demonstrate competence in these science and engineering practices* — **Activity number**

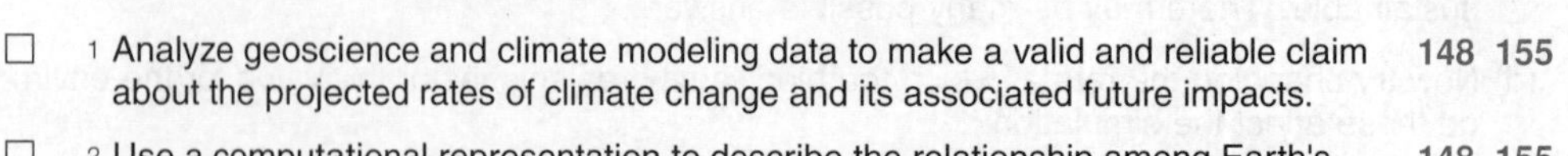

- ☐ 1 Analyze geoscience and climate modeling data to make a valid and reliable claim about the projected rates of climate change and its associated future impacts. **148 155**
- ☐ 2 Use a computational representation to describe the relationship among Earth's systems and show how these relationships are being modified by human activity. **148 155**
- ☐ 3 Evaluate design solutions to reduce the impact of climate change on Earth's systems. **152**

# 146 What is Climate Modeling?

**Key Idea**: Climate models are mathematical representations of the Earth's climate systems. Climate modeling allows scientists to predict long term climate patterns.

### What is a climate model?

- **Climate** refers to the prevailing weather conditions at a location over a long period of time, usually measured many years. The Earth's climate is the result of interactions among the Earth's systems. The complexity of these interactions makes climate difficult to understand and predict.
- Scientists use models to break the Earth's climate systems into components that can be more easily studied and understood. As the body of knowledge about a system grows, more components can be added so that it more closely represents the real system.
- Models used to understand climate are called **climate models**. They are mathematical representations and are often complicated because many different factors affect the climate.
- Changes in temperature, humidity, atmospheric pressure, wind, and precipitation (below) are some of the factors contributing to climate.

Rain storm

### How is climate data used?

- The factors affecting the climate are interconnected. A change in one factor has an effect on another. Scientists manipulate the various components of the model and see what the outcome is.
- Climate models can be developed on different scales, i.e. for a particular region or globally.
- Climate models have several purposes:
  - To understand the present climate, including the factors contributing to it.
  - To project climatic conditions into the future.
  - To investigate how natural processes or human activity may affect climate.
- Climate data is utilized by a wide range of people. Farmers may use the data to plan ahead for droughts by planting drought tolerant crops or reducing livestock numbers to cope with reduced rainfall. Towns and cities can launch water management plans so water supplies can be managed through periods of low rainfall. The energy sector uses the data to forecast energy consumption. For example, energy consumption increases in a heat wave as people use fans and air conditioning to keep cool. Energy production can be increased to meet demand. In the long term, if global warming continues, more electricity capacity may be needed to meet demand.
- By seeing what effect certain activities have, scientists can recommend changes to help prevent or slow down further climate change.

1. What is climate? ______________________________

2. (a) What is climate modeling? ______________________________

   (b) What are the three main purposes of climate modeling? ______________________________

3. How is climate modeling data used? ______________________________


**ISBN: 978-1-927309-37-7**

# 147 The History of Climate Modeling

**Key Idea**: Climate models have become more complicated and sophisticated, allowing scientists to better predict climate change.

## Climate models have become more sophisticated over time

- Climate models have been in use since the 1950s, but these very early versions really only modeled the weather in a particular region.
- The sophistication and accuracy of climate models has increased over time (below). This is because our knowledge about factors contributing to climate has increased and also because developments in computing and mathematics have allowed the more accurate prediction of more complicated scenarios.
- In 1988, the Intergovernmental Panel on Climate Change (IPCC) was established. Its role is to analyze published climate data and inform the international community about their findings.

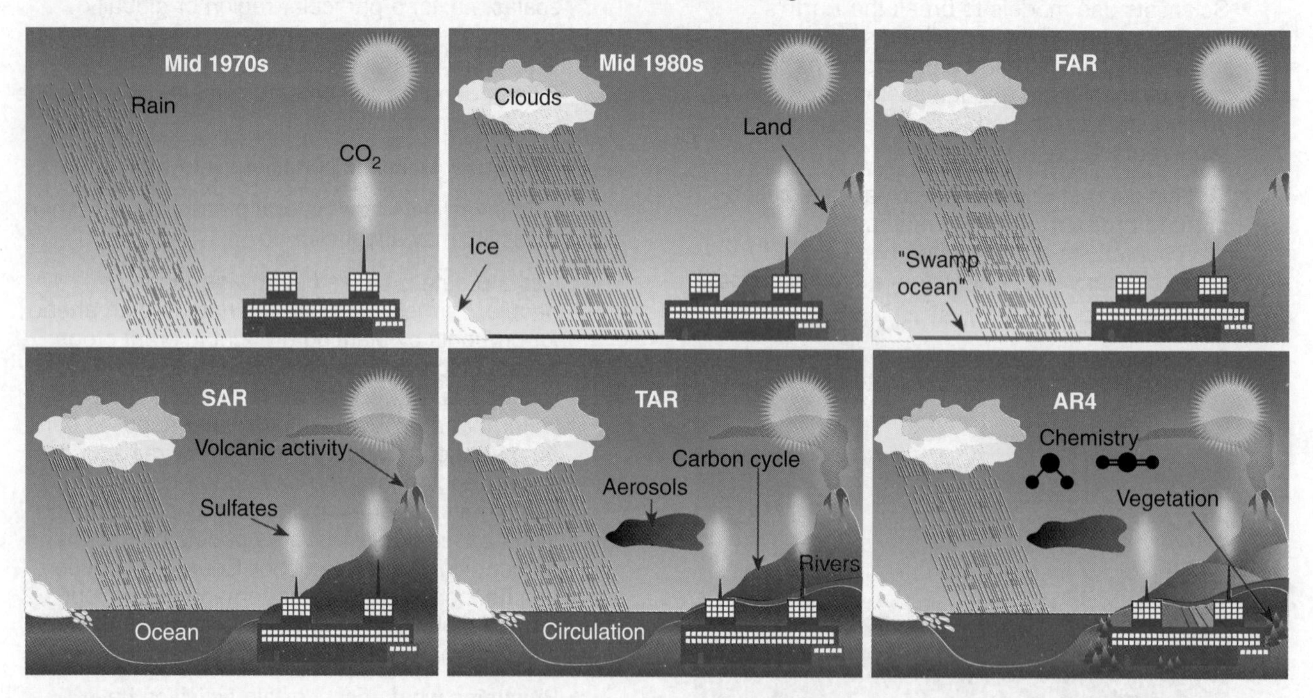


The diagrams above show how the sophistication of climate models has changed over time. Note how the complexity has increased as more elements are incorporated into the models. Early models in the 1970s were very simple and factored in only a few components (incoming sunlight, rainfall, and $CO_2$ concentration). By the 1980s, the models were becoming more complex and other features were added such as clouds, land surface features, and ice. After the establishment of the IPCC, several climate models were developed in relatively quick succession. The First Assessment Report (FAR) in the early 1990s, the Second Assessment Report (SAR) in 1995, the Third Assessment Report (TAR) in 2001, and the Fourth Assessment Report (AR4) in 2007. FAR included the ocean's effect for the first time, and subsequent models became more sophisticated, including adding the effect of atmospheric constituents such as sulfates and aerosols, the role of the carbon cycle, atmospheric chemistry, and vegetation.

### How are climate models tested?

To see how well models work, scientists enter past data and see how accurately they predict the climate changes that have already occurred. If the models recreate historical trends accurately, we can have confidence that they will also accurately predict future trends in climate change.

The graph on the right shows an example of how climate models are tested. The gray band represents data from 14 models and 58 different simulations. The black line represents the average of all 58 simulations. The gray line represents the average actual (observed) data for the same period. The gray vertical lines represent large volcanic eruptions during the period.

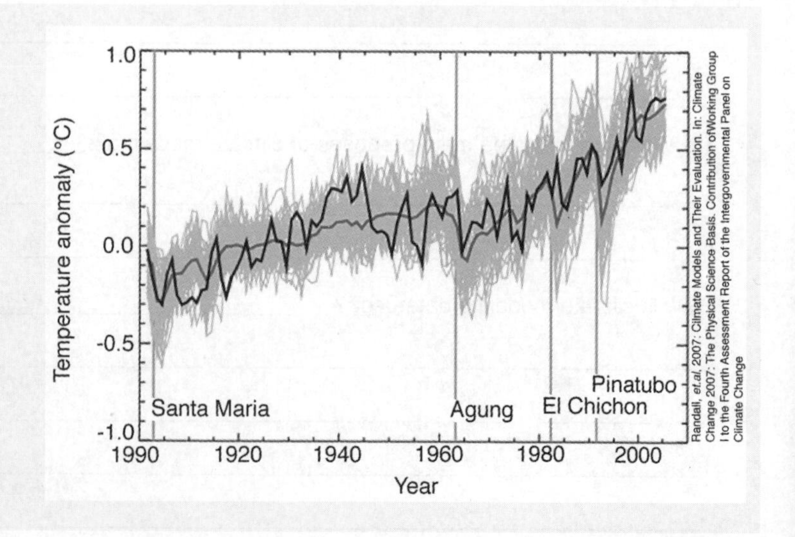


Randall, *et al.* 2007: Climate Models and Their Evaluation. In: Climate Change 2007: The Physical Science Basis. Contribution of Working Group I to the Fourth Assessment Report of the Intergovernmental Panel on Climate Change




**ISBN: 978-1-927309-37-7**

## What should a climate model include?

Climate models predict climate change more accurately when the model incorporates all the factors contributing to climate change. Some components influencing climate (e.g. the ocean and atmosphere) have their own models to better understand how the individual components can be influenced. Data from these separate models can provide more detailed information about the climate model as a whole. As we have already seen, climate models have become more complicated over time. Most now incorporate the following components:

- **Atmosphere**: This includes cloud cover, air pressure, water vapor, and gas concentrations.
- **Oceans**: Oceans have a key role in climate regulation. They help to buffer (neutralize) the effects of increasing levels of greenhouse gases in the atmosphere by acting as a carbon sink. They also act as a heat store, preventing rapid rises in global atmospheric temperature.
- **Ice sheets and sea ice (the cryosphere)**: These factors influence how much of the Sun's heat is reflected or retained. Increased ice levels reflect more heat away from Earth. Less ice allows more heat to be retained.
- **Biogeochemical cycles**: Levels of some atmospheric compounds can greatly influence climate change. Carbon is the most significant, but others such as nitrogen, phosphorus, and sulfur can also influence climate.
- **Biosphere**: The level of plant cover on Earth has a significant impact on the amount of carbon in the atmosphere. During photosynthesis, plants utilize carbon dioxide from the atmosphere to produce carbohydrates, effectively removing a major greenhouse gas from the atmosphere.
- **Human activity**: Human activity has increased the rate of global warming, especially through the actions of deforestation and carbon emissions into the atmosphere. The addition of greenhouse gases into the atmosphere through human activity is driving current climate change.
- **External influences**: These include energy variations from the Sun (e.g. through sunspot cycles) and levels of carbon dioxide and other aerosols released during volcanic eruptions.

1. (a) How has the complexity of climate models changed over time? ______

   (b) What has been the significance of this? ______

2. (a) How do scientists check the accuracy of their models? ______

   (b) Why is it important that they do this? ______

   (c) Study the testing results on the previous page. Do you think the average data from the models accurately reflects the historical data? Why or why not?

   ______

3. (a) Working in pairs or small groups, select one component of a climate model and research its significance to climate change. Summarize your findings and report back to the class.

   (b) Once all the presentations have been made, determine if any factor(s) has a larger influence than another.


**ISBN: 978-1-927309-37-7**

# 148 Models of Climate Change

**Key Idea**: Climate change models provide best-case and worst-case scenarios. The models can be used to predict the effect on Earth's systems.

## Using climate models to predict change

- There are elements of uncertainty in even well tested models. The major source is human activity and, in particular, how will consumption of fossil fuels change in the future? The level of greenhouse gases in the atmosphere will have a significant impact on future climate change.
- The IPCC often run a number of different scenarios to predict climate change. Between them, the results provide a best-case and worst-case scenario.
- The major scenarios are presented below, but there are subcategories (e.g. A1B) to help make them more accurate:
  - **A1** assumes rapid economic and technological growth, a low rate of population growth, and a very high level of energy use. Differences between "rich" and "poor" countries narrow.
  - **A2** assumes high population growth, slower technological change and economic growth, and a larger difference between countries and regions than in other scenarios. Energy use is high.
  - **B1** assumes a high level of environmental and social consciousness and sustainable development. There is low population growth, high economic and technological advancement, and low energy use. The area devoted to agriculture decreases and reforestation increases.
  - **B2** has similar assumptions to B1. However, there are more disparities between industrialized and developing nations, technological and economic growth is slower than in B1, and population growth is greater (but less than A2). Energy use is midway between B1 and A2. Changes in land use are less dramatic than in B1.

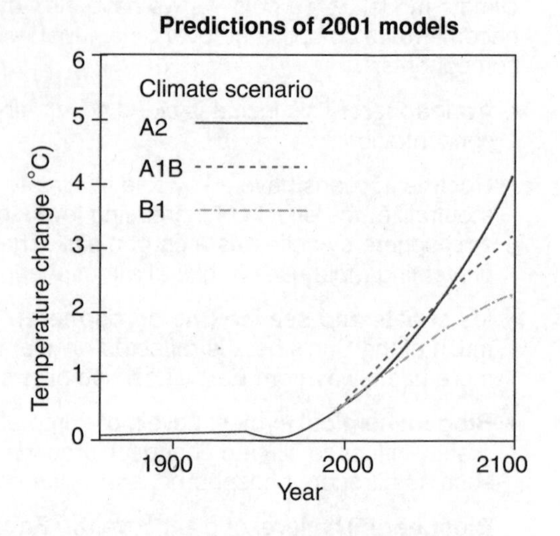


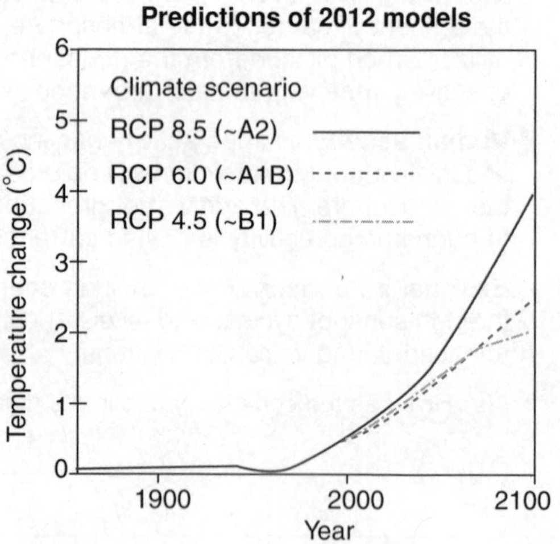


1. Why do scientists simulate a number of different scenarios when they run a climate model? ____________________

2. Study the 2001 and 2012 models of climate change predictions (above).

   (a) In the 2001 model, identify which scenario was predicted to produce the highest temperature change by 2100: ____________________

   (b) What factors are likely to contribute to this? ____________________

   (c) Why would scenario B1 produce the lowest temperature increase? ____________________

   (d) How do the predictions between the 2001 and 2012 models differ? ____________________

PRACTICES


**ISBN: 978-1-927309-37-7**

## What causes sea level rise?

The increase in global temperature is linked to a rise in global sea level. Sea level rise occurs because of two main factors, thermal expansion and melting ice. When water heats up it expands and takes up more space. Around 24 million km$^3$ of water is stored in permanent snow, glaciers, and ice caps. When these melt, they add to the volume of water in the oceans. Sea level rise will not only affect people living in coastal communities, but also Earth's systems. Many models have been developed to predict sea level rise under different scenarios in order to determine its effect.

NASA

## What effect will sea level rise have in the US?

In 2010, around 39% of the US population (around 123 million people) lived in counties directly on the shoreline. This population is expected to increase by 8% by 2020. Rising sea levels therefore represent a significant hazard to the US. Large cities such as New York (right) are in danger of becoming inundated (flooded) as sea levels rise. Other large cities, such as San Francisco and Los Angeles, are at sea level, or close to it. In New York, a sea level rise of only a few meters would inundate thousands of hectares of highly developed land. Airports, ports, railroads, housing developments, highways, factories and industry would be damaged.

3. (a) Describe the causes of sea level rise: ______________________

(b) Why is the US so vulnerable to a rise in sea level? ______________________

4. Some calculations estimate global mean sea level has increased between 10-20 cm over the last 100 years. However, for the last 20 years the rate of sea level rise has been around 3.2 mm per year.

(a) Calculate the average rate of sea level change per year for the last 100 years: ______________________

(b) How does this compare to the mean sea level change over the last 20 years: ______________________

(c) What factor could be contributing to the change observed in 4(b)? ______________________

5. (a) Study the graph on the right. What does it show?

______________________

(b) What is the worst case scenario? ______________________

(c) What is the best case scenario? ______________________

Predicted global sea level rise under a number of scenarios

Observed | Scenarios

Global sea level rise (cm above 1992): -40, 0, 40, 80, 120, 160, 200

Year: 1900, 1950, 2000, 2050, 2100

Global Sea Level Rise Scenarios for the United States National Climate Assessment, NOAA (2012)


**ISBN: 978-1-927309-37-7**

# 149 Ocean Acidification

**Key Idea**: The increasing amount of carbon dioxide in the atmosphere is affecting the pH of the ocean. Carbon dioxide reacts with water reducing water's pH.

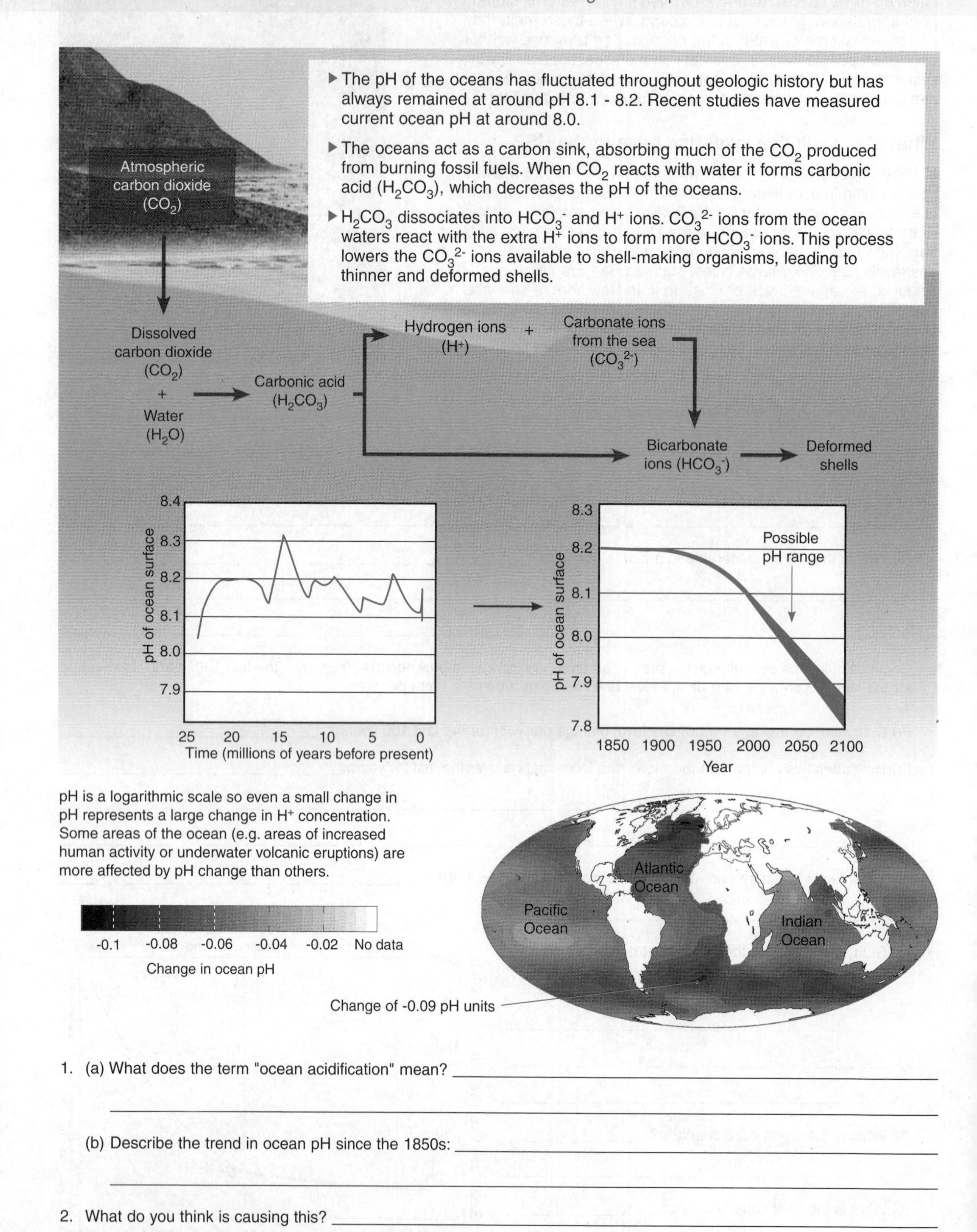


1. (a) What does the term "ocean acidification" mean? ______________________

(b) Describe the trend in ocean pH since the 1850s: ______________________

2. What do you think is causing this? ______________________




ISBN: 978-1-927309-37-7

# 150 Biodiversity and Climate Change

**Key Idea**: Global warming is causing shifts in the distribution, behavior, and even viability of plant and animal species.

- Global warming is changing the habitats of organisms and this may have profound effects on the biodiversity of specific regions as well as on the planet overall. As temperatures rise, organisms may be forced to move to areas better suited to their temperature tolerances. Those that cannot move or tolerate the temperature change may face extinction (loss of all individuals).
- Changes in precipitation as a result of climate change will also affect where organisms can live. Long term changes in climate will ultimately result in a shift in vegetation zones as some habitats contract and others expand.

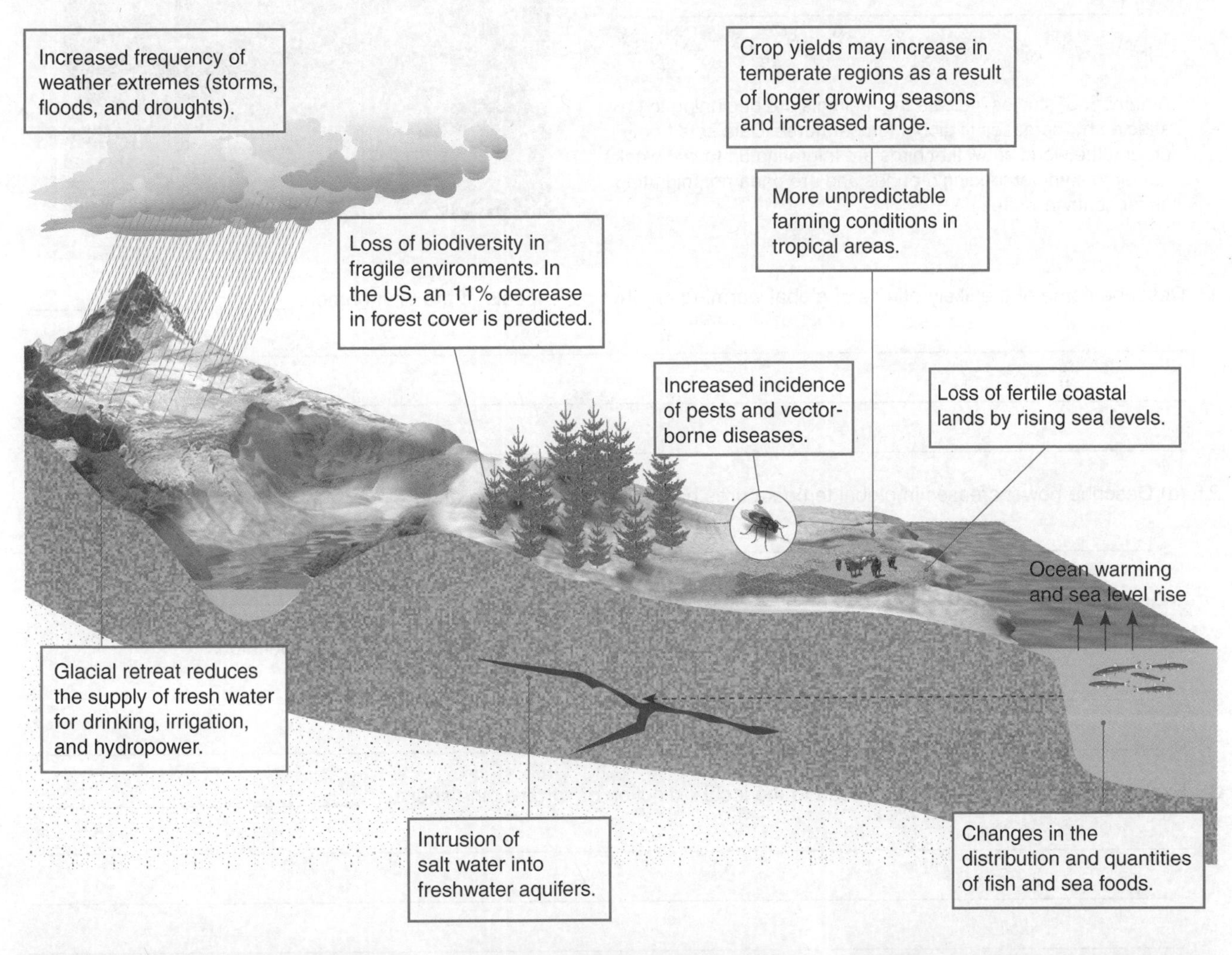

Studies of the distributions of butterfly species in many countries show their populations are shifting. Surveys of Edith's checkerspot butterfly (*Euphydryas editha*) in western North America have shown it to be moving north and to higher altitudes.

Sex ratios of reptiles are affected by the temperature. In turtles, males are produced at low incubation temperatures with females being produced at higher temperatures. Any rises in global temperatures could significantly affect these reptile populations.

An Australian study in 2004 found the center of distribution for the AdhS gene in *Drosophila*, which helps survival in hot and dry conditions, had shifted 400 kilometers south in the last twenty years.


**ISBN: 978-1-927309-37-7**

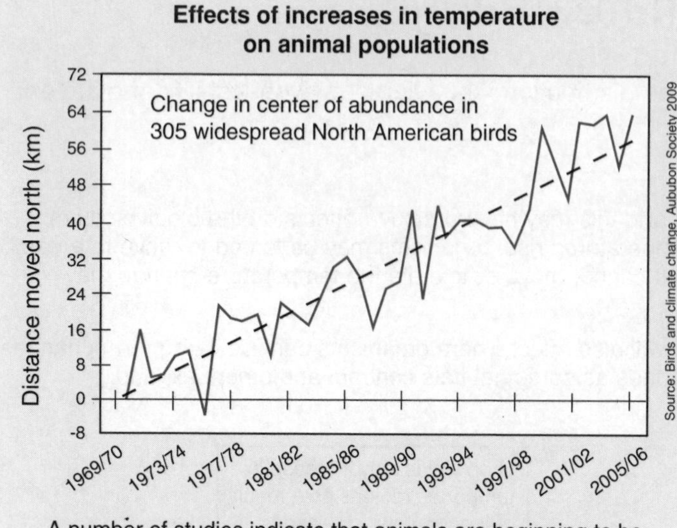


A number of studies indicate that animals are beginning to be affected by increases in global temperatures. Data sets from around the world show that birds are migrating up to two weeks earlier to summer feeding grounds and are often not migrating as far south in winter.

Animals living at altitude are also affected by warming climates and are being forced to shift their normal range. As temperatures increase, the snow line increases in altitude pushing alpine animals to higher altitudes. In some areas of North America this has resulted in the local extinction of the North American pika (*Ochotona princeps*).

1. Describe some of the likely effects of global warming on physical aspects of the environment: ______

2. (a) Describe how increases in global temperatures have affected some migratory birds: ______

(b) Explain how these changes in migratory patterns might affect food availability for these populations: ______

3. Explain how global warming could lead to the local extinction of some alpine species: ______

4. Describe the effects of climate change on three named animal examples:

(a) ______

(b) ______

(c) ______


**ISBN: 978-1-927309-37-7**

# 151 Climate Change and Agriculture

**Key Idea**: A warming climate will have significant effects on where crops can be grown and the impact of pests and diseases on them.

- The impacts of climate change on agriculture and horticulture will vary around the globe because of local climate and geography. In some regions, temperature changes will increase the growing season for existing crops or enable a wider variety of crops to be grown.
- Changes in temperature or precipitation patterns may benefit some crops, but have negative effects on others. Increasing atmospheric $CO_2$ levels will enhance the growth of some crops, but rising nighttime temperatures may affect seed set and fruit production.

### Effects of increases in temperature on crop yields

Studies on the grain production of rice have shown that maximum daytime temperatures have little effect on crop yield. However, minimum night time temperatures lower crop yield by as much as 5% for every 0.5°C increase in temperature.

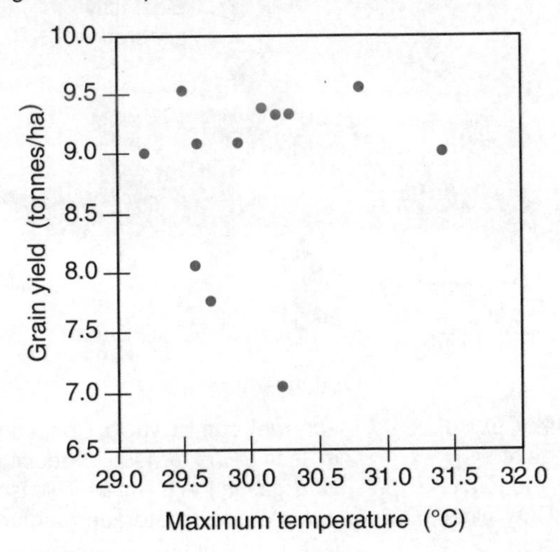


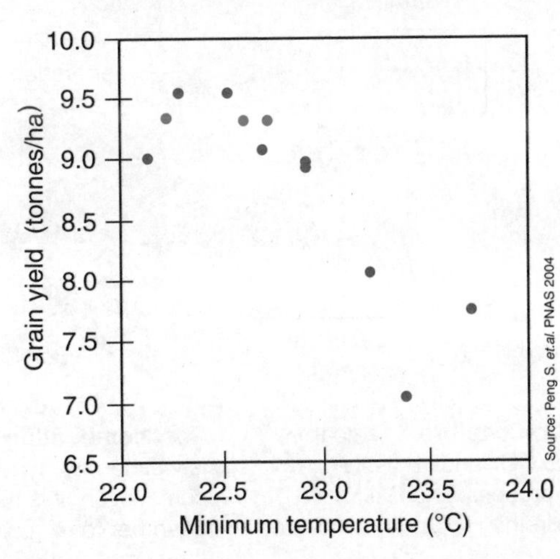


### Possible effects of increases in temperature on crop damage

The fossil record shows that global temperatures rose sharply around 56 million years ago (the Paleocene-Eocene Thermal Maximum (PETM)). Studies of fossil leaves with insect browse damage indicate that leaf damage peaked at the same time. This gives some historical evidence that, as temperatures rise, plant damage caused by insects also rises. This could have implications for agricultural crops.

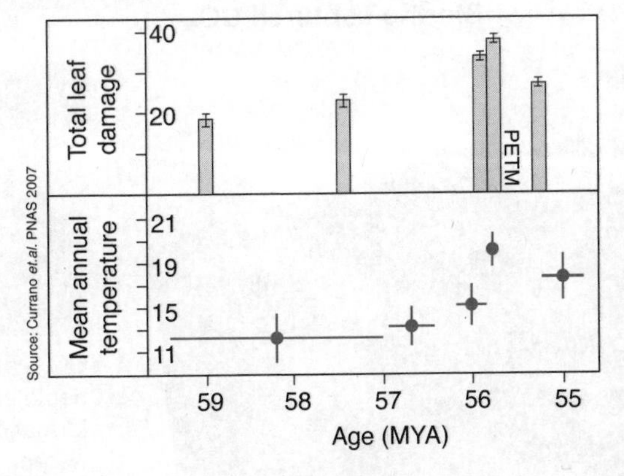


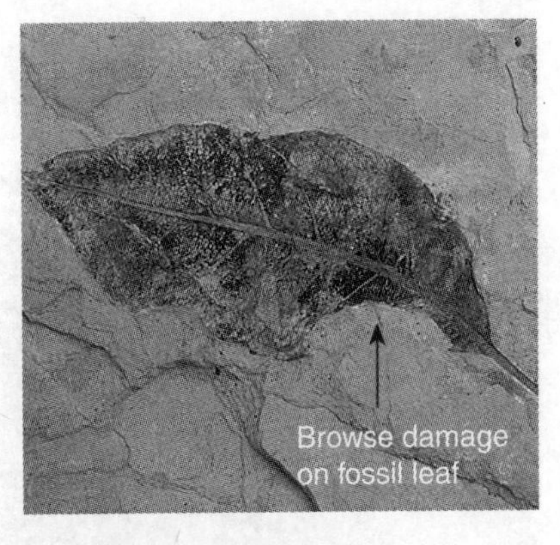


1. What minimum temperature produces the best grain yield in rice? ____________________

2. Suggest why global warming might benefit some agricultural crops, while disadvantaging others: ____________________

3. What evidence is there that global warming might influence the distribution or number of crop pests and so affect agriculture?


**ISBN: 978-1-927309-37-7**

# 152 Technological Solutions to Climate Change

**Key Idea**: New technologies are aiming to reduce carbon dioxide emissions and so help slow climate change.

- Burning fossil fuels in power stations for electricity accounts for about 40% of global carbon dioxide emissions. The transport industry accounts for at least another 30%. Even power stations using high quality coal and oil release huge volumes of $CO_2$. Systems that capture the $CO_2$ produced so that it can be stored or used for other purposes are beginning address this problem (below).
- Another important source of $CO_2$ emissions is often overlooked, probably because it is so common we don't stop to consider its effects. The manufacture of cement and concrete account for 5-10% of all $CO_2$ emissions. New types of cement and techniques for manufacture are aiming to reduce this amount.

## Schematics of possible carbon capture systems

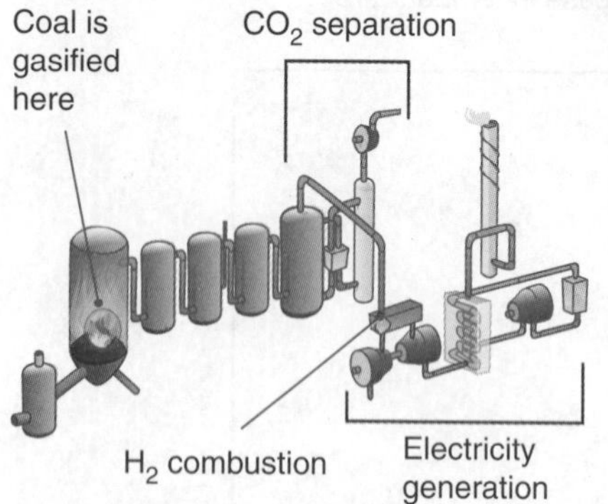


**Pre-combustion capture**: The coal is converted to $CO_2$ and $H_2$ using a gasification process. The $CO_2$ is recovered while the $H_2$ gas is combusted.

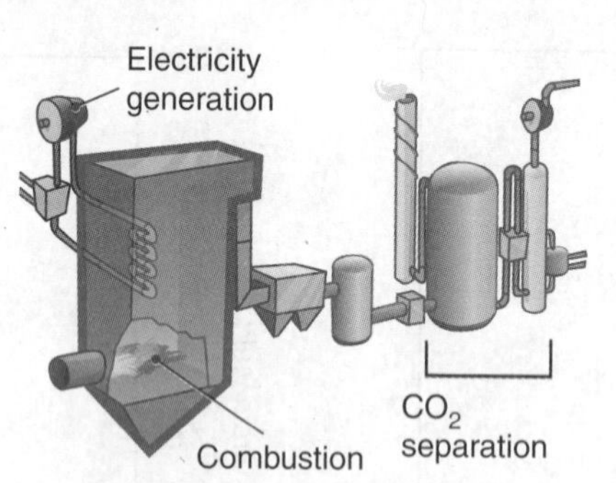


**Post combustion capture**: $CO_2$ is washed from the flue gas after combustion. It is then passed to a desorber to re-gasify the $CO_2$, where it is then compressed for storage.

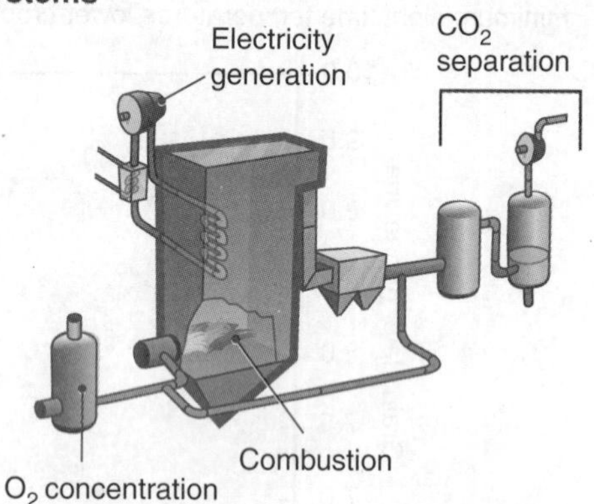


**Oxyfuel combustion**: Concentrated $O_2$ is used in the furnace, producing only $CO_2$ gas in the flue gas. This is then compressed for storage. Compressed $CO_2$ is useful as a inexpensive, nonflammable pressurized gas, e.g. for inflation and for carbonated water.

## Storing captured $CO_2$

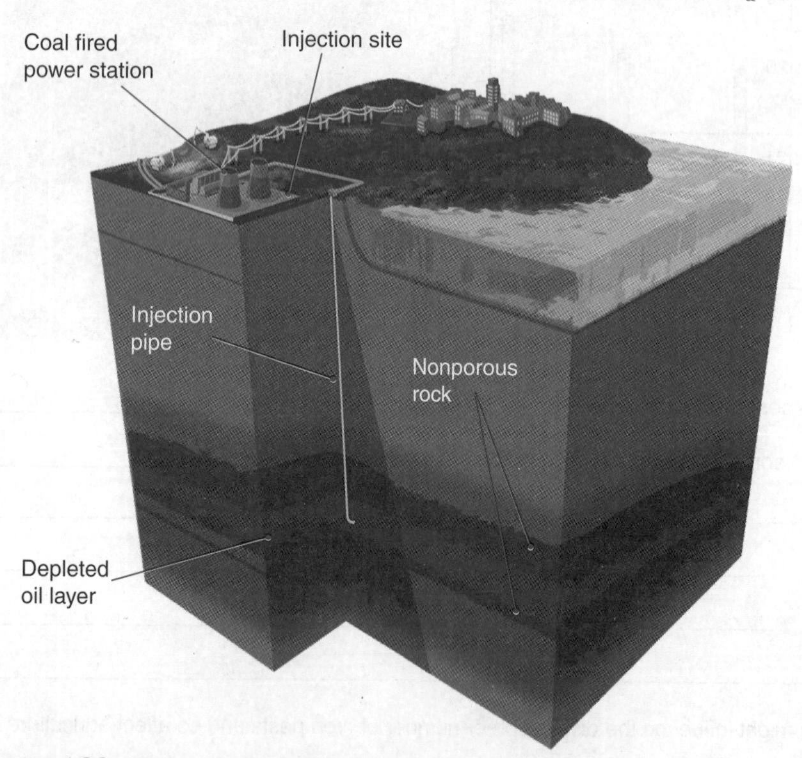


Captured $CO_2$ can be injected into porous strata between nonporous layers. Power stations near to injection sites can pipe the recovered $CO_2$ to the injected well. Other stations will need to transport the $CO_2$ to the site. The transportation of the $CO_2$ will produce less $CO_2$ than is captured by the power station, making the option viable.

$CO_2$ can be stored by injecting it into depleted oil wells or other deep geological formations, releasing it into deep ocean waters, or reacting it with minerals to form solid carbonates. The $CO_2$ can also be used as a starting point for the production of synthetic fuels.

Deep ocean storage of $CO_2$ risks lowering ocean pH, and storing $CO_2$ in geological formations risks sudden release of large quantities of $CO_2$ if the rock proves unstable. The sudden release of $CO_2$ can kill animal life in the area (above).

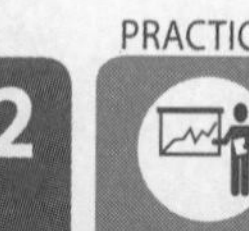

**ISBN: 978-1-927309-37-7**

## Lowering emissions in the cement industry

- Cement and concrete (the final cured product of cement mixed with water and gravel) are essential to the building industry and the global economy. 4.1 billion tonnes of cement were produced in 2015. This is expected to increase to 4.8 billion tonnes by 2030. This is important to the global climate because producing one tonne of cement also produces about one tonne of $CO_2$ (and uses the equivalent of 200 kg of coal).
- The most common cement used is called Portland cement, which is very strong when set. Its manufacture is highly energy intensive. 40% of the $CO_2$ emissions come from the burning of fossil fuels to heat limestone ($CaCO_3$) and other minerals to around 1400°C. At this temperature, the limestone degrades and releases $CO_2$. This step accounts for about 50% of the $CO_2$ emissions. Portland cement reabsorbs about half of this $CO_2$ as it hardens and over the life time of the cement.
- Reducing the $CO_2$ emissions for cement manufacture can be done at three steps in the process: Reducing the amount of fossil fuels needed to heat the raw materials, reducing the amount of $CO_2$ released by the raw materials, and increasing the amount of $CO_2$ absorbed when setting.
- New types of cement using magnesium silicates instead of limestone are under trial. These do not need to be heated to such high temperatures and also release no $CO_2$ when heated. This leads to total $CO_2$ emissions of up to 0.5 tonnes of $CO_2$ per tonne of cement produced (half that of Portland cement). Additionally, during setting, $CO_2$ is absorbed at a greater rate than in traditional cement (about 1.1 tonnes per tonne of cement produced). This type of cement is often called carbon negative cement as it actually absorbs more $CO_2$ than is produced in its manufacture (about 0.6 tonnes of $CO_2$ absorbed for every tonne of cement produced).

1. Describe the differences and similarities in the three types of carbon dioxide capture systems:

2. Describe how captured carbon dioxide might be used or stored:

3. Discuss some of the potential problems with capturing and storing carbon dioxide:

4. (a) Approximately how many tonnes of $CO_2$ were produced by the cement industry in 2015?

   (b) Where is this $CO_2$ produced in the manufacture of cement?

   (c) Explain why carbon negative cement is carbon negative:

   (d) Based on the 2015 figures, how much carbon would carbon negative cement absorb?


**ISBN: 978-1-927309-37-7**

# 153 Chapter Review

Summarize what you know about this topic so far under the headings provided. You can draw diagrams or mind maps, or write short notes to organize your thoughts. Use the introduction and the hints included to help you:

Climate modeling

HINT: What factors are important in climate models? What is the use of climate modeling? How have models changed over time?

Climate change

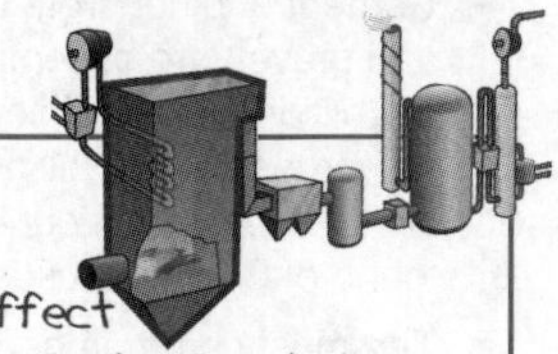

HINT: How will climate change affect humans and biodiversity? How can technology help reduce the effects of climate change?


**ISBN: 978-1-927309-37-7**

# 154 KEY TERMS AND IDEAS: Did You Get It?

1. Test your vocabulary by matching each term to its definition, as identified by its preceding letter code.

|  | |  |
|---|---|---|
| atmosphere | A | The change in distribution of weather patterns over a long period of time. |
| biosphere | B | Any gas in the atmosphere that causes the retention of heat in the Earth's atmosphere |
| climate change | C | The retention of solar energy in the Earth's atmosphere by gases that absorb heat and prevent it from being released back into space. |
| climate model | D | Mathematical or computation model that predicts future changes to the climate. |
| cryosphere | E | The process of the Earth's surface temperature steadily increasing in temperature (and its projection continuation). Usually attributed to the rise in gases produced from burning fossil fuels and other human activity such as deforestation. |
| geosphere | F | The combined mass, cycles, and location of water found in, on, and around the Earth. |
| global warming | G | The entire global ecosystem, including all living things, their interactions, and the places on Earth where they are found. |
| greenhouse effect | H | The solid part of the Earth including the rocks of the crust, the lithosphere, asthenosphere and all other geologic material. |
| greenhouse gas | I | The layers of gases that surround the Earth. |
| hydrosphere | J | The frozen part of the world's water, e.g. sea ice and glaciers. |

2. Use the graphs below to answer the following questions:

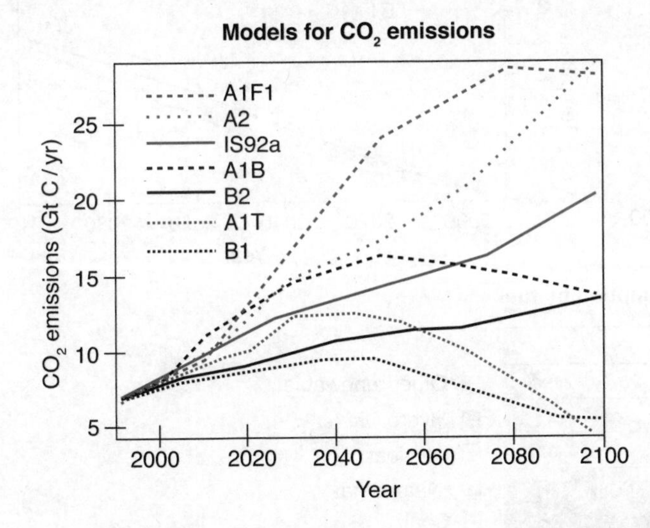


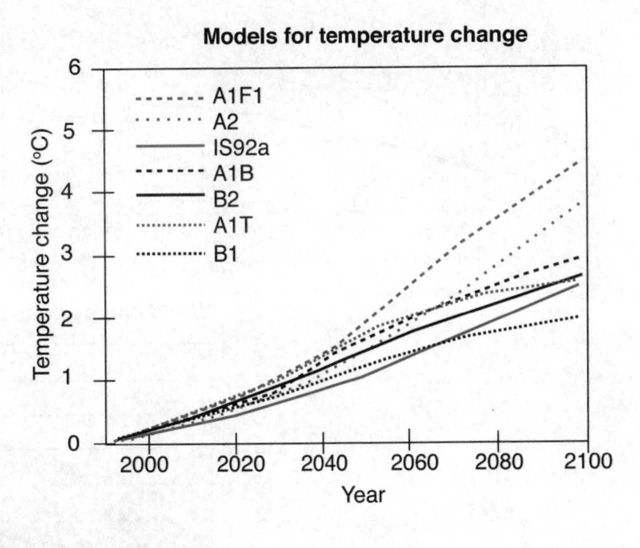


(a) Which scenario produces the highest $CO_2$ emissions by 2100? ______________________

(b) Which scenario produces the lowest $CO_2$ emissions by 2100? ______________________

(c) Which scenario produces the greatest temperature change by 2100? ______________________

(d) Which scenario produces the smallest temperature change by 2100? ______________________

(e) Explain why even in the scenarios where $CO_2$ emissions are falling, temperature still rises:

______________________

______________________

______________________

3. In groups, investigate possible technological ways to reduce carbon dioxide in the atmosphere. Technologies could include carbon scrubber "trees", algal bioreactors, or photosynthetic "frog foam". Summarize the potential benefits and problems of the system and include an evaluation of the feasibility of the technology. Staple your summary to this page.


**ISBN: 978-1-927309-37-7**

# 155 Summative Assessment

1. The graphs below show data for past and current $CO_2$ concentrations and temperature, and models for $CO_2$ and temperature for the future.

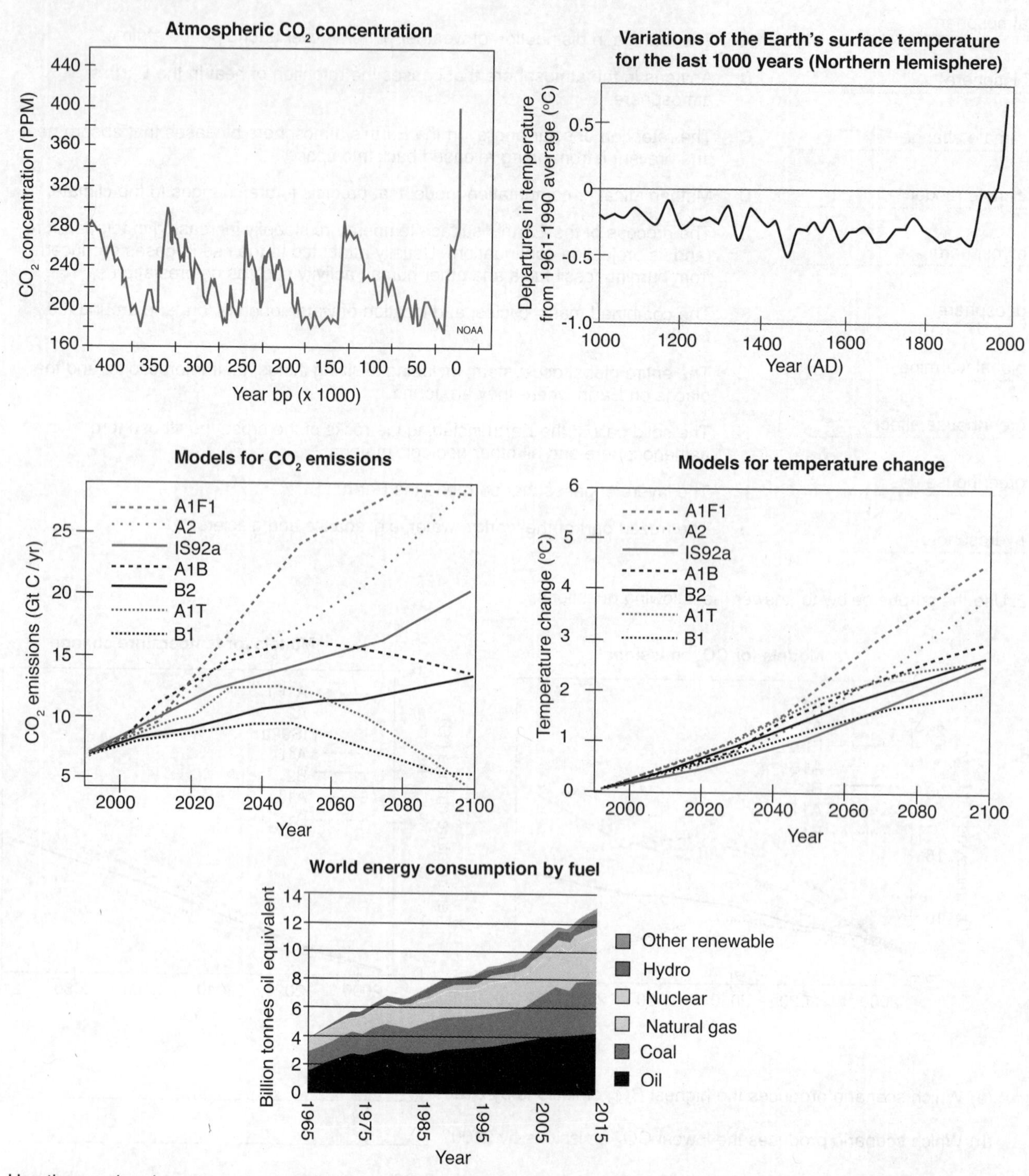


Use the graphs above to make an evidence-based forecast of the future rate of climate change, including your analysis of whether or not the rate will increase or decrease. You may use extra sources of information to further research your analysis. You may also use extra paper to write your analysis. Attach it to this page:

TEST


**ISBN: 978-1-927309-37-7**

2. The data below shows the expected temperature, precipitation changes, and grain yield for wheat grown in the Eastern Washington area over the next 65 years. The data was produced using the CCSM3 global climate model (which predicts more warming and less precipitation globally, although not necessarily locally).

| | Baseline | 2020 | 2040 | 2080 |
|---|---|---|---|---|
| Precipitation (mm) | 535.8 | 549.9 | 543.9 | 588.3 |
| Mean temperature °C | 8.5 | 10.2 | 11.2 | 12.0 |
| Yield (kg) No $CO_2$ effect | 5713 | 6022 | 5116 | 5209 |
| Yield (kg) $CO_2$ effect | 5713 | 6546 | 6034 | 7033 |

Claidio O. Stockle et al

The data below shows the percentage crop yield response of wheat and other crop plants to changes in the environment including temperature and $CO_2$ changes.

| Crop | Temperature (+1.2°C) | $CO_2$ increase (380 to 440 ppm) | Temperature, and $CO_2$ and irrigation |
|---|---|---|---|
| Wheat | -6.7 | +6.8 | +0.1 |
| Corn (midwest) | -4.0 | +1.0 | -3.0 |
| Soybean | -3.5 | +7.4 | +3.9 |
| Cotton | -5.7 | +9.2 | +3.5 |

(a) Describe the change in rainfall expected in the Eastern Washington area over the next 65 years:

(b) Describe the change in temperature expected in the Eastern Washington area over the next 65 years:

(c) Describe the effect on grain yield the change in climate will have, including effects of increased $CO_2$:

(d) Plants carry out photosynthesis, producing organic molecules. The chemical equation for photosynthesis is:

$$6CO_2 + 6H_2O \rightarrow C_6H_{12}O_6 + 6O_2$$

Why does the grain yield increase with more atmospheric $CO_2$?

(e) How will a change in temperature affect other crops grown in the USA?

(f) How will a change in $CO_2$ affect other crops grown in the USA?

(g) What overall effect might there be on crop yield due to climate change. How might this affect farmers and consumers?


**ISBN: 978-1-927309-37-7**

# Appendix

## Questioning terms

The following terms are often used when asking questions in examinations and assessments.

| | |
|---|---|
| Analyze: | Interpret data to reach stated conclusions. |
| Annotate: | Add brief notes to a diagram, drawing or graph. |
| Apply: | Use an idea, equation, principle, theory, or law in a new situation. |
| Calculate: | Find an answer using mathematical methods. Show the working unless instructed not to. |
| Compare | Show similarities between two or more items, referring to both (or all) of them throughout. |
| Construct: | Represent or develop in graphical form. |
| Contrast: | Show differences. Set in opposition. |
| Define: | Give the precise meaning of a word or phrase as concisely as possible. |
| Derive: | Manipulate a mathematical equation to give a new equation or result. |
| Describe: | Define, name, draw annotated diagrams, give characteristics of, or an account of. |
| Design: | Produce a plan, object, simulation or model. |
| Determine: | Find the only possible answer. |
| Discuss: | Show understanding by linking ideas. Where necessary, justify, relate, evaluate, compare and contrast, or analyze. |
| Distinguish: | Give the difference(s) between two or more items. |
| Draw: | Represent by means of pencil lines. Add labels unless told not to do so. |
| Estimate: | Find an approximate value for an unknown quantity, based on the information provided and application of scientific knowledge. |
| Evaluate: | Assess the implications and limitations. |
| Explain: | Provide a reason as to how or why something occurs. |
| Identify: | Find an answer from a number of possibilities. |
| Illustrate: | Give concrete examples. Explain clearly by using comparisons or examples. |
| Interpret: | Comment upon, give examples, describe relationships. Describe, then evaluate. |
| List: | Give a sequence of answers with no elaboration. |
| Measure: | Find a value for a quantity. |
| Outline: | Give a brief account or summary. |
| Predict: | Give an expected result. |
| Solve: | Obtain an answer using numerical methods. |
| State: | Give a specific name, value, or other answer. No supporting argument or calculation is necessary. |
| Suggest: | Propose a hypothesis or other possible explanation. |
| Summarize: | Give a brief, condensed account. Include conclusions and avoid unnecessary details. |


## Photo credits

We gratefully acknowledge the photographers that have made their images available through Wikimedia Commons under Creative Commons Licences 2.0, 2.5., 3.0, and 4.0: • Marlith • Coyau • Barfooz • Brocken Inaglory • Dartmouth Electron Microscope Facility (Public Domain) • Andrew Pontzen and Fabio Governato • Tobias 1984 • H. Raab • G310Luke • Valerio Pillar • PlaneMad cc 3.0 • Hcrepin • Graeme Churchard • Ashley Dace • Robert Knight • K Pryor • Meteorite Recon cc 3.0 • Mike Beauregard • Hartman and Neukum 2001 • Eugen Zibsio • Scott McDougall • Andrzej Mireck • Kochendes_wasser02 • Till Niermann • katorisi • Eurico Zimbres • Bob Embleton • Bruno de Giusti • Wknight94 talk • Tenderlok • Richard Weymouth • Hullwaren • Bob Metcalf • Sharon Loxton • Wawny • Cunningham- Saigo • United nations Logan Abassi cc2.0 • US Coastguard Public domain • Oxfam East Africa cc2.0 • Govt of Kiribati cc 3.0 • SuSanA cc 2.0 • Shahee Ilyas CC3.0 • hughe82 cc3.0 • Edwin S • Doug Kerr • Precision Seafood Harvesting • Ravedave • Paul Traytek • Chongkian

Contributors identified by coded credits are:
**BH**: Brendan Hicks (University of Waikato), **USDE**: U.S. Department of Energy, **UNFAO**: Food and Agriculture Organization of the United Nations, **NASA**: National Aeronautics and Space Administration, **JPL**: Jet Propulsion Laboratory, **NOAA**: National Oceanic and Atmospheric Administration, **ESA**: European Space Agency, **RA**: Richard Allan, **USGS**: United States Geological Survey, **USDA**: United States Department of Agriculture, **USFW**: United States Fish and Wildlife Service.

Royalty free images, purchased by Biozone International Ltd, are used throughout this book and have been obtained from the following sources: Corel Corporation from their Professional Photos CD-ROM collection; IMSI (Intl Microcomputer Software Inc.) images from IMSI's MasterClips® and MasterPhotos™ Collection, 1895 Francisco Blvd. East, San Rafael, CA 94901-5506, USA; ©1996 Digital Stock, Medicine and Health Care collection; © 2005 JupiterImages Corporation www.clipart.com; ©Hemera Technologies Inc, 1997-2001; ©Click Art, ©T/Maker Company; ©1994., ©Digital Vision; Gazelle Technologies Inc.; PhotoDisc®, Inc. USA, www.photodisc.com. • TechPool Studios, for their clipart collection of human anatomy: Copyright ©1994, TechPool Studios Corp. USA (some of these images were modified by BIOZONE) • Totem Graphics, for their clipart collection • Corel Corporation, for use of their clipart from the Corel MEGAGALLERY collection • 3D images created using Bryce, Vue 6, Poser, and Pymol • iStock images • Art Today • Dollar Photo Club.

# Index

# Index